Advances in Experimental Medicine and Biology

More information about this series at http://www.springer.com/series/5584

Christopher D. Gregory
Editor

Apoptosis in Cancer Pathogenesis and Anti-cancer Therapy

New Perspectives and Opportunities

Springer

Editor
Christopher D. Gregory
MRC Centre for Inflammation Research
University of Edinburgh Queen's Medical Research Institute
Edinburgh, UK

ISSN 0065-2598 ISSN 2214-8019 (electronic)
Advances in Experimental Medicine and Biology
ISBN 978-3-319-39404-6 ISBN 978-3-319-39406-0 (eBook)
DOI 10.1007/978-3-319-39406-0

Library of Congress Control Number: 2016947079

Printed on acid-free paper

This Springer imprint is published by Springer Nature
The registered company is Springer International Publishing AG Switzerland

Foreword

In 1993, I took up my first Chair at the University of Nottingham. Soon afterwards, my family attended my inaugural lecture, in which I described work on cell clearance by apoptosis in the resolution of inflammation. After the lecture, I was impressed by a question from my elder son, then 10 years old—"so, dad, is apoptosis a good thing?" My answer must have sounded evasive—"well it all depends…". Indeed, some years later, Valerie Fadok and I proposed that corpse clearance defines the "meaning" of cell death. Our thesis was that evolution has hijacked the phylogenetically ancient process of phagocytic clearance of cells dying by apoptosis to deliver beneficial outcomes in various states of tissue perturbation. For example, as acute inflammation resolves, phagocytes taking up apoptotic cells suppress further inflammatory responses by various mechanisms and promote tissue repair.

This superb new book, blessed with uniformly excellent contributions from contemporary leaders in the field, examines what is currently understood of the complexities of the death and clearance of cells in cancer. As one might expect, the processes are now believed to have both beneficial and deleterious consequences in tumours. What emerges for me is a surprising and enlightening concept—that cancer could be regarded "as a wound that never stops healing". If I have this right, the contributors are poised to inspire important new therapeutic approaches in cancer, building on exciting new data revealing benefit from inhibition of the PD1 pathway in T regulatory cells. Processes that are beneficial in inflammation seem to be deleterious in cancer, and their inhibition can be therapeutic. I wish I'd known that in 1993, but perhaps if I had, my son's curiosity about biology would have not have led him to read National Sciences at Cambridge, before becoming a science teacher. I hope he and his students reap the benefit of the exciting new knowledge presented here.

University of Edinburgh
Edinburgh, UK

John Savill

Preface

Apoptosis, the most widely studied and arguably best understood programmed cell death process, is the antithesis of proliferation and as such has gained much renown as an anticancer mechanism. Indeed, genes involved in suppressing apoptosis have rightly acquired the guise of oncogenes while those that promote the process, anti-oncogenes. These widely accepted, logical qualities of apoptosis, substantiated by decades of research in the context of tumour biology, are perhaps best summed up by Hanahan and Weinberg in their landmark review of the acquired characteristics of cancer cells which include the capacity to evade apoptosis.[1] Clearly, the ultimate "rogue" cell, a cell that engenders an aggressive malignant tissue that may kill its host, *must* survive and may need to evade apoptosis in order to do this. However, cells of multicellular organisms, though they have partial cell autonomous qualities, have evolved to exist in response to signals from their neighbours. Apoptotic cells are not exceptional in providing such signals and the most renowned neighbourly response to apoptosis is the phagocytic packaging and degradation of the dying cell, along with additional signalling responses from the phagocytes to yet other cells of the multicellular host. Furthermore, in development, it is well known that apoptosis can elicit proliferative responses in neighbouring cells. These examples and other consequences of apoptosis for its microenvironment have implications for the emergence, progression, invasiveness, and therapeutic responses of malignant disease, and in recent years, evidence has begun to emerge in support of these implications. This book presents a topical collection of reviews which consider the pleiotropic properties of apoptosis and indeed other forms of cell death in oncogenesis and therapy. Its primary aim is to provide a timely discussion of cell death in cancer in the broadest terms—from its widely accepted roles in prevention and cure to its potential in progression and relapse.

Chris Dillon and Doug Green begin by setting the scene with their chapter entitled "Molecular cell biology of apoptosis and necroptosis in cancer". Their review introduces the reader not only to apoptosis and its molecular mechanisms in the context of cancer but also to emerging relevant knowledge of the process of

[1] Hanahan D, Weinberg RA. The hallmarks of cancer. Cell. 2000;100:57–70.

programmed necrosis, necroptosis. They discuss the underlying physiological mechanisms, how they may become dysregulated in malignant disease, and how they may be stimulated for therapeutic gain. One of the most important take-home messages of this chapter is that cancers are likely to evolve to become tolerized to withstand cell death programs, and a pervasive view throughout this volume is that such tolerance could develop further to be of advantage to the rogue tissue.

In Chap. 2, "Clearance of dying cells by phagocytes: mechanisms and implications for disease pathogenesis", Aaron Fond and Kodi Ravichandran elegantly summarize the mechanisms by which apoptotic cells are detected, engulfed, and degraded by phagocytes. While this is the best-studied response to apoptosis and much has been learned in terms of its molecular cell biology, many questions remain, such as why is there so much variety in receptors and ligands in this process and what are the key molecules in various stages such as recognition and response? As the authors discuss, responses to apoptotic cells are far from limited to engulfment and safe degradation, and they highlight the role of apoptosis in controlling inflammation and facilitating differentiation. They not only consider normal physiology but also a variety of disease processes to which the clearance mechanisms are likely to contribute, not least cancer. Here, they highlight especially the critical pathways of dying cell clearance which connect growing tumours with anti-tumour immune responses, noting especially how the anti-immune mechanisms so typical of apoptotic cell interactions with antigen-presenting phagocytes can hold such advantage to malignant disease that can develop so effectively in the face of an effective host immune system.

In the third chapter, "Microenvironmental effects of cell death in malignant disease", my colleagues Catriona Ford and Jorine Voss and I expound on the sinister, pro-oncogenic effects of cell death in tumour tissue, mainly focusing on apoptosis but also making reference to other cell death modalities as appropriate. Our perspective is not only historical, drawing on Paget's "seed and soil" theory and on the parallels between cancer pathogenesis and wound-healing responses, but is also topical, highlighting the relationships between normal tissue regeneration and malignant disease. We rehearse the concept of the "onco-regenerative niche" (ORN), a putative cell death-driven complex of conserved tissue repair and regenerative responses that are hijacked in cancer, and propose that critical cellular responses in the ORN are those of macrophages and endothelial cells. We also propose that extracellular vesicles produced by dying cells are key biological entities that facilitate the pro-oncogenic effects of apoptosis. The concept that apoptotic mechanisms can be directed towards oncogenic processes is developed further by Yun Fan and his colleagues Catherine Dabrowska and Mingli Li in Chap. 4, "Apoptotic caspases in promoting cancer: implications from their roles in development and tissue homeostasis". Here, the authors discuss a different aspect, the utilization of central molecular mediators of the apoptosis machinery, the caspases, for functions other than cell death, especially the promotion of cell proliferation. Given the hundreds of protein substrates for apoptotic caspases, there seems little doubt that both apoptotic and non-apoptotic activities of these enzymes will prove to have a multiplicity of functions that promote neoplastic disease processes.

Chapters 5, 6, and 7 further discuss the concepts introduced in Chap. 2 surrounding the associations between apoptosis and the regulation of anti-tumour inflammatory responses and anti-tumour immunity. The review by Ian Dransfield and Sarah Farnworth (Chap. 5: "Axl and Mer receptor tyrosine kinases: distinct and nonoverlapping roles in inflammation and cancer?") considers in detail Axl and Mer, anti-inflammatory receptor tyrosine kinases known to interact with apoptotic cells via their ligands Gas6 and Protein S which bind avidly to the exposed phosphatidylserine on apoptotic cell surfaces. This receptor–ligand axis may lie at the heart of the anti-inflammatory responses of phagocytes to apoptotic cells, suggesting that it may provide a target area for anticancer drug development—though antagonism could have unwanted effects in inflammation-driven tumours! In Chap. 6 ("Immunogenic apoptotic cell death and anticancer immunity"), Peter Vandenabeele, Katrien Vandecasteele, Claus Bachert, Olga Krysko, and Dmitri Krysko discuss the particular case of immunogenic cell death, highlighting the special characteristics which make it particularly suited to anticancer treatment. The rationale here is that, if cell death processes activated by anticancer chemotherapies and radiotherapies can themselves have pro-oncogenic features, then by fostering cell death mechanisms such that the responses of the host facilitate anti-tumour immunity—for example through release of immunostimulatory adjuvants otherwise known as "DAMPs" (damage-associated molecular patterns) by dying cells—sustained tumour suppression could be achievable. This concept is considered in further depth by Udo Gaipl and colleagues Benjamin Frey, Anja Derer, Heike Scheithauer, Roland Wunderlich, and Rainer Fietkau in Chap. 7, "Cancer cell death-inducing radiotherapy: impact on local tumour control, tumour cell proliferation and induction of systemic anti-tumour immunity", especially in the context of radiation therapy. This chapter further provides a tour de force of the responses of cells to irradiation and their application in anticancer therapies.

Extending the theme of apoptosis-inducing anticancer therapy, Klaus-Michael Debatin and his colleagues Mike-Andrew Westhoff and Nicolas Marschall consider in detail in the penultimate chapter (Chap. 8: "Novel approaches to apoptosis-inducing therapies") the fundamental problem of the evolution of tumours in response to cell death-inducing therapies that lead to their resistance to apoptosis. The authors discuss the underlying mechanisms in depth and suggest solutions, notably using small molecule inhibitors of IAPs (inhibitor of apoptosis proteins) and BH3 mimetics (apoptosis-promoting small molecules), as well as by inhibiting cell survival signalling such as the PI3K/Akt/mToR and MEK/ERK pathways. To conclude the volume, and continuing new therapeutic approaches centred around apoptosis, Andreas Weigert, Javier Mora, Divya Sekar, Shahzad Syed, and Bernhard Brüne discuss the tumour-associated macrophage (TAM) as a critical cell death-responding component of the tumour microenvironment (Chap. 9: "Killing is not enough: How apoptosis hijacks tumor-associated macrophages to promote cancer progression"). Indeed, the macrophage is a cell which has the potential to induce tumour cell death as a natural anti-tumour host entity. Problematically, TAMs accumulate in response to cell death and other signals in tumours, and the activation state of TAMs is usually multi-functionally *pro*-oncogenic. The extent to which the pro-tumour functions of TAMs is driven by apoptosis requires clarification in multiple tumour types, but TAMs and their responses to cell death have real potential in anti-tumour therapeutic targeting. As Brüne and colleagues propose, effective anticancer therapies should be designed not only to induce death of tumour cells but also to target the pro-tumour TAM responses that ensue in response to that cell death.

On behalf of all contributing authors, I hope that this volume not only brings to the fore many of the varied known biological effects of cell death in tumours—both as part of their pathogenesis and of their responses to therapy—but also stimulates further discussion and research. At worst, the induction of cell death by anticancer therapies could play critical roles in relapse and evolution of more aggressive disease. At best, shifting the balance in tumour cell population dynamics from proliferation to death provides a route to disease-free survival. It seems likely that longevity in the latter will depend not only on inducing tumour cell death but also on inhibiting the consequent and perhaps inevitable pro-tumour responses. Undoubtedly, more targeted work in this area is required, and we hope that this volume will provide some inspiration towards this and that more effective cancer therapies that are applicable to a broad range of tumour types will ultimately emerge as a result.

Edinburgh, UK Christopher D. Gregory

Acknowledgements

Sincere thanks go to Karen Colvin without whose sterling administrative help this book would not have been possible.

Research in the Gregory laboratory is supported by Bloodwise and the Medical Research Council (UK).

Contents

Contributors

Claus Bachert The Upper Airway Research Laboratory, Department of Oto-Rhino-Laryngology, Ghent University Hospital, Ghent, Belgium

Bernhard Brüne Faculty of Medicine, Institute of Biochemistry I, Goethe-University Frankfurt, Frankfurt, Germany

Catherine Dabrowska School of Biosciences, University of Birmingham, Edgbaston, Birmingham, UK

Klaus-Michael Debatin Department of Pediatrics and Adolescent Medicine, University Medical Center Ulm, Ulm, Germany

Anja Derer Department of Radiation Oncology, Universitätsklinikum Erlangen, Friedrich-Alexander-Universität Erlangen-Nürnberg (FAU), Erlangen, Germany

Christopher P. Dillon Department of Immunology, St. Jude Children's Research Hospital, Memphis, TN, USA

Ian Dransfield MRC and University of Edinburgh Centre for Inflammation Research, Queen's Medical Research Institute, Edinburgh, UK

Yun Fan School of Biosciences, University of Birmingham, Edgbaston, Birmingham, UK

Sarah Farnworth MRC and University of Edinburgh Centre for Inflammation Research, Queen's Medical Research Institute, Edinburgh, UK

Rainer Fietkau Department of Radiation Oncology, Universitätsklinikum Erlangen, Friedrich-Alexander-Universität Erlangen-Nürnberg (FAU), Erlangen, Germany

Aaron M. Fond Center for Cell Clearance, and the Department of Microbiology, Immunology, and Cancer Biology, University of Virginia, Charlottesville, VA, USA

Catriona A. Ford MRC Centre for Inflammation Research, University of Edinburgh Queen's Medical Research Institute, Edinburgh, UK

Benjamin Frey Department of Radiation Oncology, Universitätsklinikum Erlangen, Friedrich-Alexander-Universität Erlangen-Nürnberg (FAU), Erlangen, Germany

Udo S. Gaipl Department of Radiation Oncology, Universitätsklinikum Erlangen, Friedrich-Alexander-Universität Erlangen-Nürnberg (FAU), Erlangen, Germany

Douglas R. Green Department of Immunology, St. Jude Children's Research Hospital, Memphis, TN, USA

Christopher D. Gregory MRC Centre for Inflammation Research, University of Edinburgh Queen's Medical Research Institute, Edinburgh, UK

Olga Krysko The Upper Airway Research Laboratory, Department of Oto-Rhino-Laryngology, Ghent University Hospital, Ghent, Belgium

Dmitri V. Krysko Molecular Signalling and Cell Death Unit, Inflammation Research Center, VIB, Ghent, Belgium

Deparment of Biomedical Molecular Biology, Ghent University, Ghent, Belgium

Mingli Li School of Biosciences, University of Birmingham, Edgbaston, Birmingham, UK

Nicolas Marschall Department of Pediatrics and Adolescent Medicine, University Medical Center Ulm, Ulm, Germany

Javier Mora Faculty of Medicine, Institute of Biochemistry I, Goethe-University Frankfurt, Frankfurt, Germany

Kodi S. Ravichandran Center for Cell Clearance, and the Department of Microbiology, Immunology, and Cancer Biology, University of Virginia, Charlottesville, VA, USA

Heike Scheithauer Department of Radiotherapy and Radiation Oncology, Ludwig Maximilian University Munich, Munich, Germany

Divya Sekar Faculty of Medicine, Institute of Biochemistry I, Goethe-University Frankfurt, Frankfurt, Germany

Shahzad Syed Faculty of Medicine, Institute of Biochemistry I, Goethe-University Frankfurt, Frankfurt, Germany

Katrien Vandecasteele Department of Radiation Oncology and Experimental Cancer Research, Ghent University, Ghent, Belgium

Peter Vandenabeele Molecular Signalling and Cell Death Unit, Inflammation Research Center, VIB, Ghent, Belgium

Deparment of Biomedical Molecular Biology, Ghent University, Ghent, Belgium

Methusalem program, Ghent University, Ghent, Belgium

Jorine J.L.P. Voss MRC Centre for Inflammation Research, University of Edinburgh Queen's Medical Research Institute, Edinburgh, UK

Andreas Weigert Faculty of Medicine, Institute of Biochemistry I, Goethe-University Frankfurt, Frankfurt, Germany

Mike-Andrew Westhoff Department of Pediatrics and Adolescent Medicine, University Medical Center Ulm, Ulm, Germany

Roland Wunderlich Research Unit Radiation Cytogenetics, Helmholtz Centre Munich, Neuherberg, Germany

Chapter 1
Molecular Cell Biology of Apoptosis and Necroptosis in Cancer

Christopher P. Dillon and Douglas R. Green

Abstract Cell death is a major mechanism to eliminate cells in which DNA is damaged, organelles are stressed, or oncogenes are overexpressed, all events that would otherwise predispose cells to oncogenic transformation. The pathways that initiate and execute cell death are complex, genetically encoded, and subject to significant regulation. Consequently, while these pathways are often mutated in malignancy, there is considerable interest in inducing cell death in tumor cells as therapy. This chapter addresses our current understanding of molecular mechanisms contributing to two cell death pathways, apoptotic cell death and necroptosis, a regulated form of necrotic cell death. Apoptosis can be induced by a wide variety of signals, leading to protease activation that dismantles the cell. We discuss the physiological importance of each apoptosis pathway and summarize their known roles in cancer suppression and the current efforts at targeting each pathway therapeutically. The intricate mechanistic link between death receptor-mediated apoptosis and necroptosis is described, as well as the potential opportunities for utilizing necroptosis in the treatment of malignancy.

Keywords Cell death • Apoptosis • Necroptosis • RIPK3 • BCL-2 • Caspases • Cancer • Inflammation • Immunogenic

1.1 Introduction

Cell death plays an important role in the maintenance and regulation of homeostasis in multicellular organisms. It is also a major mechanism by which the immune system can eliminate cells infected by pathogens or with potentially carcinogenic mutations. The pathways that initiate and execute cell death are complex, genetically encoded,

---

C.P. Dillon (✉) • D.R. Green (✉)
Department of Immunology, St. Jude Children's Research Hospital,
262 Danny Thomas Place, Memphis, TN 38105, USA
e-mail: christopher.dillon@stjude.org; douglas.green@stjude.org

C.D. Gregory (ed.), *Apoptosis in Cancer Pathogenesis and Anti-cancer Therapy*,
Advances in Experimental Medicine and Biology 930,
DOI 10.1007/978-3-319-39406-0_1

and subject to significant regulation. Mutations in genes that regulate cell death are quite common, resulting in either the elimination of pro-death proteins or the amplification of antideath proteins, a key step in the progression to cancer. This chapter will address our current understanding of molecular mechanisms contributing to two cell death pathways, apoptotic cell death and necroptosis, a regulated necrotic cell death. We discuss their physiological importance, their role in cancer suppression, and their potential use as targets for cancer therapy.

1.2 Role of Cell Death Pathways in Cancer Suppression

Cancer is believed to result from a series of accumulating mutations that result in the unregulated expansion of cells [1]. In landmark papers by Hanahan and Weinberg [2, 3], the multiple "hits" required to disable the essential cell processes that restrict proliferation and cause tumorigenesis are termed the "hallmarks" of cancer. One important hallmark is the evasion of apoptosis, a form of programmed cell death, to be discussed in detail later (Fig. 1.1). Evasion of apoptosis is often

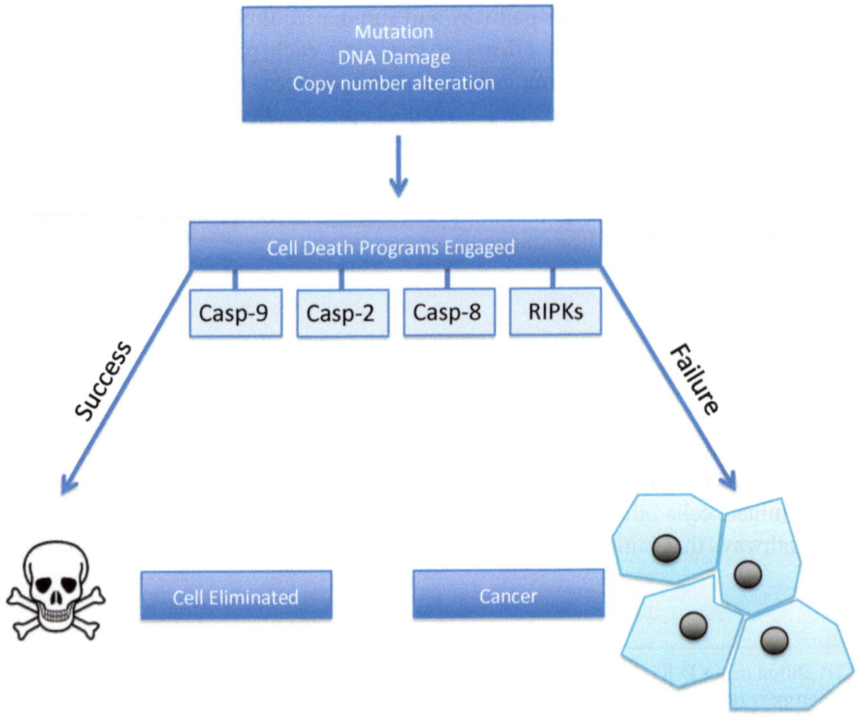

Fig. 1.1 Role of cell death programs in cancer. Various cellular insults, such as DNA damage, mutation, or changes in copy number drive activation of cell death programs, including cell death mediated by caspases-9,-2, and -8, as well as RIP kinases. Failure to engage or execute these pathways leads to cancer

construed as the complete elimination of cell death pathways in cancerous cells. But conceptually, a cell need only disrupt a death pathway specifically engaged by a tumor suppressor mechanism to elude that checkpoint. Indeed, merely dampening apoptotic responses is often sufficient to allow tumor progression, suggesting that cancer cells not only retain functionality in their cell death signaling pathways, but may be more "primed for death" than their normal counterparts. It is therefore possible that tumor cells could be driven to death by reversing the mechanisms that have dampened the death response and reactivating these death functions. On the other hand, a number of pathways of cell death with necrotic-like phenotypes have recently been described, and these might provide a new avenue to target to bypass the tumor cell's attempt to evade elimination. Here, we will address one such pathway, RIPK3-mediated necrosis, or necroptosis, and how it might be harnessed for therapy.

1.3 Apoptosis

One of the most studied and widely characterized forms of cell death is apoptosis. Known commonly as programmed cell death, apoptosis is the packaging of dying cells into fragments that are easily consumed and eliminated by phagocytes without disturbing the normal function of surrounding tissues [4]. This process is mediated by cysteine aspartate proteases known as caspases, has a stereotypical morphological appearance, and generally does not elicit an immune response. Apoptosis can be initiated through two separate pathways, the intrinsic or mitochondrial pathway and the extrinsic or death receptor pathway. Both pathways converge and utilize caspases as their downstream effectors of death, as these proteases have thousands of targets and orchestrate the final stages of apoptosis.

1.3.1 Caspases and Cell Death

The cysteine aspartate protease (caspase) family is split into multiple distinct subsets based on their specialized roles. Two subsets, known as effector and initiator caspases, are key regulators of apoptosis [5]. Effector caspases (such as caspase-3, -6, and -7 in mice and humans) are the mediators of cell destruction through the proteolysis of thousands of cellular substrates [6]. Cleavage of the inactive monomer forms of these caspases causes them to form active dimers and initiate cell death cascades [5]. Effector caspases are cleaved by the initiator caspases (caspase-2, -8, and -9 in rodents, with an additional caspase-10 in other vertebrates, including humans). Unlike effector caspases, initiator caspases are activated through conformational changes rather than cleavage [7]. Upon the integration of multiple upstream signals, initiator caspases are recruited via their long pro-domain to large macromolecular signaling platforms where they are dimerized and thereby activated [8]. Different signals and platforms regulate the activation of each individual initiator caspase, thus allowing each caspase to be specialized for different cellular tasks. Thus, the deleterious effects of caspase inhibition or deletion can vary significantly based on tissue type or pathogenic insults.

1.3.2 Intrinsic Cell Death

The intrinsic pathway of apoptosis is activated through a variety of intracellular signals ranging from DNA damage to oncogenic stress. Intrinsic cell death is most commonly associated with mitochondrial outer membrane permeabilization (MOMP) resulting in the activation of caspase-9 (Fig. 1.2) [9].

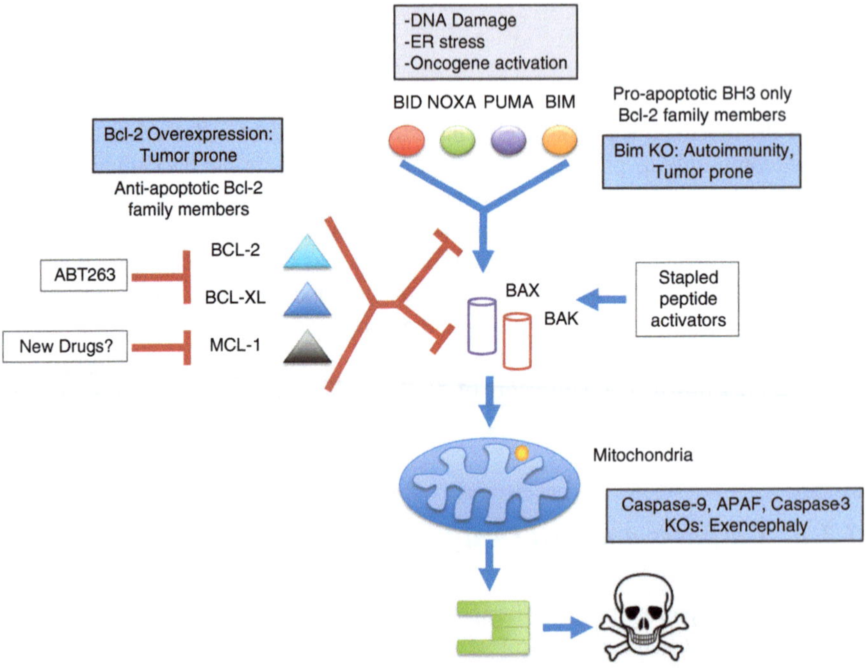

Fig. 1.2 Complex regulation of mitochondrial membrane integrity by BCL-2 family members. Intrinsic cell death driven by DNA damage, ER stress, or oncogene activation leads to the loss of mitochondria membrane permeability, release of cytochrome c, and the assembly of the apoptosome. The apoptosome, consisting of APAF1, cytochrome c, and caspase-9, serves as the activation platform for caspase-9, which in turn cleaves effector caspases and leads to the destruction and packaging of the cells for removal. BCL-2 family members regulate the mitochondrial membrane, with proapoptotic BH3-only family members being inhibited by antiapoptotics such as BCL-2, BCL-XL, and MCL-1. These antiapoptotics also directly inhibit the pore-forming proteins BAK and BAX. Therapeutics such as ABT263 target BCL-2 and BCL-XL and remove their inhibition of the pathway, while stapled peptides have been shown capable of directly binding and activating BAK and BAX

1.3.2.1 Mechanisms of Caspase-9-Mediated Apoptosis

Assembly of the caspase-9 activation platform, consisting of APAF1 and cytochrome c, is driven by intracellular signals such as DNA damage and oncogenic stress. These intracellular signals activate BAX and BAK, two BCL-2 family proteins, to form pores in the outer mitochondrial membrane, disrupting membrane integrity and releasing cytochrome c [10]. Once released from the mitochondrial intermembrane space, cytochrome c binds to APAF1 to promote the formation of a multimeric complex, termed the apoptosome, which mediates caspase-9 activation [9].

The activation of BAX and BAK, and thus MOMP, is regulated by the balance of proapoptotic and antiapoptotic BCL-2 family proteins [11]. The proapoptotic BH3-only proteins (e.g., BID, BIM, PUMA, and NOXA) contain one of four BCL-2 homology domains, and some of these can directly activate the pore-forming proteins, BAX and BAK [11]. The antiapoptotic proteins (BCL-2, BCL-XL, A1, and MCL-1) bind to and antagonize the BH-3-only proteins as well as directly act to inhibit active BAX and BAK (Fig. 1.2) [12]. Since different cellular insults induce specific BH3-only responses, and each antiapoptotic protein displays a differential regulation of each BH3-only protein family member, it may be important to target specific Bcl-2 proteins in certain contexts, such as restricting the effect of therapy to specific tissues, while activators or inhibitors of BAX and BAK may have a more universal effect [9], thus limiting their therapeutic window.

1.3.2.2 Physiological Functions of the Mitochondrial Apoptotic Pathway

Programmed cell death plays a role in the formation of the developing embryo. The importance of this process can be seen in mice deficient for components of the apoptotic pathway. Caspase-9, caspase-3, and APAF1 knockout mice have similar embryonic phenotypes, developing exencephaly and cranioschesis from a failure to properly close the neural tube during development in some genetic backgrounds [13–17]. In contrast, the combined loss of the pore-forming proteins BAX and BAK leads to several developmental defects, including persistence of interdigital webs, an imperforate vaginal canal, and excess cell accumulation in both the central nervous and hematopoietic systems [18]. Specific deletion of BAX and BAK from B cells or hematopoietic stem cells results in the development of aggressive autoimmune disorders in mice, suggesting that these proteins also play an important role in regulation of the immune system [19, 20].

Germline ablation of BIM, a proapoptotic BH3-only BCL-2 family member, leads to similar expansion of lymphoid tissues as seen in BAX and BAK deletion [21], while ablation of BID results in a myeloproliferative disorder in aging animals [22]. NOXA and PUMA knockout mice appear largely normal; however, cells from these animals placed under some stress conditions, such as DNA damage, have a survival advantage compared to WT controls [23–25].

Germline deletion of antiapoptotic BCL-2 results in widespread apoptosis and lethality shortly after birth, whereas both BCL-XL and MCL-1 knockouts are embryonic lethal [26–28]. Conditional mouse models show that MCL-1 is vital for the survival of all hematopoietic cell types with the exception of monocytes during development [29–31]. Conversely, the roles of BCL-XL and BCL-2 in hematopoiesis are more limited, with BCL-XL functioning primarily in megakaryocytes and erythrocytes and BCL-2 in lymphoid cell types [32, 33]. The variation of phenotypes seen in both these germline and conditional knockout animals suggests that diverse tumor types may be differentially responsive to therapeutics targeted at the Bcl-2 family.

1.3.2.3 Mitochondrial Apoptotic Pathway in Cancer and Cancer Therapy

In the formation of hematological tumors, antiapoptotic BCL-2 proteins have been shown to synergize with expression of the MYC oncogene [34–36]. Mice with T cells deficient in BAX and BAK develop an aggressive form of acute lymphoblastic leukemia [37] which correlates with evidence that BAX is mutated in a number of human hematological malignancies [38]. MCL-1 contributes to the development of acute myeloid leukemia [39–41], which is consistent with somatic copy-number alterations analysis, which found amplifications of both MCL-1 and BCL-2 in many cancer types [42]. Overexpression of BCL-2, BCL-XL, and MCL-1 has been found in breast, lung, prostate, colorectal, gastric, renal, hepatocellular, and pancreatic cancer [43–45], and transgenic expression of BCL-2 or BCL-XL synergizes with oncogenes to drive solid tumors in mice [46–48]. These results suggest that targeting of BCL-2 family members could prove effective for treatment for a wide variety of tumors [49].

How can potential therapies drive cell death in tumor cells? When a cell encounters unregulated oncogene expression, the signals of transformation generally trigger an apoptotic response; however, if a cell is incapable of undergoing apoptosis, due to the high expression of antiapoptotic proteins, the cell does not die. Importantly, these tumor cells are "primed to die" because the antiapoptotic proteins are occupied with proapoptotic proteins such as BIM [50]. Therefore, although tumorigenic, these cells can be easily targeted with conventional chemotherapy and exhibit an enhanced sensitivity toward death. In fact, patient tumor cells can be screened in vitro for this "priming" to predict the cytotoxic effects of chemotherapy drugs that might be most effective at reducing tumor burden [51].

A novel therapy approach that takes advantage of priming is the use "BH-3 mimetics" to directly simulate the activation of these pathways and induce apoptosis [52]. Using a structure–activity relationship design paradigm to target protein–protein interactions rather than enzymatic active sites, the drug ABT-263 (navitoclax), a BH3 mimetic, has had great success in early clinical trials [53–55], including a recent Phase II trial [56]. However, ABT-263 is not specific, binding to BCL-2, BCL-XL, and BCL-W (but not MCL-1) and patients treated with this drug develop thrombocytopenia as a result of the loss of BCL-XL-dependent platelets (Fig. 1.2) [57, 58]. A newer version of this compound (ABT-199, venetoclax) eliminates thrombocytopenia by specifically targeting BCL-2 [59]. Upregulation of MCL-1 is a common mechanism of acquired resistance to therapy in tumors [60], which may limit the use of these BH3-mimetic compounds as single agents [61]. Despite early specificity challenges [62], recent efforts to directly target MCL-1 have yielded compounds that may work effectively as single agents or in combination with ABT-263 [63, 64]. However, the broad requirement for MCL-1 across hematopoietic cells could render it difficult to find a therapeutic window for MCL-1 inhibitors [65].

Another approach to targeting the BCL-2 family is "stapled" peptides, short peptides capable of retaining their secondary structure, which are being investigated as small molecule vehicles for therapy [66]. Not only are they effective for activating BAX [67, 68], but stapled peptides also appear to bypass MCL-1 dependent acquired resistance in leukemic cells [69, 70].

BH-3 mimetics have been used to kill a broad variety of tumor cells in vitro, including chronic lymphocytic leukemia, lymphoma, lung, breast, and colorectal cancer cells [45, 71, 72]. Although single agent use of BH3-mimetics appears most effective in the hematological tumor cells, combinatorial treatment of solid tumors with existing anticancer agents worked well [73]. To date, the clinical trials of ABT-263 and ABT-199 have focused on both single agent and combination therapy for a variety of neoplasms, including a Phase 3 trial of ABT-199 in combination with rituximab for chronic lymphocytic leukemia [49]. The progression of these drugs to clinical trials has accelerated the development of new therapeutics targeting the mitochondrial pathway [74], which hopefully will expand their usage quickly to solid tumors in the clinic.

1.3.2.4 Mechanisms of Caspase-2-Mediated Apoptosis

While caspase-2, the most evolutionarily conserved caspase, is thought to function in apoptosis [75, 76], its mechanism of activation, downstream targets, and physiological role is still a matter of debate. Caspase-2 appears to play a role in cell death following DNA damage [77]. Insults causing double-stranded DNA breaks lead to activation of ATM/ATR, then p53, and ultimately cell death (Fig. 1.3) [78]. PIDD (p53-induced death domain) is a transcriptional target of p53 and a key member of the PIDDosome signaling complex that activates caspase-2 [78]. In the PIDDosome, PIDD recruits RAIDD, which in turn recruits caspase-2 through a protein–protein interaction domain on each protein known as a CARD domain [78]. Binding to the CARD domain of RAIDD puts caspase-2 into the proper conformation for activation [79–81]. In response to heat shock or endoplasmic reticulum stress, caspase-2 cleaves a number of downstream effectors, including BID, which permeabilizes the mitochondrial outer membrane [82, 83]. Recently, it has been suggested that caspase-2 plays a role in regulating genomic stability [84] and that the PIDDosome determines cell fate by integrating signals from DNA damage and mitotic checkpoints to activate caspase-2 [85].

The mechanism for activation of caspase-2 is still a matter of ongoing investigation. Recent evidence suggests that PIDD and RAIDD may not be required for activation of caspase-2 in all contexts [86, 87], yet the alternative platform for its activation in

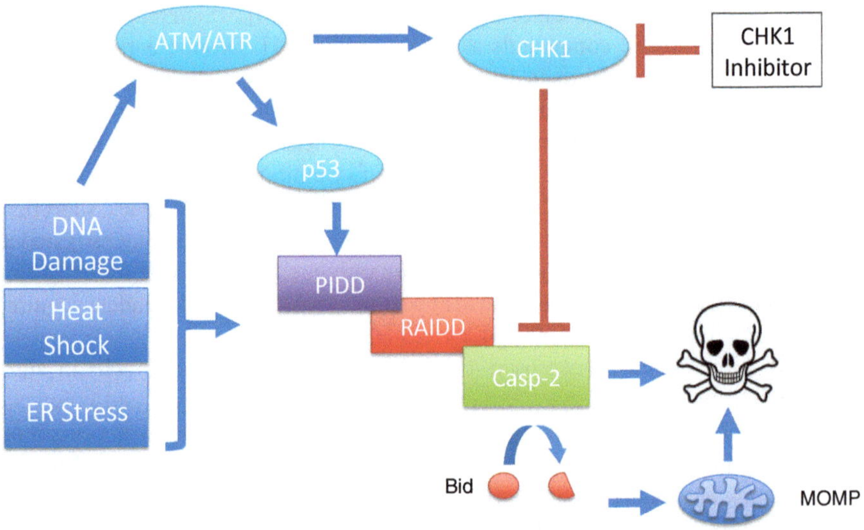

Fig. 1.3 DNA damage, heat shock, or ER stress can lead to activation of caspase-2-mediated cell death. DNA damage activates ATM/ATR, stabilizes p53, causing the upregulation of PIDD. PIDD in turn recruits RAIDD forming the activation platform for caspase-2. ATM/ATR signaling can drive caspase-2 activation in the absence of p53, but only when Chk1 is inhibited. The link between heat shock and ER stress and the assembly of the caspase-2 activation platform is unclear, and in at least some instances activation of caspase-2 may be PIDD or RAIDD independent. Targeting Chk via inhibitors might prove a therapeutically useful way to activate this pathway in cancer

these contexts is still unknown. In the absence of p53, the inactivation of Chk1 activates caspase-2 resulting in cell death [88]; however, reports conflict as to whether this signaling is dependent on PIDD [89].

1.3.2.5 Physiological Functions of the Caspase-2 Apoptosis Pathway

Initial investigations found no gross phenotype in murine knockouts of caspase-2, RAIDD, or PIDD [80, 86, 90]. Recent studies observed premature aging and oxidative stress in caspase-2 deficient animals [91–93]. Caspase-2 also appears to aid in the maintenance of bone homeostasis through elimination of damaged osteoclasts [94]. Together, these findings suggest that caspase-2 may play an important role in the elimination of transformed cells [95].

1.3.2.6 Caspase-2-Mediated Apoptosis in Cancer and Cancer Therapy

The role of caspase-2 in malignancy is still ambiguous. While the additional loss of caspase-2 does not affect the progression of tumors in p53 knockout mice [96, 97], caspase-2 does seem to influence disease development in other tumor models.

Caspase-2, ATM double knockout animals develop tumors more rapidly than ATM knockout mice alone [96, 98, 99]. Mice bearing the Eμ-Myc transgene together with the ablation of caspase-2 also develop spontaneous tumors faster than animals that only express the Eμ-Myc transgene. In acute lymphoblastic leukemia, decreased levels of caspase-2 are associated with a poorer clinical prognosis [100]. Caspase-2 loss has also been shown to accelerate the development of mammary tumors in mice driven by MMTV/c-neu, as well as lung tumors driven by K-ras [101, 102]. Interestingly, caspase-2 ablation appears to delay the onset of TH/N-MYC-driven neuroblastoma, an observation at odds with its role in other tumor types [103]. The tumor suppressive effects of caspase-2 may be apoptotic in nature [104], as mutating the catalytic sites of caspase-2 eliminates this phenotype.

Activation of caspase-2 in a therapeutic setting might be achieved through either targeting of ATM/ATR or Chk-1 in the DNA damage response (DDR) (Fig. 1.3). Recent efforts have focused on Chk-1 inhibitors as either single agents or part of a combination therapy [105, 106]. While there is a significant interest in ATM/ATR small molecule inhibitors as they move from preclinical to clinical testing phases, enhancing the DDR to kill off transformed cells could prove more effective if the caspase-2-mediated apoptosis pathway could be selectively activated [107].

1.3.3 Extrinsic Cell Death

Extracellular cues, often acting through cell surface receptors in the TNFR superfamily (such as TNFR, CD95/FAS, and TRAIL, receptors DR4 and DR5), activate the extrinsic apoptotic pathway, leading to the recruitment and activation of caspase-8.

Caspase-8 (and caspase-10, a homolog in humans) is activated by extracellular signals through the ligation of a family of proteins known as death receptors. Death receptors are a subset of the tumor necrosis factor receptor (TNFR) superfamily and include CD95, DR4, DR5, and TNFR [108]. When engaged by their respective ligands (CD95L, also known as FAS ligand for CD95; TRAIL for DR4 and DR5; and TNF for TNFR), these death receptors recruit the adaptor molecule Fas-Associated protein with Death Domain (FADD) (Fig. 1.4) [108]. Once FADD is combined with the death receptor (or in the case of TNFR after both FADD and a second adaptor molecule, TRADD, are recruited), inactive monomers of caspase-8 are recruited and dimerized into active caspase-8 homodimers [109]. The activating platform formed by the combination of the death receptor, adaptor(s), and caspase-8 is referred to as the death-inducting signaling complex (DISC) [110]. Once activated, the caspase-8 homodimer is cleaved, releasing it from the DISC, and this can cleave and thereby activate caspase-3 and -7 to promote apoptosis [109, 111]. While cleavage alone does not activate caspase-8 [112, 113], noncleavable versions of caspase-8 do not initiate apoptosis through the cleavage of caspase-3 [109]. Caspase-8 can also form a heterodimer with its catalytically inactive homolog c-FLIP, which competitively blocks the formation of the caspase-8 homodimer and

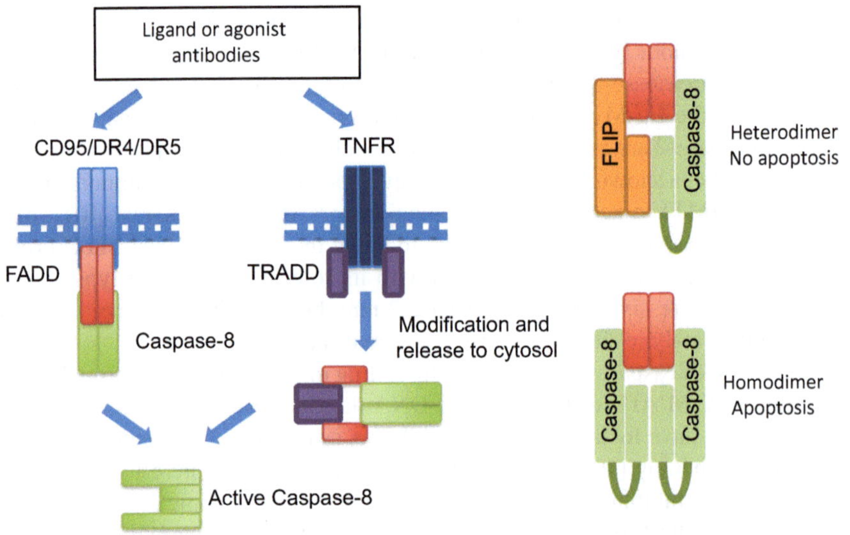

Fig. 1.4 Extrinsic cell death is mediated by caspase-8. Ligation of the death receptor by its ligand leads to receptors trimerization and recruitment of adaptor molecules FADD, and in the case of TNFR, TRADD. These in turn recruit caspase-8, putting it into the proper conformation for activation. In the case of CD95 and DR4/5, caspase-8 self-cleaves releasing a homodimer of caspase-8 from the receptor to the cytosol allowing it to cleave its downstream effector caspases and induce cell death. For TNFR, caspase-8 is recruited directly to a cytoplasmic complex for activation. The catalytically inactive caspase-8 homolog FLIP binds together with caspase-8 (heterodimer) at the receptor preventing apoptosis from occurring. Agonist antibodies have been used to induce caspase-8-mediated cell death in therapeutic settings

the resultant activation of caspase-3 [114]. As caspase-8 in the heterodimer with FLIP retains catalytic activity, it remains unclear why the heterodimer is incapable of inducing apoptosis [114].

1.3.3.1 Physiological Functions of the Caspase-8 Apoptosis Pathway

Animals ablated for caspase-8 perish early in embryogenesis [115], and therefore it has been difficult to elucidate the role of caspase-8 in normal tissue homeostasis. While the cause of this early lethality is now understood (as discussed later), the exact tissues in which caspase-8 is important for homeostatic regulation remain unclear and is confounded by its role in regulating necroptosis. Conditional ablation of caspase-8 in the gut or skin results in animals that die as juveniles [116, 117], while deletion of caspase-8 from the liver, heart, muscle, or brain does not appear to have major effects on development [118–120]. Deletion of caspase-8 from the liver, however, does protect against lethality induced by death ligands such as CD95L, which in WT animals causes rapid death due to apoptotic loss of hepatocytes and significant liver damage [118]. Deletion of caspase-8 does not appear to affect development of lymphoid cells such as T and B cells, but does significantly affect their lifespan in the periphery after activation [121–123].

1.3.3.2 Caspase-8-Mediated Apoptosis in Cancer and Cancer Therapy

The role of caspase-8 in tumor pathology appears to vary significantly by tumor type. In some models, loss of caspase-8 appears to promote oncogenesis. In the TH/N-MYC model of neuroblastoma, caspase-8 loss appears to accelerate tumor progression, an effect that was attributed to a potential role for caspase-8 in inhibiting cell motility [124, 125]. Caspase-8 is also frequently deleted in small cell lung carcinoma, medulloblastoma, glioma, gastric, and hepatocellular carcinomas, providing further support of its tumor limiting function [126–131]. Conversely, mutations in the caspase-8 promoter that reduce caspase-8 expression have been associated with decreased incidence of lung, esophageal, gastric, colorectal, cervical, and breast cancers, suggesting that caspase-8 could function as a tumor promoter in these contexts [132, 133]. These paradoxical and opposing roles for caspase-8 in tumorigenesis are confusing, but may now make sense in light of recent work showing a role for caspase-8 in limiting necroptosis (discussed later).

If an appropriate therapeutic window can be identified, caspase-8 could function as a therapeutic target in two possible ways. In those tissues where reduction of caspase-8 has been associated with reduced tumor progression, caspase inhibitors could have valuable clinical applications if they can be targeted appropriately, as second-generation caspase inhibitors have very few side effects [134]. However, in situations where the engagement of death receptor-mediated apoptosis does play an important role in limiting tumor growth, the application of caspase inhibitors could be detrimental and instead we would want to administer therapeutics that engage the death receptors, possibly in combination with drugs that could also reduce FLIP expression or stability. The use of agonistic drugs that target death receptors is complicated by the pro-survival roles of these pathways. In fact, there is evidence that tonic low levels of TNF or CD95 signaling may be important for the survival of certain cancer cell types (Fig. 1.4) [135, 136]. Despite these complications, clinical trials have utilized agonists to target TRAIL with initial success in preferentially killing malignant cells over normal cells, suggesting the possibility of a therapeutic window for the use of such agonists [137]. New work linking the activation of apoptotic pathways to both programmed necrosis and inflammatory signaling further complicates the targeting of these death receptors for clinical applications.

1.3.4 RIPK-Mediated Necroptosis: A New Target for Tumor Therapy?

The early embryonic death and pale phenotype of caspase-8 null mouse embryos puzzled the programmed cell death field for many years [115], since there was no evidence of excess cell number resulting from blocking apoptosis as predicted. Early evidence to explain this phenotype came from the observation that inhibition of caspase-8 increased levels of cell death in certain types of cells in vitro [138]. It was later demonstrated that instead of apoptosis, the observed death had features of necrosis and could be inhibited by necrostatin-1, a drug that targets

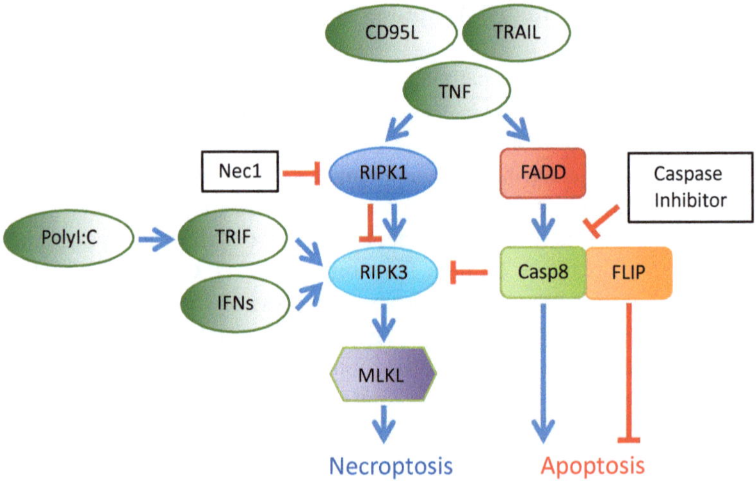

Fig. 1.5 Necroptosis is a programmed form of necrotic cell death. RIPK3, through its downstream target, MLKL is capable of inducing necrotic cell death. RIPK3 can be activated by RIPK1 via the ligation of death receptors by TNFα, CD95L, or TRAIL. Alternatively, innate immune signaling via TRIF or interferons can activate RIPK3. A heterodimer of caspase-8 and FLIP negatively regulates RIPK3 activation. Necroptosis may be induced therapeutically via a combination of death receptor agonists and caspase inhibitors. Because necroptosis is a more inflammatory form of cell death than apoptosis, induction of necroptosis may be used as an adjuvant to enhance antitumor immune responses

Receptor-Interacting serine/threonine-Protein Kinase 1 (RIPK1) [139, 140]. These findings led to the recognition of a new form of programmed necrotic death mediated by RIP kinases called necroptosis. Importantly, it was further discovered that the extrinsic apoptotic pathway was intricately connected to the necroptotic pathway, with TNFR signaling and caspase 8 activation at the forefront of signaling events driving or inhibiting necroptosis.

1.3.4.1 Mechanisms of Necroptosis

Necroptosis can be actively inhibited by components of the death receptor-mediated apoptotic pathway, including caspase-8, FLIP, and FADD. Similar to extrinsic apoptosis, it is induced through ligation of a death receptor such as TNFR1 (Fig. 1.5). Upon its ligation, RIPK1 is recruited to the death receptor as part of complex 1. RIPK1 is deubiquitinated and then released from complex 1 into the cytosol where it activates its homolog RIPK3 by binding through a protein–protein interaction domain known as the RIP Homotypic Interaction Motif (RHIM) domain [141]. Active RIPK3 subsequently recruits and activates the pseudokinase MLKL through phosphorylation. MLKL is then targeted to the plasma membrane where it appears to generate pores, causing the rapid death of the cell [142–148]. Necroptosis is blocked by caspase-8 activity, although the exact mechanism by which this occurs is not clear. RIPK3 activation is counteracted by the caspase-8: FLIP heterodimer, which is known to have catalytic activity [149]. It is plausible that the inhibition of necroptosis by caspase-8 is the result of the cleavage of RIPK1 or RIPK3 by this heterodimer; however, this remains to be formally demonstrated [114, 149, 150].

The cross talk between death receptor-mediated apoptosis and necroptosis presents difficulties for the design of therapies, as the modulation of one pathway could have unforeseen effects on the other. In particular, RIPK3 functions not only to drive necroptosis but can also activate apoptosis through a complex known as the ripoptosome [151, 152] in a manner that is independent of its kinase activity [153]. The specific cellular contexts that drive the necroptotic versus apoptotic functions of RIPK3 are still being elucidated. Additionally, several forms of programmed necrosis have now been described, and it is possible that there may be compensation and/or coregulation between these pathways [154]. For example, protection against ischemic reperfusion injury requires the inhibition of programmed necrosis mediated through both cyclophilin D and RIPK3 [155].

In addition to induction by death receptors, necroptosis can also be stimulated by a number of other signals. For example, RIPK3 can be activated directly by TRIF, an integrator of innate immune signals, presumably through its RHIM domain [156]. Interferons also activate RIPK3 through an unknown mechanism thought to depend on protein kinase R (PKR) [157]. Adding to the complexity of this pathway is the fact that RIPK1 can both activate and inhibit necroptosis depending on context, and inhibition of RIPK1 and RIPK3 may have opposing effects, suggesting that targeting necroptosis in pathophysiological contexts is extremely complicated [158].

Necroptosis as a form of cell death is distinct from apoptosis in its ability to induce inflammation. Caspase-8, RIPK1, and RIPK3 all seem to play a role in the production of cytokines either directly (in the case of RIPK1 and RIPK3) or through the inflammasome (in the case of caspase-8) [159]. Caspase-1 and caspase-8 appear to work in parallel to produce IL-1β in some contexts [160] and caspase-8 may somehow facilitate the activation of caspase-1 in response to both canonical and noncanonical activators of the NRLP3 inflammasome as well as inflammasome activation in response to *Yersinia* [161, 162]. It has been difficult, however, to ascertain whether cell death per se and the production of damage-associated molecular patterns (DAMPs), or the cytokines produced via these pathways are responsible for the observed inflammation [163]. It has also been suggested that killing cells via necroptosis may be an effective way to limit the production of DAMPs as it quickly eliminates the damaged cells [164]. The regulation of necroptosis, and how it might interact with extrinsic apoptosis, remains an area of intense research, and new knowledge will have implications for targeting this pathway in cancer.

1.3.4.2 Physiological Role of RIPK-Mediated Necroptosis

Necroptosis plays an important role in development under some conditions of genetic perturbation. Ablation of RIPK3 from caspase-8 null embryos completely rescues these animals, preventing the early lethality and resulting in grossly normal adult mice [149, 165]. The embryonic lethality in *casp8*$^{-/-}$ mice is also RIPK1 dependent, which suggests that the primary role of caspase-8 in development is to restrain necroptosis [149, 165–168]. The loss of FADD or FLIP also yields the same embryonic phenotype as *casp8*$^{-/-}$ mice [169, 170]. The lethality in FADD-deficient embryos can be rescued by concurrent ablation of RIPK1 or RIPK3; however, FLIP-deficient animals only survive when both RIPK3 and FADD are also deleted [120, 171].

Interestingly, RIPK3 and MLKL knockout mice appear grossly normal unless challenged in pathogenic contexts [145, 172]. In contrast, *ripk1*$^{-/-}$ mice can be rescued from perinatal lethality by simultaneous ablation of caspase-8 and RIPK3 [166–168, 173], suggesting that RIPK1 is actually *required* for the inhibition of some forms of RIPK3-dependent death, including necroptosis triggered through TRIF or interferons [166].

The conditional loss of caspase-8, FADD, or RIPK1 causes pathology in the development and maintenance of the endothelium, intestine, and skin [116–118, 174–176], and the loss of RIPK1 also adversely impacts hematopoiesis [177].

Necroptosis therefore appears to play an important role in the development and maintenance of barrier tissues, whose high turnover and exposure to external insults make them prone to development of malignancy.

1.3.4.3 Necroptosis in Cancer and Cancer Therapy

Necroptosis may act as a tumor suppressor in specific cells and certain types of cancers. In chronic lymphocytic leukemia, CYLD, which deubiquitinates RIPK1 to activate necroptosis, is downregulated, preventing normally elimination of these tumorigenic cells [178]. In a study examining inflammatory carcinogenesis, researchers found that RIPK3 inhibited tumor growth, although the authors believe this was due to limitation of caspase-8-induced compensatory proliferation rather than activation of necroptosis per se [179]. To date, little direct evidence for necroptosis-mediated tumor suppression has been found and functioning necroptotic pathways have also been found in some, but not all, tumor cells in vitro [180–182]. In fact, inducing necroptosis has been found effective in killing pancreatic and ovarian cancer cells [137, 183].

In cancers where the elimination or reduction of caspase-8 reduces tumor burden [132], it is possible that this effect is the result of increased necroptotic activity, a hypothesis that remains to be formally tested. In support of this hypothesis, in vitro experiments have shown that caspase inhibition primes colon cancer cells for necroptotic death (Fig. 1.5) [184]. However, another group found that caspase inhibition did not alter the cell death response in colon cancer cells upon treatment with traditional chemotherapeutics, such as doxorubicin or etoposide [185]. Cross talk between apoptosis and necroptosis could therefore vary between cancer type or even within cancers of the same tissue origin [186].

Given that at least some tumors have intact necroptosis pathways, it is possible that activating necroptosis could be a fruitful therapeutic strategy for some cancers [187], especially those where caspase-8 apoptosis has been eliminated through either downregulation of the surface death receptor or loss of caspase-8 expression [188]. Since death receptor activation is important for both apoptotic and necroptotic cell death, targeting the downstream effectors of necroptosis, such as RIPK3 or MLKL, eliminates the requirement for efficient TNFR signaling. Screens for drugs that directly activate RIPK3 or MLKL are underway.

Necroptosis could also be used as an adjuvant for tumor therapy to stimulate a more immunogenic form of cell death and therefore recruit activated immune cells to the diseased tissues. Activating necroptosis in tumor cells via polyI:C in immune-competent, but not immune-deficient, hosts substantially reduced tumor burden [189]. This could be explained by the fact that, in addition to the production of immune-stimulatory cytokines released by dying cells, RIPK1 has been shown to be essential to cross-priming of CD8 cells for an efficient immune response [190].

1.3.5 Perspectives

Our understanding of the cellular processes that contribute to cancer has increased substantially since Hanahan and Weinberg first detailed the "hallmarks" of cancer, including emerging ideas about tumor metabolism, inflammatory microenvironments, and the mechanisms behind the failure to activate an anticancer immune response. Even original hallmarks such as evasion of cell death have been reevaluated in the context of the last decade of research. Rather than broadly eliminating all cell death pathways, we now understand that tumors mutate or dampen only the pathways which are specifically engaged by tumor suppressive mechanisms, leaving intact the ability to induce cell death by unimpaired alternative pathways or to reactivate the dampened pathway. Furthermore, the identification of previously unknown mechanisms of cell death has elucidated new molecular targets for drug discovery. New work utilizing both functional profiling and deep sequencing methods has focused on accelerating the ability to determine which death pathways are intact in patient tumor cells, ushering an era of personalized medicine. Thus, one can foresee a future treatment paradigm where in vitro screening of patient tumors might be able to identify which among many pathways of cell death are still intact, and predict the response to therapies that target one or more of these pathways while minimizing the possibility of acquired resistance. Reaching this point will require continued efforts at the basic level to find and describe the molecular details of cell death pathways.

Acknowledgments We would like to thank Heather Dillon and Katherine Baran for their critical reading of this manuscript.

References

1. Armitage P, Doll R. The age distribution of cancer and a multi-stage theory of carcinogenesis. Br J Cancer. 1954;8(1):1–12.
2. Hanahan D, Weinberg RA. Hallmarks of cancer: the next generation. Cell. 2011;144(5): 646–74.
3. Hanahan D, Weinberg RA. The hallmarks of cancer. Cell. 2000;100(1):57–70.
4. Taylor RC, Cullen SP, Martin SJ. Apoptosis: controlled demolition at the cellular level. Nat Rev Mol Cell Biol. 2008;9(3):231–41.
5. Ashkenazi A, Salvesen G. Regulated cell death: signaling and mechanisms. Annu Rev Cell Dev Biol. 2014;30:337–56.
6. Stroh C, Schulze-Osthoff K. Death by a thousand cuts: an ever increasing list of caspase substrates. Cell Death Differ. 1998;5(12):997–1000.
7. MacKenzie SH, Clark AC. Death by caspase dimerization. Adv Exp Med Biol. 2012;747:55–73.
8. Mace PD, Riedl SJ. Molecular cell death platforms and assemblies. Curr Opin Cell Biol. 2010;22(6):828–36.
9. Tait SW, Green DR. Mitochondrial regulation of cell death. Cold Spring Harb Perspect Biol. 2013;5(9). pii: a008706.
10. Tait SW, Green DR. Mitochondria and cell death: outer membrane permeabilization and beyond. Nat Rev Mol Cell Biol. 2010;11(9):621–32.
11. Chipuk JE, et al. The BCL-2 family reunion. Mol Cell. 2010;37(3):299–310.
12. Llambi F, et al. A unified model of mammalian BCL-2 protein family interactions at the mitochondria. Mol Cell. 2011;44(4):517–31.
13. Honarpour N, et al. Adult Apaf-1-deficient mice exhibit male infertility. Dev Biol. 2000;218(2):248–58.
14. Houde C, et al. Caspase-7 expanded function and intrinsic expression level underlies strain-specific brain phenotype of caspase-3-null mice. J Neurosci. 2004;24(44):9977–84.
15. Leonard JR, et al. Strain-dependent neurodevelopmental abnormalities in caspase-3-deficient mice. J Neuropathol Exp Neurol. 2002;61(8):673–7.
16. Hakem R, et al. Differential requirement for caspase 9 in apoptotic pathways in vivo. Cell. 1998;94(3):339–52.
17. Woo M, et al. Essential contribution of caspase 3/CPP32 to apoptosis and its associated nuclear changes. Genes Dev. 1998;12(6):806–19.
18. Lindsten T, et al. The combined functions of proapoptotic Bcl-2 family members bak and bax are essential for normal development of multiple tissues. Mol Cell. 2000;6(6):1389–99.
19. Takeuchi O, et al. Essential role of BAX, BAK in B cell homeostasis and prevention of autoimmune disease. Proc Natl Acad Sci U S A. 2005;102(32):11272–7.
20. Rathmell JC, et al. Deficiency in Bak and Bax perturbs thymic selection and lymphoid homeostasis. Nat Immunol. 2002;3(10):932–9.
21. Bouillet P, et al. Proapoptotic Bcl-2 relative Bim required for certain apoptotic responses, leukocyte homeostasis, and to preclude autoimmunity. Science. 1999;286(5445):1735–8.
22. Zinkel SS, et al. Proapoptotic BID is required for myeloid homeostasis and tumor suppression. Genes Dev. 2003;17(2):229–39.
23. Michalak EM, et al. In several cell types tumour suppressor p53 induces apoptosis largely via Puma but Noxa can contribute. Cell Death Differ. 2008;15(6):1019–29.
24. Villunger A, et al. p53- and drug-induced apoptotic responses mediated by BH3-only proteins puma and noxa. Science. 2003;302(5647):1036–8.
25. Shibue T, et al. Integral role of Noxa in p53-mediated apoptotic response. Genes Dev. 2003;17(18):2233–8.
26. Rinkenberger JL, et al. Mcl-1 deficiency results in peri-implantation embryonic lethality. Genes Dev. 2000;14(1):23–7.
27. Veis DJ, et al. Bcl-2-deficient mice demonstrate fulminant lymphoid apoptosis, polycystic kidneys, and hypopigmented hair. Cell. 1993;75(2):229–40.

28. Motoyama N, et al. Massive cell death of immature hematopoietic cells and neurons in Bcl-x-deficient mice. Science. 1995;267(5203):1506–10.
29. Opferman JT, et al. Obligate role of anti-apoptotic MCL-1 in the survival of hematopoietic stem cells. Science. 2005;307(5712):1101–4.
30. Opferman JT, et al. Development and maintenance of B and T lymphocytes requires anti-apoptotic MCL-1. Nature. 2003;426(6967):671–6.
31. Dzhagalov I, John AS, He YW. The antiapoptotic protein Mcl-1 is essential for the survival of neutrophils but not macrophages. Blood. 2007;109(4):1620–6.
32. Perciavalle RM, Opferman JT. Delving deeper: MCL-1's contributions to normal and cancer biology. Trends Cell Biol. 2013;23(1):22–9.
33. Hildeman D, et al. Apoptosis and the homeostatic control of immune responses. Curr Opin Immunol. 2007;19(5):516–21.
34. Strasser A, et al. Novel primitive lymphoid tumours induced in transgenic mice by cooperation between myc and bcl-2. Nature. 1990;348(6299):331–3.
35. Beverly LJ, Varmus HE. MYC-induced myeloid leukemogenesis is accelerated by all six members of the antiapoptotic BCL family. Oncogene. 2009;28(9):1274–9.
36. McDonnell TJ, et al. bcl-2-immunoglobulin transgenic mice demonstrate extended B cell survival and follicular lymphoproliferation. Cell. 1989;57(1):79–88.
37. Biswas S, et al. A role for proapoptotic Bax and Bak in T-cell differentiation and transformation. Blood. 2010;116(24):5237–46.
38. Meijerink JP, et al. Hematopoietic malignancies demonstrate loss-of-function mutations of BAX. Blood. 1998;91(8):2991–7.
39. Campbell KJ, et al. Elevated Mcl-1 perturbs lymphopoiesis, promotes transformation of hematopoietic stem/progenitor cells, and enhances drug resistance. Blood. 2010;116(17):3197–207.
40. Glaser SP, et al. Anti-apoptotic Mcl-1 is essential for the development and sustained growth of acute myeloid leukemia. Genes Dev. 2012;26(2):120–5.
41. Xiang Z, et al. Mcl1 haploinsufficiency protects mice from Myc-induced acute myeloid leukemia. J Clin Invest. 2010;120(6):2109–18.
42. Beroukhim R, et al. The landscape of somatic copy-number alteration across human cancers. Nature. 2010;463(7283):899–905.
43. Kirkin V, Joos S, Zornig M. The role of Bcl-2 family members in tumorigenesis. Biochim Biophys Acta. 2004;1644(2–3):229–49.
44. Belmar J, Fesik SW. Small molecule Mcl-1 inhibitors for the treatment of cancer. Pharmacol Ther. 2015;145:76–84.
45. Delbridge AR, Strasser A. The BCL-2 protein family, BH3-mimetics and cancer therapy. Cell Death Differ. 2015;22(7):1071–80.
46. Williams MM, Cook RS. Bcl-2 family proteins in breast development and cancer: could Mcl-1 targeting overcome therapeutic resistance? Oncotarget. 2015;6(6):3519–30.
47. Fedorov LM, et al. Bcl-2 determines susceptibility to induction of lung cancer by oncogenic CRaf. Cancer Res. 2002;62(21):6297–303.
48. Pena JC, Rudin CM, Thompson CB. A Bcl-xL transgene promotes malignant conversion of chemically initiated skin papillomas. Cancer Res. 1998;58(10):2111–6.
49. Correia C, et al. Emerging understanding of Bcl-2 biology: implications for neoplastic progression and treatment. Biochim Biophys Acta. 2015;1853(7):1658–71.
50. Certo M, et al. Mitochondria primed by death signals determine cellular addiction to anti-apoptotic BCL-2 family members. Cancer Cell. 2006;9(5):351–65.
51. Montero J, et al. Drug-induced death signaling strategy rapidly predicts cancer response to chemotherapy. Cell. 2015;160(5):977–89.
52. Vela L, Marzo I. Bcl-2 family of proteins as drug targets for cancer chemotherapy: the long way of BH3 mimetics from bench to bedside. Curr Opin Pharmacol. 2015;23:74–81.
53. Gandhi L, et al. Phase I study of Navitoclax (ABT-263), a novel Bcl-2 family inhibitor, in patients with small-cell lung cancer and other solid tumors. J Clin Oncol. 2011;29(7):909–16.
54. Roberts AW, et al. Substantial susceptibility of chronic lymphocytic leukemia to BCL2 inhibition: results of a phase I study of navitoclax in patients with relapsed or refractory disease. J Clin Oncol. 2012;30(5):488–96.

55. Wilson WH, et al. Navitoclax, a targeted high-affinity inhibitor of BCL-2, in lymphoid malignancies: a phase 1 dose-escalation study of safety, pharmacokinetics, pharmacodynamics, and antitumour activity. Lancet Oncol. 2010;11(12):1149–59.
56. Kipps TJ, et al. A phase 2 study of the BH3 mimetic BCL2 inhibitor navitoclax (ABT-263) with or without rituximab, in previously untreated B-cell chronic lymphocytic leukemia. Leuk Lymphoma. 2015:56(10):2826–33.
57. Zhang H, et al. Bcl-2 family proteins are essential for platelet survival. Cell Death Differ. 2007;14(5):943–51.
58. Merino D, et al. Bcl-2, Bcl-x(L), and Bcl-w are not equivalent targets of ABT-737 and navitoclax (ABT-263) in lymphoid and leukemic cells. Blood. 2012;119(24):5807–16.
59. Souers AJ, et al. ABT-199, a potent and selective BCL-2 inhibitor, achieves antitumor activity while sparing platelets. Nat Med. 2013;19(2):202–8.
60. Yecies D, et al. Acquired resistance to ABT-737 in lymphoma cells that up-regulate MCL-1 and BFL-1. Blood. 2010;115(16):3304–13.
61. Rudin CM, et al. Phase II study of single-agent navitoclax (ABT-263) and biomarker correlates in patients with relapsed small cell lung cancer. Clin Cancer Res. 2012;18(11):3163–9.
62. Varadarajan S, et al. Evaluation and critical assessment of putative MCL-1 inhibitors. Cell Death Differ. 2013;20(11):1475–84.
63. Leverson JD, et al. Potent and selective small-molecule MCL-1 inhibitors demonstrate on-target cancer cell killing activity as single agents and in combination with ABT-263 (navitoclax). Cell Death Dis. 2015;6, e1590.
64. Varadarajan S, et al. Maritoclax and dinaciclib inhibit MCL-1 activity and induce apoptosis in both a MCL-1-dependent and -independent manner. Oncotarget. 2015;6(14):12668–81.
65. Brumatti G, Ekert PG. Seeking a MCL-1 inhibitor. Cell Death Differ. 2013;20(11):1440–1.
66. Walensky LD, Bird GH. Hydrocarbon-stapled peptides: principles, practice, and progress. J Med Chem. 2014;57(15):6275–88.
67. Walensky LD, et al. Activation of apoptosis in vivo by a hydrocarbon-stapled BH3 helix. Science. 2004;305(5689):1466–70.
68. Walensky LD, et al. A stapled BID BH3 helix directly binds and activates BAX. Mol Cell. 2006;24(2):199–210.
69. Cohen NA, et al. A competitive stapled peptide screen identifies a selective small molecule that overcomes MCL-1-dependent leukemia cell survival. Chem Biol. 2012;19(9):1175–86.
70. LaBelle JL, et al. A stapled BIM peptide overcomes apoptotic resistance in hematologic cancers. J Clin Invest. 2012;122(6):2018–31.
71. Vaillant F, et al. Targeting BCL-2 with the BH3 mimetic ABT-199 in estrogen receptor-positive breast cancer. Cancer Cell. 2013;24(1):120–9.
72. Mattoo AR, et al. Inhibition of NANOG/NANOGP8 downregulates MCL-1 in colorectal cancer cells and enhances the therapeutic efficacy of BH3 mimetics. Clin Cancer Res. 2014;20(21):5446–55.
73. Zhang L, Ming L, Yu J. BH3 mimetics to improve cancer therapy; mechanisms and examples. Drug Resist Updat. 2007;10(6):207–17.
74. Besbes S, et al. New dimension in therapeutic targeting of BCL-2 family proteins. Oncotarget. 2015;6(15):12862–71.
75. Li H, et al. Activation of caspase-2 in apoptosis. J Biol Chem. 1997;272(34):21010–7.
76. Khar A, et al. Caspase-2/NEDD-2 protease mediates execution of apoptosis in AK-5 tumor cells. Apoptosis. 1997;2(6):494–500.
77. Bouchier-Hayes L, Green DR. Caspase-2: the orphan caspase. Cell Death Differ. 2012;19(1):51–7.
78. Janssens S, Tinel A. The PIDDosome, DNA-damage-induced apoptosis and beyond. Cell Death Differ. 2012;19(1):13–20.
79. Shi M, et al. DNA-PKcs-PIDDosome: a nuclear caspase-2-activating complex with role in G2/M checkpoint maintenance. Cell. 2009;136(3):508–20.
80. Berube C, et al. Apoptosis caused by p53-induced protein with death domain (PIDD) depends on the death adapter protein RAIDD. Proc Natl Acad Sci U S A. 2005;102(40):14314–20.

81. Bouchier-Hayes L, et al. Characterization of cytoplasmic caspase-2 activation by induced proximity. Mol Cell. 2009;35(6):830–40.
82. Bonzon C, et al. Caspase-2-induced apoptosis requires bid cleavage: a physiological role for bid in heat shock-induced death. Mol Biol Cell. 2006;17(5):2150–7.
83. Upton JP, et al. Caspase-2 cleavage of BID is a critical apoptotic signal downstream of endoplasmic reticulum stress. Mol Cell Biol. 2008;28(12):3943–51.
84. Olsson M, Forsberg J, Zhivotovsky B. Caspase-2: the reinvented enzyme. Oncogene. 2015;34(15):1877–82.
85. Thompson R, et al. An Inhibitor of PIDDosome Formation. Mol Cell. 2015;58(5):767–79.
86. Manzl C, et al. Caspase-2 activation in the absence of PIDDosome formation. J Cell Biol. 2009;185(2):291–303.
87. Peintner L, et al. The tumor-modulatory effects of Caspase-2 and Pidd1 do not require the scaffold protein Raidd. Cell Death Differ. 2015;22:1803–11.
88. Sidi S, et al. Chk1 suppresses a caspase-2 apoptotic response to DNA damage that bypasses p53, Bcl-2, and caspase-3. Cell. 2008;133(5):864–77.
89. Manzl C, et al. Death of p53-defective cells triggered by forced mitotic entry in the presence of DNA damage is not uniquely dependent on Caspase-2 or the PIDDosome. Cell Death Dis. 2013;4, e942.
90. Bergeron L, et al. Defects in regulation of apoptosis in caspase-2-deficient mice. Genes Dev. 1998;12(9):1304–14.
91. Zhang Y, et al. Caspase-2 deficiency enhances aging-related traits in mice. Mech Ageing Dev. 2007;128(2):213–21.
92. Wilson CH, et al. Age-related proteostasis and metabolic alterations in Caspase-2-deficient mice. Cell Death Dis. 2015;6, e1597.
93. Shalini S, et al. Caspase-2 protects against oxidative stress in vivo. Oncogene. 2015;34: 4995–5002.
94. Sharma R, et al. Caspase-2 maintains bone homeostasis by inducing apoptosis of oxidatively-damaged osteoclasts. PLoS One. 2014;9(4), e93696.
95. Kumar S. Caspase function in programmed cell death. Cell Death Differ. 2007;14(1):32–43.
96. Manzl C, et al. PIDDosome-independent tumor suppression by Caspase-2. Cell Death Differ. 2012;19(10):1722–32.
97. Dorstyn L, et al. Caspase-2 deficiency promotes aberrant DNA-damage response and genetic instability. Cell Death Differ. 2012;19(8):1288–98.
98. Puccini J, et al. Loss of caspase-2 augments lymphomagenesis and enhances genomic instability in Atm-deficient mice. Proc Natl Acad Sci U S A. 2013;110(49):19920–5.
99. Ho LH, et al. A tumor suppressor function for caspase-2. Proc Natl Acad Sci U S A. 2009;106(13):5336–41.
100. Holleman A, et al. Decreased PARP and procaspase-2 protein levels are associated with cellular drug resistance in childhood acute lymphoblastic leukemia. Blood. 2005;106(5): 1817–23.
101. Terry MR, et al. Caspase-2 impacts lung tumorigenesis and chemotherapy response in vivo. Cell Death Differ. 2015;22(5):719–30.
102. Parsons MJ, et al. Genetic deletion of caspase-2 accelerates MMTV/c-neu-driven mammary carcinogenesis in mice. Cell Death Differ. 2013;20(9):1174–82.
103. Dorstyn L, et al. An unexpected role for caspase-2 in neuroblastoma. Cell Death Dis. 2014;5, e1383.
104. Ren K, et al. Tumor-suppressing function of caspase-2 requires catalytic site Cys-320 and site Ser-139 in mice. J Biol Chem. 2012;287(18):14792–802.
105. Thompson R, Eastman A. The cancer therapeutic potential of Chk1 inhibitors: how mechanistic studies impact on clinical trial design. Br J Clin Pharmacol. 2013;76(3):358–69.
106. Dent P, et al. CHK1 inhibitors in combination chemotherapy: thinking beyond the cell cycle. Mol Interv. 2011;11(2):133–40.
107. Weber AM, Ryan AJ. ATM and ATR as therapeutic targets in cancer. Pharmacol Ther. 2015;149:124–38.

108. Aggarwal BB. Signalling pathways of the TNF superfamily: a double-edged sword. Nat Rev Immunol. 2003;3(9):745–56.
109. Oberst A, et al. Inducible dimerization and inducible cleavage reveal a requirement for both processes in caspase-8 activation. J Biol Chem. 2010;285(22):16632–42.
110. Walczak H, Sprick MR. Biochemistry and function of the DISC. Trends Biochem Sci. 2001;26(7):452–3.
111. Kallenberger SM, et al. Intra- and interdimeric caspase-8 self-cleavage controls strength and timing of CD95-induced apoptosis. Sci Signal. 2014;7(316):ra23.
112. McStay GP, Salvesen GS, Green DR. Overlapping cleavage motif selectivity of caspases: implications for analysis of apoptotic pathways. Cell Death Differ. 2008;15(2):322–31.
113. Tu S, et al. In situ trapping of activated initiator caspases reveals a role for caspase-2 in heat shock-induced apoptosis. Nat Cell Biol. 2006;8(1):72–7.
114. Pop C, et al. FLIP(L) induces caspase 8 activity in the absence of interdomain caspase 8 cleavage and alters substrate specificity. Biochem J. 2011;433(3):447–57.
115. Varfolomeev EE, et al. Targeted disruption of the mouse Caspase 8 gene ablates cell death induction by the TNF receptors, Fas/Apo1, and DR3 and is lethal prenatally. Immunity. 1998;9(2):267–76.
116. Gunther C, et al. Caspase-8 regulates TNF-alpha-induced epithelial necroptosis and terminal ileitis. Nature. 2011;477(7364):335–9.
117. Kovalenko A, et al. Caspase-8 deficiency in epidermal keratinocytes triggers an inflammatory skin disease. J Exp Med. 2009;206(10):2161–77.
118. Kang TB, et al. Caspase-8 serves both apoptotic and nonapoptotic roles. J Immunol. 2004;173(5):2976–84.
119. Krajewska M, et al. Neuronal deletion of caspase 8 protects against brain injury in mouse models of controlled cortical impact and kainic acid-induced excitotoxicity. PLoS One. 2011;6(9), e24341.
120. Dillon CP, et al. Survival function of the FADD-CASPASE-8-cFLIP(L) complex. Cell Rep. 2012;1(5):401–7.
121. Lemmers B, et al. Essential role for caspase-8 in Toll-like receptors and NFkappaB signaling. J Biol Chem. 2007;282(10):7416–23.
122. Salmena L, Hakem R. Caspase-8 deficiency in T cells leads to a lethal lymphoinfiltrative immune disorder. J Exp Med. 2005;202(6):727–32.
123. Beisner DR, et al. Cutting edge: innate immunity conferred by B cells is regulated by caspase-8. J Immunol. 2005;175(6):3469–73.
124. Teitz T, et al. Th-MYCN mice with caspase-8 deficiency develop advanced neuroblastoma with bone marrow metastasis. Cancer Res. 2013;73(13):4086–97.
125. Teitz T, et al. Caspase 8 is deleted or silenced preferentially in childhood neuroblastomas with amplification of MYCN. Nat Med. 2000;6(5):529–35.
126. Ashley DM, et al. Caspase 8 is absent or low in many ex vivo gliomas. Cancer. 2005;104(7):1487–96.
127. Soung YH, et al. CASPASE-8 gene is inactivated by somatic mutations in gastric carcinomas. Cancer Res. 2005;65(3):815–21.
128. Soung YH, et al. Caspase-8 gene is frequently inactivated by the frameshift somatic mutation 1225_1226delTG in hepatocellular carcinomas. Oncogene. 2005;24(1):141–7.
129. Shivapurkar N, et al. Differential inactivation of caspase-8 in lung cancers. Cancer Biol Ther. 2002;1(1):65–9.
130. Harada K, et al. Deregulation of caspase 8 and 10 expression in pediatric tumors and cell lines. Cancer Res. 2002;62(20):5897–901.
131. Pingoud-Meier C, et al. Loss of caspase-8 protein expression correlates with unfavorable survival outcome in childhood medulloblastoma. Clin Cancer Res. 2003;9(17):6401–9.
132. Sun T, et al. A six-nucleotide insertion-deletion polymorphism in the CASP8 promoter is associated with susceptibility to multiple cancers. Nat Genet. 2007;39(5):605–13.
133. Li C, et al. The six-nucleotide deletion/insertion variant in the CASP8 promoter region is inversely associated with risk of squamous cell carcinoma of the head and neck. Cancer Prev Res (Phila). 2010;3(2):246–53.

134. MacKenzie SH, Schipper JL, Clark AC. The potential for caspases in drug discovery. Curr Opin Drug Discov Devel. 2010;13(5):568–76.
135. Chen L, et al. CD95 promotes tumour growth. Nature. 2010;465(7297):492–6.
136. Jing Y, et al. Tumor necrosis factor-alpha promotes tumor growth by inducing vascular endothelial growth factor. Cancer Invest. 2011;29(7):485–93.
137. Zhang M, et al. The roles of ROS and caspases in TRAIL-induced apoptosis and necroptosis in human pancreatic cancer cells. PLoS One. 2015;10(5), e0127386.
138. Vercammen D, et al. Inhibition of caspases increases the sensitivity of L929 cells to necrosis mediated by tumor necrosis factor. J Exp Med. 1998;187(9):1477–85.
139. Degterev A, et al. Identification of RIP1 kinase as a specific cellular target of necrostatins. Nat Chem Biol. 2008;4(5):313–21.
140. Degterev A, et al. Chemical inhibitor of nonapoptotic cell death with therapeutic potential for ischemic brain injury. Nat Chem Biol. 2005;1(2):112–9.
141. Cho YS, et al. Phosphorylation-driven assembly of the RIP1-RIP3 complex regulates programmed necrosis and virus-induced inflammation. Cell. 2009;137(6):1112–23.
142. Dondelinger Y, et al. MLKL compromises plasma membrane integrity by binding to phosphatidylinositol phosphates. Cell Rep. 2014;7(4):971–81.
143. Galluzzi L, Kepp O, Kroemer G. MLKL regulates necrotic plasma membrane permeabilization. Cell Res. 2014;24(2):139–40.
144. Hildebrand JM, et al. Activation of the pseudokinase MLKL unleashes the four-helix bundle domain to induce membrane localization and necroptotic cell death. Proc Natl Acad Sci U S A. 2014;111(42):15072–7.
145. Murphy JM, et al. The pseudokinase MLKL mediates necroptosis via a molecular switch mechanism. Immunity. 2013;39(3):443–53.
146. Rodriguez DA, et al. Characterization of RIPK3-mediated phosphorylation of the activation loop of MLKL during necroptosis. Cell Death Differ. 2016;23:76–88.
147. Shin YK, Kim J, Yang Y. Switch for the necroptotic permeation pore. Structure. 2014;22(10):1374–6.
148. Su L, et al. A plug release mechanism for membrane permeation by MLKL. Structure. 2014;22(10):1489–500.
149. Oberst A, et al. Catalytic activity of the caspase-8-FLIP(L) complex inhibits RIPK3-dependent necrosis. Nature. 2011;471(7338):363–7.
150. Green DR, et al. RIPK-dependent necrosis and its regulation by caspases: a mystery in five acts. Mol Cell. 2011;44(1):9–16.
151. Feoktistova M, et al. cIAPs block Ripoptosome formation, a RIP1/caspase-8 containing intracellular cell death complex differentially regulated by cFLIP isoforms. Mol Cell. 2011;43(3):449–63.
152. Tenev T, et al. The Ripoptosome, a signaling platform that assembles in response to genotoxic stress and loss of IAPs. Mol Cell. 2011;43(3):432–48.
153. Mandal P, et al. RIP3 induces apoptosis independent of pronecrotic kinase activity. Mol Cell. 2014;56(4):481–95.
154. Vanden Berghe T, et al. Regulated necrosis: the expanding network of non-apoptotic cell death pathways. Nat Rev Mol Cell Biol. 2014;15(2):135–47.
155. Linkermann A, et al. Two independent pathways of regulated necrosis mediate ischemia-reperfusion injury. Proc Natl Acad Sci U S A. 2013;110(29):12024–9.
156. Kaiser WJ, et al. Toll-like receptor 3-mediated necrosis via TRIF, RIP3, and MLKL. J Biol Chem. 2013;288(43):31268–79.
157. Thapa RJ, et al. Interferon-induced RIP1/RIP3-mediated necrosis requires PKR and is licensed by FADD and caspases. Proc Natl Acad Sci U S A. 2013;110(33):E3109–18.
158. Linkermann A, et al. Dichotomy between RIP1- and RIP3-mediated necroptosis in tumor necrosis factor-alpha-induced shock. Mol Med. 2012;18:577–86.
159. Lukens JR, et al. RIP1-driven autoinflammation targets IL-1alpha independently of inflammasomes and RIP3. Nature. 2013;498(7453):224–7.

160. Lukens JR, et al. Dietary modulation of the microbiome affects autoinflammatory disease. Nature. 2014;516(7530):246–9.
161. Gurung P, et al. FADD and caspase-8 mediate priming and activation of the canonical and noncanonical Nlrp3 inflammasomes. J Immunol. 2014;192(4):1835–46.
162. Philip NH, et al. Caspase-8 mediates caspase-1 processing and innate immune defense in response to bacterial blockade of NF-kappaB and MAPK signaling. Proc Natl Acad Sci U S A. 2014;111(20):7385–90.
163. Cullen SP, et al. Diverse activators of the NLRP3 inflammasome promote IL-1beta secretion by triggering necrosis. Cell Rep. 2015;11(10):1535–48.
164. Kearney CJ, et al. Necroptosis suppresses inflammation via termination of TNF- or LPS-induced cytokine and chemokine production. Cell Death Differ. 2015;22:1313–27.
165. Kaiser WJ, et al. RIP3 mediates the embryonic lethality of caspase-8-deficient mice. Nature. 2011;471(7338):368–72.
166. Dillon CP, et al. RIPK1 blocks early postnatal lethality mediated by caspase-8 and RIPK3. Cell. 2014;157(5):1189–202.
167. Kaiser WJ, et al. RIP1 suppresses innate immune necrotic as well as apoptotic cell death during mammalian parturition. Proc Natl Acad Sci U S A. 2014;111(21):7753–8.
168. Rickard JA, et al. RIPK1 regulates RIPK3-MLKL-driven systemic inflammation and emergency hematopoiesis. Cell. 2014;157(5):1175–88.
169. Yeh WC, et al. FADD: essential for embryo development and signaling from some, but not all, inducers of apoptosis. Science. 1998;279(5358):1954–8.
170. Yeh WC, et al. Requirement for Casper (c-FLIP) in regulation of death receptor-induced apoptosis and embryonic development. Immunity. 2000;12(6):633–42.
171. Zhang H, et al. Functional complementation between FADD and RIP1 in embryos and lymphocytes. Nature. 2011;471(7338):373–6.
172. Newton K, Sun X, Dixit VM. Kinase RIP3 is dispensable for normal NF-kappa Bs, signaling by the B-cell and T-cell receptors, tumor necrosis factor receptor 1, and Toll-like receptors 2 and 4. Mol Cell Biol. 2004;24(4):1464–9.
173. Kelliher MA, et al. The death domain kinase RIP mediates the TNF-induced NF-kappaB signal. Immunity. 1998;8(3):297–303.
174. Dannappel M, et al. RIPK1 maintains epithelial homeostasis by inhibiting apoptosis and necroptosis. Nature. 2014;513(7516):90–4.
175. Lee P, et al. Dynamic expression of epidermal caspase 8 simulates a wound healing response. Nature. 2009;458(7237):519–23.
176. Takahashi N, et al. RIPK1 ensures intestinal homeostasis by protecting the epithelium against apoptosis. Nature. 2014;513(7516):95–9.
177. Roderick JE, et al. Hematopoietic RIPK1 deficiency results in bone marrow failure caused by apoptosis and RIPK3-mediated necroptosis. Proc Natl Acad Sci U S A. 2014;111(40): 14436–41.
178. Liu P, et al. Dysregulation of TNFalpha-induced necroptotic signaling in chronic lymphocytic leukemia: suppression of CYLD gene by LEF1. Leukemia. 2012;26(6): 1293–300.
179. Vucur M, et al. RIP3 inhibits inflammatory hepatocarcinogenesis but promotes cholestasis by controlling caspase-8- and JNK-dependent compensatory cell proliferation. Cell Rep. 2013;4(4):776–90.
180. Schmidt SV, et al. RIPK3 expression in cervical cancer cells is required for PolyIC-induced necroptosis, IL-1alpha release, and efficient paracrine dendritic cell activation. Oncotarget. 2015;6(11):8635–47.
181. Melo-Lima S, Celeste Lopes M, Mollinedo F. Necroptosis is associated with low procaspase-8 and active RIPK1 and -3 in human glioma cells. Oncoscience. 2014;1(10):649–64.
182. Geserick P, et al. Absence of RIPK3 predicts necroptosis resistance in malignant melanoma. Cell Death Dis. 2015;6, e1884.
183. McCabe KE, et al. Triggering necroptosis in cisplatin and IAP antagonist-resistant ovarian carcinoma. Cell Death Dis. 2014;5, e1496.

184. Oliver Metzig M, et al. Inhibition of caspases primes colon cancer cells for 5-fluorouracil-induced TNF-alpha-dependent necroptosis driven by RIP1 kinase and NF-kappaB. Oncogene. 2015 Nov 2. doi:10.1038/onc.2015.398.
185. Moriwaki K, et al. Differential roles of RIPK1 and RIPK3 in TNF-induced necroptosis and chemotherapeutic agent-induced cell death. Cell Death Dis. 2015;6, e1636.
186. Radogna F, Dicato M, Diederich M. Cancer-type-specific crosstalk between autophagy, necroptosis and apoptosis as a pharmacological target. Biochem Pharmacol. 2015;94(1):1–11.
187. Fulda S. Therapeutic exploitation of necroptosis for cancer therapy. Semin Cell Dev Biol. 2014;35:51–6.
188. Mohammad RM, et al. Broad targeting of resistance to apoptosis in cancer. Semin Cancer Biol. 2015;35:S78–103.
189. Takemura R, et al. PolyI:C-induced, TLR3/RIP3-dependent necroptosis backs up immune effector-mediated tumor elimination in vivo. Cancer Immunol Res. 2015;3(8):902–14.
190. Yatim N, et al. RIPK1 and NF-kappaB signaling in dying cells determines cross-priming of CD8(+) T cells. Science. 2015;350(6258):328–34.

Chapter 2
Clearance of Dying Cells by Phagocytes: Mechanisms and Implications for Disease Pathogenesis

Aaron M. Fond and Kodi S. Ravichandran

Abstract The efficient clearance of apoptotic cells is an evolutionarily conserved process crucial for homeostasis in multicellular organisms. The clearance involves a series of steps that ultimately facilitates the recognition of the apoptotic cell by the phagocytes and the subsequent uptake and processing of the corpse. These steps include the phagocyte sensing of "find-me" signals released by the apoptotic cell, recognizing "eat-me" signals displayed on the apoptotic cell surface, and then intracellular signaling within the phagocyte to mediate phagocytic cup formation around the corpse and corpse internalization, and the processing of the ingested contents. The engulfment of apoptotic cells by phagocytes not only eliminates debris from tissues but also produces an anti-inflammatory response that suppresses local tissue inflammation. Conversely, impaired corpse clearance can result in loss of immune tolerance and the development of various inflammation-associated disorders such as autoimmunity, atherosclerosis, and airway inflammation but can also affect cancer progression. Recent studies suggest that the clearance process can also influence antitumor immune responses. In this review, we will discuss how apoptotic cells interact with their engulfing phagocytes to generate important immune responses, and how modulation of such responses can influence pathology.

Keywords Engulfment • Cell clearance • Apoptosis • Cancer • Immune tolerance

2.1 Introduction

Phagocytes are often thought of as the "garbage collectors" of the body, eliminating pathogens, immune complexes, and dying cells. Sensing infection and coordinating an immune response is fundamental in the body's fight to prevent disease, but

---

A.M. Fond • K.S. Ravichandran (✉)
Center for Cell Clearance, and the Department of Microbiology, Immunology, and Cancer Biology, University of Virginia, Charlottesville, VA 22908, USA
e-mail: ravi@virginia.edu

C.D. Gregory (ed.), *Apoptosis in Cancer Pathogenesis and Anti-cancer Therapy*, Advances in Experimental Medicine and Biology 930,
DOI 10.1007/978-3-319-39406-0_2

discriminating between harmless debris and a true threat is equally vital. Every day, billions of cells in the body undergo apoptotic cell death as part of the normal physiology/homeostasis that is essential for healthy living [1]. The phagocytes that clear them must respond appropriately to prevent an unnecessary and unwanted immune response to such homeostatic cell turnover [2–5].

Apoptosis is often described as "immunologically silent" cell death; however, apoptotic cells are anything but unheard. The homeostatic clearance of apoptotic cells elicits critical immunosuppressive responses in the phagocytes that are often specific to apoptotic cell recognition. This includes the release of anti-inflammatory cytokines, the inhibition of proinflammatory cytokine expression, and the regulation of new immune cell production. Importantly, these are active responses to apoptotic cells and not just the lack of an inflammatory response [6–10]. Showcasing the vital role of apoptotic cell clearance in the maintenance of immune tolerance, uncleared corpses can result in inflammation [11]. This is partly due to the nature of apoptosis, which when left uncompleted by engulfment, progresses to "secondary necrosis," a state in which the dying cell loses its membrane integrity and releases some or most of its intracellular components [12]. However, while the body may not want to produce an immunogenic response to apoptotic cells that are routinely turned over as part of homeostasis, generating immune responses to tumor-derived antigens could obviously be beneficial. Recent studies suggest that the cell clearance process can also initiate antitumor immunity [13, 14].

In this review, we will discuss how phagocytes sense, engulf, and respond to apoptotic cells. Since the identification of phosphatidylserine (PtdSer) as an eat-me signal on apoptotic cells in the early 1990s, there has been a rapid expansion in our knowledge of how apoptotic cells are recognized and removed. As apoptosis can be immunologically silent only when the corpses are cleared in a timely manner, understanding the mechanisms of cell clearance has provided important insights and tools for the study of cell clearance and its relationship to disease.

2.2 Types of Apoptotic Cells and Phagocytes

It is estimated that every day, we turnover about 200–300 billion cells, or about one million cells/s. The cells can die by many different modalities, including caspase-dependent apoptosis, necroptosis, as well as necrosis [15] (see Chap. 1 for more details). Even within each modality of death, there are multiple subroutines that eventually lead to the death of the cells fated to die [15]. This chapter primarily deals with clearance of cells that undergo caspase-mediated apoptosis as this is the most common and perhaps the best understood mode of cell death in vivo and in vitro.

There are different types of phagocytes that mediate the removal of the dying cells in the various tissues [16]. Phagocytes can be broadly classified into three different types: professional phagocytes, nonprofessional phagocytes, and specialized phagocytes. The professional phagocytes include the macrophages and immature dendritic cells. Given the recent understanding of self-renewing tissue resident, and recruited macrophages, they both seem capable of engulfing apoptotic cells. The professional phagocytes are named as such because in different analyses they tend to have a large capacity to engulf, engulf the targets with faster kinetics, and can also ingest multiple corpses in succession. The nonprofessional phagocytes include cell types such as epithelial cells, fibroblasts, and other tissue-resident cells. Although these nonprofessional phagocytes often display a lower capacity to engulf and slower kinetics in vitro, they do have a numerical supremacy and due to their proximity to the apoptotic cells (e.g., neighbor), these nonprofessional phagocytes likely contribute substantially to the clearance of apoptotic cells in vivo. In fact, disruption of nonprofessional phagocyte engulfment has important immunological consequences.

The third type of phagocytes, specialized phagocytes, refers to cells that generally have a multitude of functions, one of them being phagocytosis of dying cells. The examples include the Sertoli cells of the testis (which provide nurse function for the developing male germ cells while also engulfing those with defective development) [17] and retinal pigment epithelial cells (which remove the "used" photoreceptor outer segments daily in a circadian fashion) [18, 19]. Obviously, the key difference between the three types of phagocytes is that not all of them have to express the same collection of receptors and they may not have similar postengulfment responses. However, some of the responses of professional phagocytes and nonprofessional phagocytes clearly overlap, and could be relevant in tissues where often both of them are involved in apoptotic cell clearance simultaneously.

2.3 Steps in Apoptotic Cell Engulfment

Work from a number of laboratories over the past nearly two decades has detailed a series of distinguishable steps in apoptotic cell recognition and clearance (Fig. 2.1). These have helped us understand how the phagocytes and apoptotic cells get near each other, how the phagocytes specifically recognize the targets, the types of intracellular signaling within phagocytes that leads to the corpse uptake, as well as some of the subsequent responses of phagocytes.

2.3.1 Find-Me Signals

The apoptotic cell is an active participant in its own clearance. The response of phagocytes to apoptotic cells can be influenced by several actions taken by the apoptotic cell. These include the release of find-me signals that attract phagocytes to the site of death and the exposure of eat-me signals that allow the phagocyte to distinguish the dying cell from its healthy neighbors [20].

For apoptotic cells to be rapidly cleared, which is the case in vivo, they must be rapidly "found" [21]. During homeostatic cell turnover in tissues, a single corpse

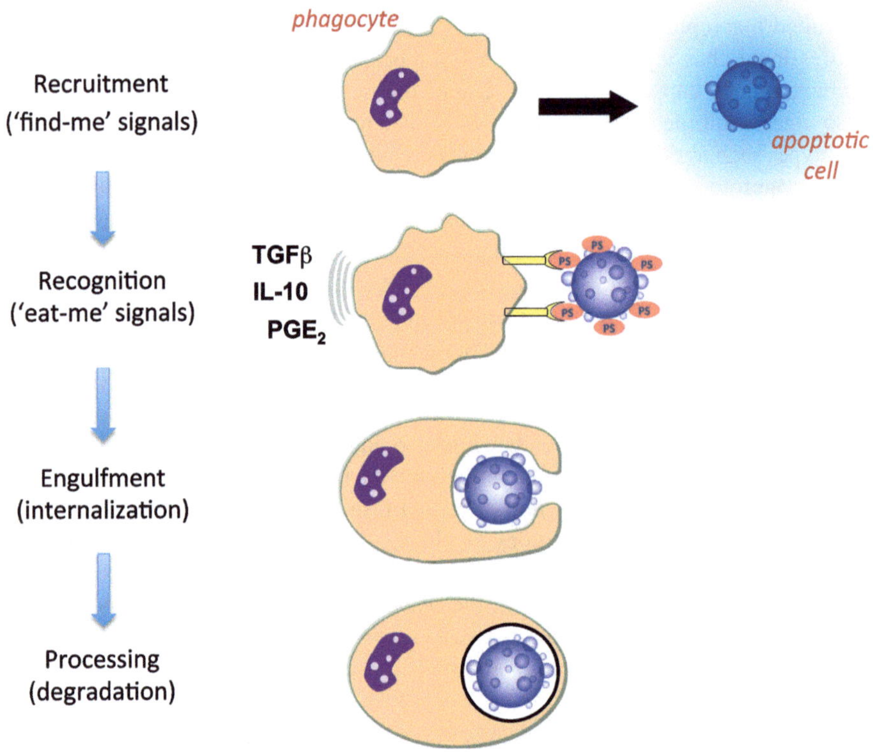

Fig. 2.1 Major steps in apoptotic cell clearance. The key events in recognition and clearance of dying cells can be broadly classified into four steps. Step 1 depicts the recruitment of motile phagocytes (such as tissue-resident macrophages) by apoptotic cells via the release of find-me signals. Step 2 is the specific recognition of the apoptotic cell via "eat-me" ligands on the dying cells engaged by the receptors on phagocytes. Often, the recognition of apoptotic cells alone (even without corpse internalization) is sufficient to trigger some of the key anti-inflammatory mediators from the phagocytes. Step 3 is the intracellular signaling that occurs within the phagocytes leading to physical corpse internalization. The fourth step is the processing/digestion of the internalized targets and the regulation of the metabolic overload within the phagocytes

might be surrounded by a vast number of healthy neighbors, and therefore "calling out" the professional phagocytes such as resident macrophages to come clear the apoptotic cell is important for the prompt removal. Several such "find-me" signals have been identified and may be differentially important depending on the situation. The first find-me signal proposed was lysophosphatidylcholine (LPC); however, the role LPC played to attract phagocytes seemed specific to both the type of apoptotic cell (the MCF-7 breast cancer line) and phagocyte (THP-1 monocyte line) used [22]. Furthermore, in vivo relevance of LPC as a find-me signal remains to be established. Later, an elegant study showed that cleavage of CX3CL1/Fractalkine (FKN) during apoptosis leads to release of a soluble fragment that induces the migration of monocytes to Burkitt lymphoma B-cells in vitro and to germinal centers in vivo [23].

This could also be relevant for attraction of monocytes and the complex interplay between macrophages and tumor cells in a tumor microenvironment. The significance of FKN has been established for locating apoptotic B-cells, but FKN per se as a universal find-me signal in other cell types is at present less defined. Finally, the triphosphate nucleotides ATP and UTP were found to be released in a regulated manner during apoptosis by the caspase-mediated cleavage of Pannexin-1 (PANX1), a transmembrane protein that forms hexameric hemichannels [24]. The nucleotides released by PANX1 cleavage are chemotactic for monocytes in vitro and in vivo by signaling through the nucleotide receptor P2Y2 [24, 25]. Although nucleotides clearly are relevant find-me signals, one of the interesting challenges with such nucleotide find-me signals is how far the nucleotide signal can travel before extracellular nucleotidases convert them into their nonchemotactic diphosphate and monophosphate forms. In addition to attracting phagocytes to the site of death, these find-me signals may also prime the phagocytes for engulfment, although this has only been shown in the case of FKN, which stimulates macrophages to produce the apoptotic cell bridging molecule milk fat globule-EGF factor 8 (MFG-E8, discussed later) [26, 27].

2.3.2 *Eat-Me Signals*

Once the phagocyte has been brought to the area of the dying cell, it must identify the specific cell that needs to be cleared, which is achieved by recognition of eat-me signals on the surface of the apoptotic cell. There are many "eat-me" markers identified to date on apoptotic cells that are linked to corpse uptake. The classic eat-me signal is the lipid phosphatidylserine (PtdSer). It had been known that aged red blood cells lose their phospholipid asymmetry, but Fadok and colleagues demonstrated that PtdSer is also exposed by thymocytes as they undergo apoptosis [28]. Furthermore, they found that apoptotic thymocyte engulfment by macrophages is inhibited by the competitive addition of PtdSer-containing liposomes. Since then, PtdSer exposure has been found to be an evolutionarily conserved general feature of apoptosis from lower organisms to man and is now commonly used to assay the apoptotic status of a cell [29, 30].

Phosphatidylserine (PtdSer) as an eat-me signal has stood the test of time due to a preponderance of evidence of its importance [31]. Exogenous incorporation of PtdSer into the outer leaflet of viable cells in some cases is sufficient to cause their engulfment by macrophages, and PtdSer liposomes alone in certain circumstances can elicit some of the responses induced in the phagocyte [32, 33]. The asymmetric distribution of PtdSer in healthy cells is maintained through flippases that actively mediate the movement of PtdSer from the outer to the inner membrane [31]. In contrast, during apoptosis induction, the flippases appear to be inactivated, while another set of enzymes called "phospholipid scramblases" become active, and the latter randomize the PtdSer levels between the outer and inner leaflets. The exposed PtdSer is then recognized by specific receptors on the phagocytes, contributing to corpse internalization [31, 34, 35]. The P4-ATPase family member ATP11C and its chaperone CDC50 have been identified as key components for the flippase function seen in healthy cells. With respect to the scramblases, members of the Xkr-family with six transmembrane domains appear to perform this role. Remarkably, both the Xkr8 scramblase and ATP11C flippase have sites that can be cleaved by apoptotic caspases [31, 34, 35]. Thus, in live cells, the flippase remains active while the scramblase is inactive, while this occurs in opposite ways after caspase-mediated cleavage of these proteins during apoptosis. Current evidence based on mutant proteins suggests that the flippase is likely more dominant in maintaining the PtdSer asymmetry and that it has to be inactivated for the scramblase to fully promote the PtdSer exposure.

While PtdSer exposure is clearly central in apoptotic cell recognition and widely studied, unfortunately that has been at the expense of thorough characterization of many other eat-me signals that have been seen in different apoptotic contexts. These include the ER resident protein calreticulin (CRT), which some studies have found to translocate to the cell surface during apoptosis [36, 37]. However, CRT has also been shown to play a role on the surface of the phagocyte in interacting with Mannose Binding Lectin (MBL) and complement C1q bound to the surface of apoptotic cells [38]. In addition to CRT and PtdSer, many modifications to the apoptotic cell surface have been implicated, such as the presentation of oxidized Low Density Lipoprotein (oxLDL)-like sites, changes to glycosylation such as the capping of CD43, the exposure of Annexin I, and the expression of ICAM3 [39–42]. Although these are less well characterized than PtdSer, they indicate that the apoptotic cell has many ways to make itself known to the phagocyte. Recently, DD1α, a p53-inducible protein that mediates homotypic interaction between apoptotic cells and phagocytes has been defined and has been linked to establishment of immune responses to cancer cells [43, 44].

2.3.3 The Engulfment Receptors and Bridging Molecules

Eat-me signals on the surface of apoptotic cells are not useful without cognate receptors on phagocytes to recognize the eat-me signals. This is the role of various engulfment receptors on phagocytes and other soluble bridging molecules. Due to the importance of phosphatidylserine, much work has been done to identify its receptors. Although a PtdSer recognizing membrane receptor (simply termed “PSR”) was first identified using an antibody that blocked apoptotic cell engulfment, this is no longer considered a PtdSer recognition receptor as the knockout of the gene in mice did not impact engulfment and PSR is now thought to be a nuclear protein [45]. Since then, multiple receptors that directly or indirectly bind PtdSer have been identified and play a role in engulfment (Fig. 2.2). In 2007, Brain Angiogenesis Inhibitor 1 (BAI1), T-cell immunoglobulin domain-containing 4 (TIM4), and Stabilin-2 (Stab2) were all identified as receptors that can bind directly to PtdSer [46–48]. Modifying BAI1, TIM4, or Stab2 levels altered the engulfment

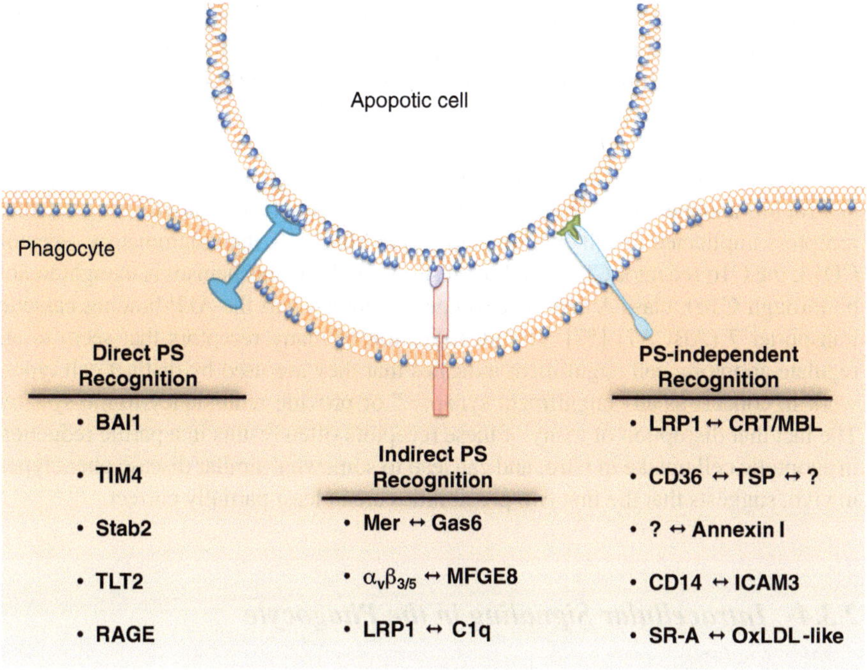

Fig. 2.2 Recognition of apoptotic cells by phagocytes. Phosphatidylserine, one of the key eat-me signals on apoptotic cells, can be recognized either directly via phagocytic receptors or indirectly through bridging molecules. Although phosphatidylserine is a key recognition entity, a number of other eat-me markers can also participate to different degrees in the recognition and uptake of apoptotic cells

capacity of phagocytes in vitro. In the years since the identification of these receptors, other receptors have been proposed, including triggering receptor expressed on myeloid cells-like protein 2 (TLT2) and the receptor for advanced glycation end-products (RAGE) [49, 50]. The relative role of each of these engulfment receptors to apoptotic cell engulfment, either independent of the others or in cooperation, still needs to be fully elucidated. Some preliminary studies suggest that TIM-4 and MER proteins can cooperate in the clearance of apoptotic targets by peritoneal macrophages [51]. Moreover, MER homolog appears to phosphorylate ELMO1 proteins that also function downstream of BAI1 [52].

PtdSer can also be recognized indirectly by phagocytic receptors via bridging molecules. One of the first engulfment receptors identified was $\alpha_v\beta_3$ integrin, which has since been shown to bind apoptotic cells via the PtdSer-dependent bridging molecule MFG-E8 [8, 53]. Later, the receptor tyrosine kinase Mer (as well as its homologs Tyro3 and Axl, part of a family of receptors termed TAM receptors) was found to mediate corpse clearance [54, 55]. Mer functions by binding to growth arrest-specific gene 6 (Gas6) or protein S, which recognizes PtdSer [54] (see Chap. 6 for more details).

Other less well understood bridging molecules include C1q, MBL, and Thrombospondin-1 (TSP-1) [9, 38]. C1q is thought to opsonize late apoptotic and early necrotic cells [56]. It has multiple receptors, but its role in cell engulfment is thought to be through LDL-related receptor protein 1 (LRP1 or CD91), a multifunctional receptor that has also been found to mediate engulfment through CRT and MBL [36, 38]. Finally, TSP-1 was found to increase macrophage binding to apoptotic neutrophils and mediate their engulfment via the phagocytic receptor CD36 [9]. Other receptors implicated in cell engulfment include MEGF10, the inflammatory receptor CD14, the C1q receptor CD93 (although surprisingly its mechanism is thought to not be through C1q), class A and B scavenger receptors, and the ATP-binding cassette transporter 7 (ABCA7) [57]. The identification of many receptors that seem to all regulate apoptotic cell engulfment indicates that they are used by distinct cell types, work in concert as an "engulfment synapse," or provide redundancy to the system. The fact that disruption of many of these receptors often results in a partial reduction in apoptotic cell uptake in vitro, and can lead to somewhat similar disease phenotypes in vivo, suggests that the first two possibilities are at least partially correct.

2.3.4 Intracellular Signaling in the Phagocyte

Once a phagocyte recognizes an apoptotic cell, signaling occurs to rearrange the cytoskeleton and engulf the target. In *C. elegans*, where much of the early work was done to identify some of the relevant engulfment genes, two phagocytic signaling pathways were discovered that share homology with mammalian engulfment pathways. In both nematodes and mammals, the pathways converge on the Rho family GTPase CED10/Rac1, which in turn signals through WAVE to Arp2/3, initiating actin nucleation and cytoskeletal rearrangement [58, 59]. Actin polymerization forms the phagocytic cup around the apoptotic cell and mediates the physical act of engulfment. The intracellular signals will be discussed briefly here, but more in-depth reviews can be found elsewhere [60–64].

The first evolutionarily conserved engulfment pathway contains the nemotode genes cell death defective-1 (CED-1), CED-6, and CED-7. In mammals, the orthologous pathway members, respectively, are LRP1 or MEGF10, engulfment adaptor GULP1, and ABCA1 or ABCA7 [61, 62]. In mammals, GULP1 has been shown to be downstream of LRP1 as well as Stab2, whereas the direct functions of ABCA1 and ABCA7 in engulfment have been controversial and may not play the same role as CED-7 in the nematode [65, 66]. Although this pathway requires Rac1 for engulfment, the mechanism by which GULP connects to Rac1 is currently unknown.

The second pathway that is shared between nematodes and mammals is the CED-2, CED-5, CED-12 pathway, corresponding to the mammalian proteins CrkII, Dock180, and engulfment and cell motility (ELMO), respectively. In this pathway, ELMO and Dock180 act together as a bipartite guanine nucleotide-exchange factor (GEF) for Rac1 activation [67]. The phagocytic receptor BAI1 has been shown to signal directly to ELMO, but other unknown receptors may feed in to the pathway through the TRIO protein activating the GTPase RhoG, which can activate ELMO to promote engulfment [47, 68]. In fact, it has been shown that the TAM family receptor MER can lead to phosphorylation of ELMO1. Finally, while CED-2/CrkII is associated with this pathway and has been found in complex with ELMO/Dock180 [69], its actual role is unclear, as ELMO and Dock180 can act without binding to CrkII [70]. This suggests that there may be subpathways within this group of genes regulating engulfment.

In addition to these known canonical signaling pathways, there are alternate mediators of cell engulfment. For example, it was proposed that RAGE acts as an engulfment receptor in mice by activating Rac1 through Diaphenous-1 (mDia1) [50]. The signaling pathways for many of the other phagocytic receptors linked to apoptotic cell clearance remain to be defined. TIM4 has a very short intracellular domain, which has been shown to be dispensable for its function in engulfment [71]. TIM4 appears to function cooperatively with the MER receptor in clearance of apoptotic cells in the peritoneum. Others have suggested that TIM4 works as a tether in conjunction with $\alpha_v\beta_3$/MFG-E8 to mediate engulfment; however, TIM4-mediated engulfment in the peritoneum is benefited by sequestration of MFG-E8 to prevent its binding to apoptotic cells, suggesting that TIM4 and $\alpha_v\beta_3$/MFG-E8 do not work in a single engulfment pathway in vivo [72, 73]. Interestingly, the sequestration of MFG-E8 is accomplished by the oxidation of PE on the surface of resident noninflammatory macrophages by 12/15-lipoxygenase as a way to prevent inflammatory infiltrating monocytes from recognizing and clearing the apoptotic cells and initiating an unintended immune response [73]. This highlights the importance of the correct engulfment signals occurring in the correct cell type for the right output, which is the generalized suppression of inflammation characteristic of apoptotic cell engulfment.

2.4 Effects of Apoptotic Cell Clearance

2.4.1 Induction of an Anti-inflammatory Program

One of the first recognized effects of apoptotic cells was their ability to induce the production of anti-inflammatory cytokines by engulfing phagocytes [6]. While the phagocyte encounter with bacterial lipopolysaccharide (LPS) induces production of inflammatory cytokines, phagocyte interaction with apoptotic cells instead stimulates the release of anti-inflammatory cytokines [6]. This finding changed the assumption that apoptotic cells were "immunologically inert" and that it is not just the lack of proinflammatory signals, rather apoptotic cells carry ligands that can actively induce an anti-inflammatory signaling within the phagocytes (Fig. 2.3). It was also shown that engulfment of apoptotic cells uniquely caused the release of anti-inflammatory signals compared to other forms of uptake, such as engulfment of zymosan or IgG-opsonized apoptotic cells [74]. Although it has been reported that PtdSer signaling alone is sufficient to induce these signals, such as the anti-inflammatory effects of

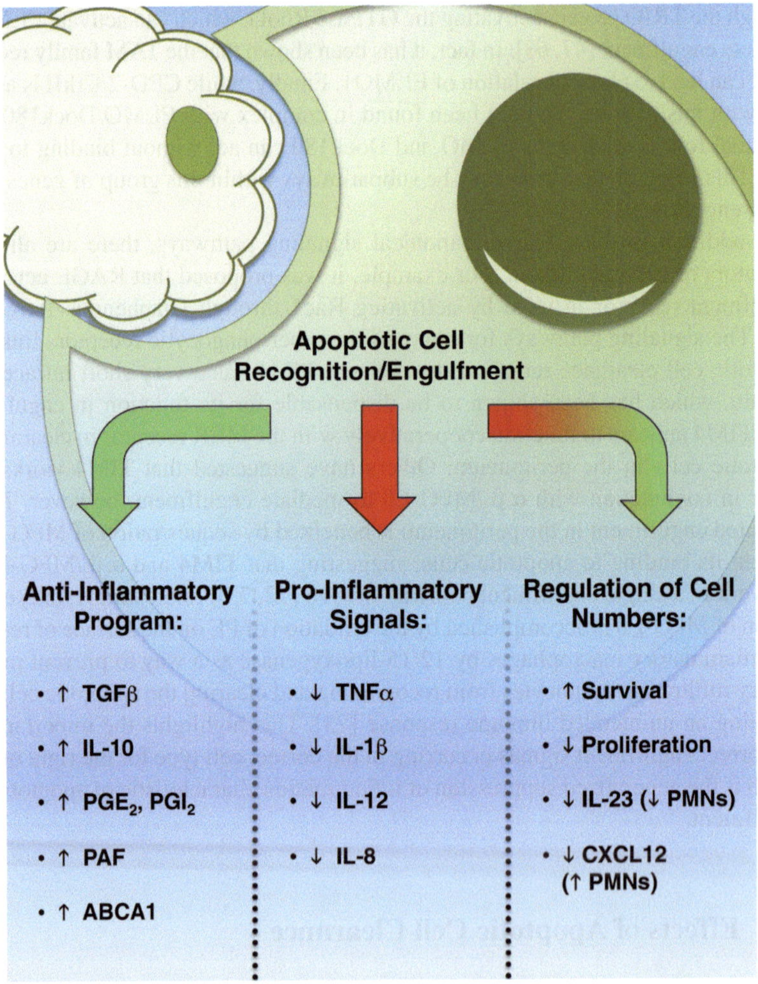

Fig. 2.3 Responses of phagocytes. When phagocytes (both professional phagocytes and nonprofessional phagocytes) engage and engulf apoptotic cells, they produce anti-inflammatory mediators such as TGF-β, interleukin-10 (IL-10), platelet activating factor (PAF), prostaglandin E2, as well as the membrane protein ABCA1 that induces anti-inflammatory effects by a yet to be defined mechanism. Apoptotic cell recognition also suppresses the release of proinflammatory cytokines such as TNFα, interleukin-1β, interleukin-12, and interleukin-8. The apoptotic cell recognition process also regulates the numbers of neutrophils and hematopoietic precursor cells via the cytokine interleukin-23 (IL-23) and the chemokine CXCL12

PtdSer liposomes [33, 75], or the administration of PtdSer liposomes in mouse models of inflammation to reduce disease [76], the effect of isolated PtdSer liposomes is very variable and they are never as potent as whole apoptotic cells. Either the conformation of the PtdSer exposure on the apoptotic cells, or more likely, one or more additional signals on the apoptotic cells are necessary for the full elicitation of the anti-inflammatory responses from phagocytes.

Transforming Growth Factor-β (TGF-β) is a classic example of an anti-inflammatory signal released as a consequence of apoptotic cell engulfment. It is transcriptionally upregulated by apoptotic cell recognition in a p38 MAPK, JNK, and ERK-dependent process [77]. The apoptotic cell recognition-dependent transcription, translation, and release of TGF-β after PtdSer recognition is not unique to professional phagocytes such as macrophages but also observed in bronchial epithelial cell lines and vascular smooth muscle cells [78, 79]. Similar to TGF-β, the transcriptional upregulation of IL-10 in phagocytes requires p38 MAPK, and the pharmacological inhibition of p38 completely abrogates the IL-10 response to apoptotic cells [80]. Besides TGF-β and IL-10, other anti-inflammatory mediators are produced via different means. Several eicosanoids are made in response to apoptotic cell engulfment, including prostaglandins E_2 and I_2 (PGE_2, PGI_2) [74, 75]. The increased production of these is due to upregulation of important enzymes in their synthetic pathways, including COX-2, PGES, and PGIS. However, this response was found to be dependent on TGF-β autocrine and paracrine signaling to the phagocyte, indicating that TGF-β is a master regulator of the cell's anti-inflammatory response after the engulfment of corpses [75].

Phagocytes also respond to apoptotic cells by upregulating the cholesterol transporter ABCA1 [81, 82]. This upregulation is transcriptional and induced by recognition of PtdSer [82]. Recent studies show that the upregulation of ABCA1 downstream of apoptotic cell recognition occurs via a novel membrane-initiated pathway [83]. This induction of ABCA1 transcription was rapid, and at least in part, involves the upstream phagocytic receptor BAI1, as well as the downstream signaling intermediates ELMO1 and Rac1. Importantly, this ABCA1 upregulation does not involve the classic LXR-dependent ABCA1 upregulation that is normally seen with increase in intra cytoplasmic oxysterol levels.

The upregulation of ABCA1 can have several effects important for the anti-inflammatory tone of the cells. First, the upregulation of cholesterol transporters unsurprisingly increases cholesterol efflux from the phagocyte. The presence of cholesterol in membranes affects the signaling of cell surface receptors as well as the function of transporters, and therefore the ability to efflux excess cholesterol is an important response to engulfed cells to prevent cholesterol loading [84]. The inability to maintain cholesterol homeostasis can lead to activation of Toll-like receptors (TLRs) and inflammatory signaling [85]. However, ABCA1 (and ABCG1) have other effects as the absence of ABCA1 causes an inflammatory cell death of the phagocyte mediated by sustained JNK activation [86]. Finally, ABCA1 itself can signal as an anti-inflammatory surface protein by signals through ABCA1 to Jak2 and STAT3 [87]. Whether ABCA1 is truly a significant anti-inflammatory molecule in phagocytes immediately after apoptotic cell engulfment remains to be seen, but it is clear that ABCA1 can play important functions in the response to apoptotic cells.

The increase in anti-inflammatory receptors, intracellular signals, and secreted mediators by the phagocyte are important for the dampening of immune responses/local inflammation within a tissue context. In addition to just acting in the engulfing phagocyte, the secreted immunosuppressive cytokines can signal back on the phagocyte as well as to neighboring cells to suppress their proinflammatory responses.

2.4.2 Suppression of Proinflammatory Signals

The suppression of proinflammatory signals is subtly different from the active initiation of an anti-inflammatory program, although the two are connected (Fig. 2.3). When peripheral blood mononuclear cells (PBMCs) were preincubated with apoptotic cells before being stimulated with LPS, they produced less of the inflammatory cytokines TNFα, IL-1β, and IL-12 [6]. The suppression of TNFα, IL-8, GM-CSF, and proinflammatory eicosanoids like leukotriene C_4 is mediated by the autocrine signaling from released TGFβ, Platelet Activating Factor (PAF), and PGE_2 [74]. Although some of the immunosuppressive effects of apoptotic cells can be traced back to the release of anti-inflammatory cytokines by phagocytes, this is not the only mechanism as LPS-induced production of IL-12 gets suppressed by treatment with apoptotic cells independent of TGFβ or IL-10 [88]. The suppression of IL-12 expression was found to be through GC-binding protein (GC-BP), a zinc finger nuclear factor that binds to the IL-12 promoter after apoptotic cell recognition [88]. Others have shown that LXR is crucial for the apoptotic cell-dependent suppression of IL-12 expression; however, LXR also controls a positive feedback loop whereby apoptotic cells signal through LXR to upregulate apoptotic cell receptors [89]. Sorting out the precise signaling pathways downstream of apoptotic cell recognition, the specific transcription factors activated, which genes they activate, and how they link to upregulation of anti-inflammatory cytokines and suppression of proinflammatory signals is a key challenge in the field.

2.5 Defective Apoptotic Cell Clearance and Inflammatory Disease

2.5.1 Autoimmunity

Considering the anti-inflammatory effects of apoptotic cell clearance, it is no surprise that defects in the engulfment of apoptotic cells often result in systemic inflammation. The cause of the inflammation likely depends on the cause of the engulfment defect, but the presence of uncleared apoptotic cells which then undergo secondary necrosis provides a sufficient inflammatory stimulus to cause autoimmunity [3, 5, 90]. Apoptotic cells in the absence of clearance naturally progress to secondarily necrotic cells and lose their membrane integrity, resulting in the release of inflammatory intracellular contents [12, 91]. Some cells during apoptosis may even release proinflammatory cytokines over time [92]. This stresses the importance of efficient clearance early in the apoptotic program as a way to diffuse a ticking time bomb of inflammation and prevent homeostatic cell turnover from inducing a detrimental response. Before much of the apoptotic cell receptors or intracellular signals important for clearance were established, it was observed that excess apoptotic cells injected into mice were immunogenic, inducing the production of antibodies indicative of autoimmunity such as antinuclear and anticardiolipin antibodies [11]. Similarly, blocking the uptake of endogenous apoptotic cells in mice by masking PtdSer to prevent its recognition by phagocytes results in autoimmunity [93].

Now that some of the opsonins of apoptotic cells as well as engulfment receptors have been discovered, autoimmunity has been observed as a frequent result of their deletion in mice. Deficiency of the apoptotic cell bridging molecules C1q or MFG-E8 both result in marked autoimmunity as measured by autoantibodies and the development of glomerulonephritis [94, 95]. Consequently, disrupting the ability of phagocytes to bind MFG-E8 by deleting α_v integrin also results in autoimmunity [96]. Receptors for apoptotic cells using other bridging molecules are not able to compensate for this loss to maintain immune tolerance, and in fact deleting a component of seemingly independent engulfment pathways all result in autoimmunity. For example, preventing signaling by Mer, a receptor for the apoptotic cell bridging molecule Gas6, also results in lupus-like autoimmunity in mice [55, 97]. If direct recognition of PtdSer on apoptotic cells by TIM4 is eliminated by genetic deletion, again autoimmunity results [98, 99].

In all of these mouse models in which autoimmunity occurs, there are also lingering uncleared apoptotic cells in vivo, making it difficult to uncouple these two features. However, DNaseII-deficient macrophages are able to phagocytose apoptotic cells but lack the ability to degrade the engulfed nuclei, resulting in an abundance of DNA-containing bodies throughout the embryo [100, 101]. Although global deletion of DNaseII in mice is embryonic lethal, the induced deletion of DNaseII in adult mice causes the development of autoimmune polyarthritis similar to rheumatoid arthritis [102]. The macrophages that are still able to efficiently engulf apoptotic cells are unable to process the corpses and subsequently have an inflammatory phenotype, including high TNFα expression [102]. These findings suggest that although cells undergoing secondary necrosis can be sufficiently immunogenic, the anti-inflammatory signaling in the phagocytes is also important to prevent autoimmunity.

In patients with systemic lupus erythematosus (SLE), a prototypical autoimmune disease characterized by chronic systemic inflammation, there is an increase in uncleared apoptotic cells in lymph node germinal centers, suggesting failed clearance [103]. The engulfment capacity of peripheral blood mononuclear cells (PBMCs) isolated from SLE patients is markedly decreased compared to PBMCs from normal healthy donors [104]. Whether impaired clearance is a cause or an effect of autoimmunity in humans is yet to be definitively established.

2.5.2 Airway Inflammation

The lung is an interesting model organ to examine how apoptotic cells interact with multiple cell types, including professional phagocytes such as macrophages and nonprofessional phagocytes such as neighboring epithelial cells. The phagocytic responsibilities of these populations could vary based on the different circumstances. For example, during acute injury when there is increased apoptosis of epithelial cells as well as the infiltrating neutrophils, the professional phagocytes are likely the most capable cell type to handle the increased load. This is seen when excessive apoptotic cells are instilled into the lungs of mice. Although multiple cell types are able to clear them, only CD103+ dendritic cells are able to traffic the engulfed apoptotic cells to the lung-draining lymph node to present antigen [105]. In an acute lung injury model in mice, MFG-E8 deficiency exacerbated the effects, suggesting that engulfment is protective in the lung [106].

Mice in which Rac1 is deleted specifically in the epithelial cells are more susceptible to house dust mite (HDM) or ovalbumin-induced airway inflammation [79]. Interestingly, airway epithelial cells rather than the macrophages from these mice have impaired engulfment of apoptotic cells, suggesting that engulfment by the epithelial cells helps to maintain tolerance to allergens. In this context, the cytokines secreted by the engulfing epithelial cells, especially IL-10, appear to be important for the prevention of disease [79]. Although Rac1 is linked to many functions outside of apoptotic cell engulfment, the in vivo and in vitro evidence implicates Rac1 as important in the initiation of airway inflammation. In support of the importance of epithelial cells as engulfers of apoptotic cells, it has been shown that epithelial cells from cystic fibrosis patients, who have increased numbers of free apoptotic cells in their sputum, are defective in their ability to engulf apoptotic cells due to increased levels of RhoA, a GTPase with an inhibitory role in engulfment [107].

In other diseases of the lung, such as emphysema and chronic obstructive pulmonary disease (COPD), apoptotic cell clearance is beneficial. The porcine pancreatic elastase model of emphysema in mice is driven partially by apoptosis in the lung. Cotreatment of these mice with Annexin V to block apoptotic cell uptake worsened the emphysema [108]. Furthermore, smoking has been linked to impaired engulfment of apoptotic cells [109]. The alveolar macrophages from patients with COPD exhibit defective engulfment capacity in vitro, corresponding to the increase in free apoptotic cells in the lungs of these patients [110].

2.5.3 *Inflammatory Colitis*

An example of an inflammatory disease that may be affected by the clearance of apoptotic cells is inflammatory bowel disease. In mouse models of colitis, mice treated with dextran sulfate sodium (DSS) develop reversible acute intestinal inflammation associated with massive apoptosis. When DSS is given to mice lacking MFG-E8, the colitis is more severe than in treated wild-type mice [111]. Mirroring the effect of MFG-E8, mice with conditional knockout of α_v integrin in hematopoietic cells develop spontaneous systemic inflammation, including severe colitis [96]. A causative link between impaired engulfment and the extent of colitis in mice has not been made, and in fact it has been suggested that the effects of MFG-E8 are through alternate anti-inflammatory mechanisms [112]. Recent studies suggest that the levels of the phagocytic receptor BAI1 are altered during DSS-induced colitis and also in human patients with ulcerative colitis [113]. Mice deficient in BAI1 have a much more pronounced disease with large numbers of uncleared apoptotic cells. Perhaps most telling of the relevance of apoptotic cell clearance in this disease model, mice with transgenic overexpression of BAI1 show attenuated disease, with reduction in proinflammatory cytokines and overall reduced inflammation. Interestingly, BAI1 expression in intestinal epithelial cells (rather than the myeloid cells) was critical for the beneficial effect, once again highlighting the key role played by nonprofessional phagocytes during inflammation [113].

2.6 Phagocytes, Anti-inflammatory Responses, and Cancer Context

2.6.1 Phagocytic Signaling Pathways in a Malignant Tissue

It is well known that apoptotic cells are found in many types of solid tumors along with mononuclear phagocytes, particularly, tumor-associated macrophages (TAMs). The current evidence based on in vitro studies, and profiling of the TAMs suggests that the collection of engulfment proteins expressed in these TAMs is not vastly different from conventional tissue-resident macrophages. Therefore, the types of signaling pathways activated during apoptotic cell clearance in the tumor are likely not vastly different from the normal uptake mechanisms. But the real challenge is determining the interplay between the apoptotic tumor cells and the TAMs (see Chap. 3 for more discussion). Some of these challenges include: the type of responses initiated within the phagocytes upon contact with apoptotic and live tumor cells within the tumor tissue; the nature of death within the tumor; hypoxia versus normoxia (depending on the extent of vascularization of the tumor tissue) conditions when the phagocytes engulf; the effect of factors secreted by the healthy and dying tumor cells on the TAMs; the rates of proliferation within the tumor; and finally, the cross talk between the normal cells, tumor cells, and the TAMs and how they may influence the response of the engulfing phagocytes. Currently, there are significant efforts to induce immune responses to tumors via immunogenic cell death (see Chap. 7).

Interestingly, a number of engulfment genes that are linked to apoptotic cell clearance are also linked to tumor development in different tissues. These have been identified either via large-scale screening approaches, individual candidate gene approaches, or via altered expression profiles. We have chosen four examples below to highlight the link between engulfment genes and cancer.

2.6.2 BAI1

The gene encoding the phagocytic receptor BAI1 was initially identified as a p53-inducible gene whose expression was severely reduced or lost in glioblastomas (GBM) [114, 115]. Moreover, it was shown that the thrombospondin repeats of

BAI1 (which also binds PtdSer) could function to inhibit angiogenesis [115]. Given the high degree of vascularization of GBM, it was initially hypothesized that the loss of BAI1 might influence the state of vascularization within the tumors [116–119]. However, given the recent studies directly linking BAI1 to apoptotic cell clearance, this needs to be revisited. GBMs are also known to have a number of apoptotic and necrotic cells, with characteristic "palisade necrosis" seen within the advanced stages of GBM (often lacking BAI1 expression). Whether this is due to defective clearance or defective angiogenesis or both remains to be determined.

2.6.3 ELMO1

The engulfment adapter protein ELMO1 has been linked to different types of cancers and is likely involved in apoptotic cell clearance and its role in tumor cell migration [120–124]. While macrophages express one of the highest levels of ELMO1, ELMO1 is widely expressed and mutations in ELMO1 that confer activation have been linked to head and neck tumors. Since ELMO1 associates with Dock180 to act as a guanine nucleotide exchange factor to activate Rac1 during cell migration [121], ELMO1 has been linked to metastasis of tumor cells in different contexts.

2.6.4 Pannexin Channels

In addition to the known role of Panx1 in the release of find-me signals from apoptotic cells, recently a mutant variant form of pannexin has been identified in highly metastatic human breast cancer cells [125]. This variant has been linked to extrusion of tumor cells from small vessels and metastasis. Although the precise mechanism by which the short fragment of Panx1 mediates this response of tumor cells is unclear, because Panx1 is a hexameric channel, it is possible that there is interference with the "holo channel." Work in the context of apoptosis already suggests that Panx1 can influence cell shape and the cytoskeleton [126], and the short fragment of Panx1 might influence or augment the natural function of Panx1.

2.6.5 MER Tyrosine Kinase

The tyrosine kinase phagocytic receptor MER-TK has been linked to many different types of cancers [127, 128]. Besides its high expression in macrophages, and the induction of MER-TK expression by different stimuli and conditions, MER-TK is also expressed on certain epithelial cells [113]. Moreover, in gene expression profiling studies MER-TK expression has been shown to be both upregulated and downregulated [129]. Given the anti-inflammatory properties associated with MER-TK expression on macrophages [130], there could be complex scenarios depending on MER-TK expression in the macrophages versus tumor cells (see Chap. 6 for more details).

2.7 Concluding Thoughts

Our understanding of apoptotic cell clearance has expanded greatly in the past few decades, both mechanistically and in its physiological roles. There is also much work left to do to parse the complex signaling networks involved in clearance. There are inevitably more components of the engulfment machinery that remain to be identified, whether eat-me signals, engulfment receptors or coreceptors, or intracellular signaling pathways. Many of the players already known do not yet have clearly defined roles, which offer the opportunity to make new progress in this area. Furthermore, it is still unclear how different cell types utilize different engulfment pathways to clear cells and how these differences affect the response to apoptotic cells. A key aspect of tissue homeostasis is the loss/disposal of used/aged cells while replacing them with newer cells/regeneration. This is particularly important for cell types such as neutrophils that are produced in very large numbers, have a relatively short lifespan, and whose production has to be tightly coordinated. Since the loss of neutrophil numbers is a critical complication in many cancer therapies, defining the specific molecular details how the neutrophil death/engulfment/production cycle is coordinated in healthy and treated patients could be highly therapeutically relevant. Another of the beautiful challenges ahead in the context of cancer and cell death is determining under which contexts the apoptotic cell clearance is altered in tumors, whether the anti-inflammatory properties of apoptotic cell recognition are dampening an immune response or modifying the macrophages, how the tumor cells (via engulfment) might be made immunogenic, and how manipulation of the engulfment machinery can be of benefit. In addition, with recent evidence that apoptotic cell clearance can be augmented in vivo [113], the next level challenge is boosting cell clearance via small molecules and to target the myriad of inflammatory diseases.

References

1. Arandjelovic S, Ravichandran KS. Phagocytosis of apoptotic cells in homeostasis. Nat Immunol. 2015;16:907–17.
2. Gregory CD, Pound JD. Cell death in the neighbourhood: direct microenvironmental effects of apoptosis in normal and neoplastic tissues. J Pathol. 2011;223:177–94.
3. Henson PM. Dampening inflammation. Nat Immunol. 2005;6:1179–81.
4. Ravichandran KS, Lorenz U. Engulfment of apoptotic cells: signals for a good meal. Nat Rev Immunol. 2007;7:964–74.

5. Savill J, Dransfield I, Gregory C, Haslett C. A blast from the past: clearance of apoptotic cells regulates immune responses. Nat Rev Immunol. 2002;2:965–75.
6. Voll RE, Herrmann M, Roth EA, Stach C, Kalden JR, Girkontaite I. Immunosuppressive effects of apoptotic cells. Nature. 1997;390:350–1.
7. Haslett C, Savill JS, Whyte MK, Stern M, Dransfield I, Meagher LC. Granulocyte apoptosis and the control of inflammation. Philos Trans R Soc Lond B Biol Sci. 1994;345:327–33.
8. Savill J, Dransfield I, Hogg N, Haslett C. Vitronectin receptor-mediated phagocytosis of cells undergoing apoptosis. Nature. 1990;343:170–3.
9. Savill J, Hogg N, Ren Y, Haslett C. Thrombospondin cooperates with CD36 and the vitronectin receptor in macrophage recognition of neutrophils undergoing apoptosis. J Clin Invest. 1992;90:1513–22.
10. Savill JS, Henson PM, Haslett C. Phagocytosis of aged human neutrophils by macrophages is mediated by a novel "charge-sensitive" recognition mechanism. J Clin Invest. 1989;84:1518–27.
11. Mevorach D, Zhou JL, Song X, Elkon KB. Systemic exposure to irradiated apoptotic cells induces autoantibody production. J Exp Med. 1998;188:387–92.
12. Silva MT. Secondary necrosis: the natural outcome of the complete apoptotic program. FEBS Lett. 2010;584:4491–9.
13. Green DR, Ferguson T, Zitvogel L, Kroemer G. Immunogenic and tolerogenic cell death. Nat Rev Immunol. 2009;9:353–63.
14. Kroemer G, Senovilla L, Galluzzi L, Andre F, Zitvogel L. Natural and therapy-induced immunosurveillance in breast cancer. Nat Med. 2015;21:1128–38.
15. Galluzzi L, Bravo-Sanpedro JM, Vitale I, Aaronson SA, Abrams JM, Adam D, Alnemri ES, Altucci L, Andrews D, Annicchiarico-Petruzzelli M, Baehrecke EH, Bazan NG, Bertrand MJ, Bianchi K, Blagosklonny MV, Blomgren K, Borner C, Bredesen DE, Brenner C, Campanella M, Candi E, Cecconi F, Chan FK, Chandel NS, Cheng EH, Chipuk JE, Cidlowski JA, Ciechanover A, Dawson TM, Dawson VL, de Laurenzi V, de Maria R, Debatin KM, di Daniele N, Dixit VM, Dynlacht BD, el-Deiry WS, Fimia GM, Flavell RA, Fulda S, Garrido C, Gougeon ML, Green DR, Gronemeyer H, Hajnoczky G, Hardwick JM, Hengartner MO, Ichijo H, Joseph B, Jost PJ, Kaufmann T, Kepp O, Klionsky DJ, Knight RA, Kumar S, Lemasters JJ, Levine B, Linkermann A, Lipton SA, Lockshin RA, Lopez-Otin C, Lugli E, Madeo F, Malorni W, Marine JC, Martin SJ, Martinou JC, Medema JP, Meier P, Melino S, Mizushima N, Moll U, Munoz-Pinedo C, Nunez G, Oberst A, Panaretakis T, Penninger JM, Peter ME, Piacentini M, Pinton P, Prehn JH, Puthalakath H, Rabinovich GA, Ravichandran KS, Rizzuto R, Rodrigues CM, Rubinsztein DC, Rudel T, Shi Y, Simon HU, Stockwell BR, Szabadkai G, Tait SW, Tang HL, Tavernarakis N, Tsujimoto Y, vanden Berghe T, Vandenabeele P, Villunger A, Wagner EF, et al. Essential versus accessory aspects of cell death: recommendations of the NCCD 2015. Cell Death Differ. 2015;22:58–73.
16. Arandjelovic S, Ravichandran KS. A MERry response after myocardial infarction. Circ Res. 2013;113:949–51.
17. Elliott MR, Ravichandran KS. ELMO1 signaling in apoptotic germ cell clearance and spermatogenesis. Ann N Y Acad Sci. 2010;1209:30–6.
18. Burstyn-Cohen T, Lew ED, Traves PG, Burrola PG, Hash JC, Lemke G. Genetic dissection of TAM receptor-ligand interaction in retinal pigment epithelial cell phagocytosis. Neuron. 2012;76:1123–32.
19. Prasad D, Rothlin CV, Burrola P, Burstyn-Cohen T, Lu Q, Garcia De Frutos P, Lemke G. TAM receptor function in the retinal pigment epithelium. Mol Cell Neurosci. 2006;33:96–108.
20. Poon IK, Lucas CD, Rossi AG, Ravichandran KS. Apoptotic cell clearance: basic biology and therapeutic potential. Nat Rev Immunol. 2014;14:166–80.
21. Ravichandran KS. Beginnings of a good apoptotic meal: the find-me and eat-me signaling pathways. Immunity. 2011;35:445–55.
22. Lauber K, Bohn E, Kröber SM, Xiao Y-J, Blumenthal SG, Lindemann RK, Marini P, Wiedig C, Zobywalski A, Baksh S, Xu Y, Autenrieth IB, Schulze-Osthoff K, Belka C, Stuhler G,

Wesselborg S. Apoptotic cells induce migration of phagocytes via caspase-3-mediated release of a lipid attraction signal. Cell. 2003;113:717–30.
23. Truman LA, Ford CA, Pasikowska M, Pound JD, Wilkinson SJ, Dumitriu IE, Melville L, Melrose LA, Ogden CA, Nibbs R, Graham G, Combadiere C, Gregory CD. CX3CL1/fractalkine is released from apoptotic lymphocytes to stimulate macrophage chemotaxis. Blood. 2008;112:5026–36.
24. Chekeni FB, Elliott MR, Sandilos JK, Walk SF, Kinchen JM, Lazarowski ER, Armstrong AJ, Penuela S, Laird DW, Salvesen GS, Isakson BE, Bayliss DA, Ravichandran KS. Pannexin 1 channels mediate 'find-me' signal release and membrane permeability during apoptosis. Nature. 2010;467:863–7.
25. Elliott MR, Chekeni FB, Trampont PC, Lazarowski ER, Kadl A, Walk SF, Park D, Woodson RI, Ostankovich M, Sharma P, Lysiak JJ, Harden TK, Leitinger N, Ravichandran KS. Nucleotides released by apoptotic cells act as a find-me signal to promote phagocytic clearance. Nature. 2009;461:282–6.
26. Leonardi-Essmann F, Emig M, Kitamura Y, Spanagel R, Gebicke-Haerter PJ. Fractalkine-upregulated milk-fat globule EGF factor-8 protein in cultured rat microglia. J Neuroimmunol. 2005;160:92–101.
27. Miksa M, Amin D, Wu R, Ravikumar TS, Wang P. Fractalkine-induced MFG-E8 leads to enhanced apoptotic cell clearance by macrophages. Mol Med. 2007;13:553–60.
28. Fadok VA, Voelker DR, Campbell PA, Cohen JJ, Bratton DL, Henson PM. Exposure of phosphatidylserine on the surface of apoptotic lymphocytes triggers specific recognition and removal by macrophages. J Immunol. 1992;148:2207–16.
29. Martin SJ, Reutelingsperger CP, Mcgahon AJ, Rader JA, van Schie RC, Laface DM, Green DR. Early redistribution of plasma membrane phosphatidylserine is a general feature of apoptosis regardless of the initiating stimulus: inhibition by overexpression of Bcl-2 and Abl. J Exp Med. 1995;182:1545–56.
30. Vermes I, Haanen C, Steffens-Nakken H, Reutelingsperger C. A novel assay for apoptosis. Flow cytometric detection of phosphatidylserine expression on early apoptotic cells using fluorescein labelled Annexin V. J Immunol Methods. 1995;184:39–51.
31. Segawa K, Nagata S. An apoptotic 'eat me' signal: phosphatidylserine exposure. Trends Cell Biol. 2015;25:639–50.
32. Borisenko GG, Matsura T, Liu S-X, Tyurin VA, Jianfei J, Serinkan FB, Kagan VE. Macrophage recognition of externalized phosphatidylserine and phagocytosis of apoptotic Jurkat cells—existence of a threshold. Arch Biochem Biophys. 2003;413:41–52.
33. Huynh M-LN, Fadok VA, Henson PM. Phosphatidylserine-dependent ingestion of apoptotic cells promotes TGF-beta1 secretion and the resolution of inflammation. J Clin Invest. 2002;109:41–50.
34. Segawa K, Kurata S, Yanagihashi Y, Brummelkamp TR, Matsuda F, Nagata S. Caspase-mediated cleavage of phospholipid flippase for apoptotic phosphatidylserine exposure. Science. 2014;344:1164–8.
35. Suzuki J, Fujii T, Imao T, Ishihara K, Kuba H, Nagata S. Calcium-dependent phospholipid scramblase activity of TMEM16 protein family members. J Biol Chem. 2013;288: 13305–16.
36. Gardai SJ, Mcphillips KA, Frasch SC, Janssen WJ, Starefeldt A, Murphy-Ullrich JE, Bratton DL, Oldenborg P-A, Michalak M, Henson PM. Cell-surface calreticulin initiates clearance of viable or apoptotic cells through trans-activation of LRP on the phagocyte. Cell. 2005;123:321–34.
37. Obeid M, Tesniere A, Ghiringhelli F, Fimia GM, Apetoh L, Perfettini JL, Castedo M, Mignot G, Panaretakis T, Casares N, Metivier D, Larochette N, van Endert P, Ciccosanti F, Piacentini M, Zitvogel L, Kroemer G. Calreticulin exposure dictates the immunogenicity of cancer cell death. Nat Med. 2007;13:54–61.
38. Ogden CA, Decathelineau A, Hoffmann PR, Bratton D, Ghebrehiwet B, Fadok VA, Henson PM. C1q and mannose binding lectin engagement of cell surface calreticulin and CD91 initiates macropinocytosis and uptake of apoptotic cells. J Exp Med. 2001;194:781–95.

39. Arur S, Uche UE, Rezaul K, Fong M, Scranton V, Cowan AE, Mohler W, Han DK. Annexin I is an endogenous ligand that mediates apoptotic cell engulfment. Dev Cell. 2003;4:587–98.
40. Eda S, Yamanaka M, Beppu M. Carbohydrate-mediated phagocytic recognition of early apoptotic cells undergoing transient capping of CD43 glycoprotein. J Biol Chem. 2004;279:5967–74.
41. Gregory CD, Devitt A. The macrophage and the apoptotic cell: an innate immune interaction viewed simplistically? Immunology. 2004;113:1–14.
42. Moffatt OD, Devitt A, Bell ED, Simmons DL, Gregory CD. Macrophage recognition of ICAM-3 on apoptotic leukocytes. J Immunol. 1999;162:6800–10.
43. Yoon KW, Byun S, Kwon E, Hwang SY, Chu K, Hiraki M, Jo SH, Weins A, Hakroush S, Cebulla A, Sykes DB, Greka A, Mundel P, Fisher DE, Mandinova A, Lee SW. Control of signaling-mediated clearance of apoptotic cells by the tumor suppressor p53. Science. 2015;349:1261669.
44. Zitvogel L, Kroemer G. CANCER. A p53-regulated immune checkpoint relevant to cancer. Science. 2015;349:476–7.
45. Fadok VA, Bratton DL, Rose DM, Pearson A, Ezekewitz RA, Henson PM. A receptor for phosphatidylserine-specific clearance of apoptotic cells. Nature. 2000;405:85–90.
46. Miyanishi M, Tada K, Koike M, Uchiyama Y, Kitamura T, Nagata S. Identification of Tim4 as a phosphatidylserine receptor. Nature. 2007;450:435–9.
47. Park D, Tosello-Trampont A-C, Elliott MR, Lu M, Haney LB, Ma Z, Klibanov AL, Mandell JW, Ravichandran KS. BAI1 is an engulfment receptor for apoptotic cells upstream of the ELMO/Dock180/Rac module. Nature. 2007;450:430–4.
48. Park S-Y, Jung M-Y, Kim H-J, Lee S-J, Kim S-Y, Lee B-H, Kwon T-H, Park R-W, Kim I-S. Rapid cell corpse clearance by stabilin-2, a membrane phosphatidylserine receptor. Cell Death Differ. 2008;15:192–201.
49. de Freitas A, Banerjee S, Xie N, Cui H, Davis KI, Friggeri A, Fu M, Abraham E, Liu G. Identification of TLT2 as an engulfment receptor for apoptotic cells. J Immunol. 2012;188:6381–8.
50. He M, Kubo H, Morimoto K, Fujino N, Suzuki T, Takahasi T, Yamada M, Yamaya M, Maekawa T, Yamamoto Y, Yamamoto H. Receptor for advanced glycation end products binds to phosphatidylserine and assists in the clearance of apoptotic cells. EMBO Rep. 2011;12:358–64.
51. Nishi C, Toda S, Segawa K, Nagata S. Tim4- and MerTK-mediated engulfment of apoptotic cells by mouse resident peritoneal macrophages. Mol Cell Biol. 2014;34:1512–20.
52. Abu-Thuraia A, Gauthier R, Chidiac R, Fukui Y, Screaton RA, Gratton JP, Cote JF. Axl phosphorylates Elmo scaffold proteins to promote Rac activation and cell invasion. Mol Cell Biol. 2015;35:76–87.
53. Hanayama R, Tanaka M, Miwa K, Shinohara A, Iwamatsu A, Nagata S. Identification of a factor that links apoptotic cells to phagocytes. Nature. 2002;417:182–7.
54. Nakano T, Ishimoto Y, Kishino J, Umeda M, Inoue K, Nagata K, Ohashi K, Mizuno K, Arita H. Cell adhesion to phosphatidylserine mediated by a product of growth arrest-specific gene 6. J Biol Chem. 1997;272:29411–4.
55. Scott RS, Mcmahon EJ, Pop SM, Reap EA, Caricchio R, Cohen PL, Earp HS, Matsushima GK. Phagocytosis and clearance of apoptotic cells is mediated by MER. Nature. 2001;411:207–11.
56. Gaipl US, Kuenkele S, Voll RE, Beyer TD, Kolowos W, Heyder P, Kalden JR, Herrmann M. Complement binding is an early feature of necrotic and a rather late event during apoptotic cell death. Cell Death Differ. 2001;8:327–34.
57. Hamon Y, Trompier D, Ma Z, Venegas V, Pophillat M, Mignotte V, Zhou Z, Chimini G. Cooperation between engulfment receptors: the case of ABCA1 and MEGF10. PLoS One. 2006;1, e120.
58. Jaffe AB, Hall A. Rho GTPases: biochemistry and biology. Annu Rev Cell Dev Biol. 2005;21:247–69.
59. Kinchen JM, Cabello J, Klingele D, Wong K, Feichtinger R, Schnabel H, Schnabel R, Hengartner MO. Two pathways converge at CED-10 to mediate actin rearrangement and corpse removal in C. elegans. Nature. 2005;434:93–9.

60. He B, Lu N, Zhou Z. Cellular and nuclear degradation during apoptosis. Curr Opin Cell Biol. 2009;21:900–12.
61. Kinchen JM, Doukoumetzidis K, Almendinger J, Stergiou L, Tosello-Trampont A, Sifri CD, Hengartner MO, Ravichandran KS. A pathway for phagosome maturation during engulfment of apoptotic cells. Nat Cell Biol. 2008;10:556–66.
62. Kinchen JM, Hengartner MO. Tales of cannibalism, suicide, and murder: programmed cell death in C. elegans. Curr Top Dev Biol. 2005;65:1–45.
63. Kinchen JM, Ravichandran KS. Journey to the grave: signaling events regulating removal of apoptotic cells. J Cell Sci. 2007;120:2143–9.
64. Zhou Z, Yu X. Phagosome maturation during the removal of apoptotic cells: receptors lead the way. Trends Cell Biol. 2008;18:474–85.
65. Park S-Y, Kang K-B, Thapa N, Kim S-Y, Lee S-J, Kim I-S. Requirement of adaptor protein GULP during stabilin-2-mediated cell corpse engulfment. J Biol Chem. 2008;283:10593–600.
66. Su HP, Nakada-Tsukui K, Tosello-Trampont A-C, Li Y, Bu G, Henson PM, Ravichandran KS. Interaction of CED-6/GULP, an adapter protein involved in engulfment of apoptotic cells with CED-1 and CD91/low density lipoprotein receptor-related protein (LRP). J Biol Chem. 2002;277:11772–9.
67. Brugnera E, Haney L, Grimsley C, Lu M, Walk SF, Tosello-Trampont A-C, Macara IG, Madhani H, Fink GR, Ravichandran KS. Unconventional Rac-GEF activity is mediated through the Dock180-ELMO complex. Nat Cell Biol. 2002;4:574–82.
68. Debakker CD, Haney LB, Kinchen JM, Grimsley C, Lu M, Klingele D, Hsu P-K, Chou B-K, Cheng L-C, Blangy A, Sondek J, Hengartner MO, Wu Y-C, Ravichandran KS. Phagocytosis of apoptotic cells is regulated by a UNC-73/TRIO-MIG-2/RhoG signaling module and armadillo repeats of CED-12/ELMO. Curr Biol. 2004;14:2208–16.
69. Gumienny TL, Brugnera E, Tosello-Trampont AC, Kinchen JM, Haney LB, Nishiwaki K, Walk SF, Nemergut ME, Macara IG, Francis R, Schedl T, Qin Y, van Aelst L, Hengartner MO, Ravichandran KS. CED-12/ELMO, a novel member of the CrkII/Dock180/Rac pathway, is required for phagocytosis and cell migration. Cell. 2001;107:27–41.
70. Tosello-Trampont A-C, Kinchen JM, Brugnera E, Haney LB, Hengartner MO, Ravichandran KS. Identification of two signaling submodules within the CrkII/ELMO/Dock180 pathway regulating engulfment of apoptotic cells. Cell Death Differ. 2007;14:963–72.
71. Park D, Hochreiter-Hufford A, Ravichandran KS. The phosphatidylserine receptor TIM-4 does not mediate direct signaling. Curr Biol. 2009;19:346–51.
72. Toda S, Hanayama R, Nagata S. Two-step engulfment of apoptotic cells. Mol Cell Biol. 2012;32:118–25.
73. Uderhardt S, Herrmann M, Oskolkova OV, Aschermann S, Bicker W, Ipseiz N, Sarter K, Frey B, Rothe T, Voll R, Nimmerjahn F, Bochkov VN, Schett G, Krönke G. 12/15-lipoxygenase orchestrates the clearance of apoptotic cells and maintains immunologic tolerance. Immunity. 2012;36:834–46.
74. Fadok VA, Bratton DL, Konowal A, Freed PW, Westcott JY, Henson PM. Macrophages that have ingested apoptotic cells in vitro inhibit proinflammatory cytokine production through autocrine/paracrine mechanisms involving TGF-beta, PGE2, and PAF. J Clin Invest. 1998;101:890–8.
75. Freire-De-Lima CG, Xiao YQ, Gardai SJ, Bratton DL, Schiemann WP, Henson PM. Apoptotic cells, through transforming growth factor-beta, coordinately induce anti-inflammatory and suppress pro-inflammatory eicosanoid and NO synthesis in murine macrophages. J Biol Chem. 2006;281:38376–84.
76. Ramos GC, Fernandes D, Charão CT, Souza DG, Teixeira MM, Assreuy J. Apoptotic mimicry: phosphatidylserine liposomes reduce inflammation through activation of peroxisome proliferator-activated receptors (PPARs) in vivo. Br J Pharmacol. 2007;151: 844–50.
77. Xiao YQ, Freire-De-Lima CG, Schiemann WP, Bratton DL, Vandivier RW, Henson PM. Transcriptional and translational regulation of TGF-beta production in response to apoptotic cells. J Immunol. 2008;181:3575–85.

78. Fries DM, Lightfoot R, Koval M, Ischiropoulos H. Autologous apoptotic cell engulfment stimulates chemokine secretion by vascular smooth muscle cells. Am J Pathol. 2005; 167:345–53.
79. Juncadella IJ, Kadl A, Sharma AK, Shim YM, Hochreiter-Hufford A, Borish L, Ravichandran KS. Apoptotic cell clearance by bronchial epithelial cells critically influences airway inflammation. Nature. 2013;493:547–51.
80. Chung EY, Liu J, Homma Y, Zhang Y, Brendolan A, Saggese M, Han J, Silverstein R, Selleri L, Ma X. Interleukin-10 expression in macrophages during phagocytosis of apoptotic cells is mediated by homeodomain proteins Pbx1 and Prep-1. Immunity. 2007;27:952–64.
81. Gerbod-Giannone M-C, Li Y, Holleboom A, Han S, Hsu L-C, Tabas I, Tall AR. TNFalpha induces ABCA1 through NF-kappaB in macrophages and in phagocytes ingesting apoptotic cells. Proc Natl Acad Sci U S A. 2006;103:3112–7.
82. Kiss RS, Elliott MR, Ma Z, Marcel YL, Ravichandran KS. Apoptotic cells induce a phosphatidylserine-dependent homeostatic response from phagocytes. Curr Biol. 2006;16:2252–8.
83. Fond AM, Lee CS, Schulman IG, Kiss RS, Ravichandran KS. Apoptotic cells trigger a membrane-initiated pathway to increase ABCA1. J Clin Invest. 2015;125:2748–58.
84. Lee AG. How lipids affect the activities of integral membrane proteins. Biochim Biophys Acta. 2004;1666:62–87.
85. Sun Y, Ishibashi M, Seimon T, Lee M, Sharma SM, Fitzgerald KA, Samokhin AO, Wang Y, Sayers S, Aikawa M, Jerome WG, Ostrowski MC, Bromme D, Libby P, Tabas IA, Welch CL, Tall AR. Free cholesterol accumulation in macrophage membranes activates toll-like receptors and p38 mitogen-activated protein kinase and induces cathepsin K. Circ Res. 2009;104:455–65.
86. Yvan-Charvet L, Pagler TA, Seimon TA, Thorp E, Welch CL, Witztum JL, Tabas I, Tall AR. ABCA1 and ABCG1 protect against oxidative stress-induced macrophage apoptosis during efferocytosis. Circ Res. 2010;106:1861–9.
87. Tang C, Liu Y, Kessler PS, Vaughan AM, Oram JF. The macrophage cholesterol exporter ABCA1 functions as an anti-inflammatory receptor. J Biol Chem. 2009;284:32336–43.
88. Kim S, Elkon KB, Ma X. Transcriptional suppression of interleukin-12 gene expression following phagocytosis of apoptotic cells. Immunity. 2004;21:643–53.
89. A-Gonzalez N, Bensinger SJ, Hong C, Beceiro S, Bradley MN, Zelcer N, Deniz J, Ramirez C, Díaz M, Gallardo G, de Galarreta CR, Salazar J, Lopez F, Edwards P, Parks J, Andujar M, Tontonoz P, Castrillo A. Apoptotic cells promote their own clearance and immune tolerance through activation of the nuclear receptor LXR. Immunity. 2009;31:245–58.
90. Nagata S, Hanayama R, Kawane K. Autoimmunity and the clearance of dead cells. Cell. 2010;140:619–30.
91. Elliott MR, Ravichandran KS. Clearance of apoptotic cells: implications in health and disease. J Cell Biol. 2010;189:1059–70.
92. Cullen SP, Henry CM, Kearney CJ, Logue SE, Feoktistova M, Tynan GA, Lavelle EC, Leverkus M, Martin SJ. Fas/CD95-induced chemokines can serve as "find-me" signals for apoptotic cells. Mol Cell. 2013;49:1034–48.
93. Asano K, Miwa M, Miwa K, Hanayama R, Nagase H, Nagata S, Tanaka M. Masking of phosphatidylserine inhibits apoptotic cell engulfment and induces autoantibody production in mice. J Exp Med. 2004;200:459–67.
94. Botto M, Dell'agnola C, Bygrave AE, Thompson EM, Cook HT, Petry F, Loos M, Pandolfi PP, Walport MJ. Homozygous C1q deficiency causes glomerulonephritis associated with multiple apoptotic bodies. Nat Genet. 1998;19:56–9.
95. Hanayama R, Tanaka M, Miyasaka K, Aozasa K, Koike M, Uchiyama Y, Nagata S. Autoimmune disease and impaired uptake of apoptotic cells in MFG-E8-deficient mice. Science. 2004;304:1147–50.
96. Lacy-Hulbert A, Smith AM, Tissire H, Barry M, Crowley D, Bronson RT, Roes JT, Savill JS, Hynes RO. Ulcerative colitis and autoimmunity induced by loss of myeloid alphav integrins. Proc Natl Acad Sci U S A. 2007;104:15823–8.

97. Cohen PL, Caricchio R, Abraham V, Camenisch TD, Jennette JC, Roubey RAS, Earp HS, Matsushima G, Reap EA. Delayed apoptotic cell clearance and lupus-like autoimmunity in mice lacking the c-mer membrane tyrosine kinase. J Exp Med. 2002;196:135–40.
98. Rodriguez-Manzanet R, Sanjuan MA, Wu HY, Quintana FJ, Xiao S, Anderson AC, Weiner HL, Green DR, Kuchroo VK. T and B cell hyperactivity and autoimmunity associated with niche-specific defects in apoptotic body clearance in TIM-4-deficient mice. Proc Natl Acad Sci. 2010;107:8706–11.
99. Wong K, Valdez PA, Tan C, Yeh S, Hongo J-A, Ouyang W. Phosphatidylserine receptor Tim-4 is essential for the maintenance of the homeostatic state of resident peritoneal macrophages. Proc Natl Acad Sci U S A. 2010;107:8712–7.
100. Kawane K, Fukuyama H, Yoshida H, Nagase H, Ohsawa Y, Uchiyama Y, Okada K, Iida T, Nagata S. Impaired thymic development in mouse embryos deficient in apoptotic DNA degradation. Nat Immunol. 2003;4:138–44.
101. Krieser RJ, Maclea KS, Longnecker DS, Fields JL, Fiering S, Eastman A. Deoxyribonuclease IIalpha is required during the phagocytic phase of apoptosis and its loss causes perinatal lethality. Cell Death Differ. 2002;9:956–62.
102. Kawane K, Ohtani M, Miwa K, Kizawa T, Kanbara Y, Yoshioka Y, Yoshikawa H, Nagata S. Chronic polyarthritis caused by mammalian DNA that escapes from degradation in macrophages. Nature. 2006;443:998–1002.
103. Baumann I, Kolowos W, Voll RE, Manger B, Gaipl U, Neuhuber WL, Kirchner T, Kalden JR, Herrmann M. Impaired uptake of apoptotic cells into tingible body macrophages in germinal centers of patients with systemic lupus erythematosus. Arthritis Rheum. 2002;46:191–201.
104. Herrmann M, Voll RE, Zoller OM, Hagenhofer M, Ponner BB, Kalden JR. Impaired phagocytosis of apoptotic cell material by monocyte-derived macrophages from patients with systemic lupus erythematosus. Arthritis Rheum. 1998;41:1241–50.
105. Desch AN, Randolph GJ, Murphy K, Gautier EL, Kedl RM, Lahoud MH, Caminschi I, Shortman K, Henson PM, Jakubzick CV. CD103+ pulmonary dendritic cells preferentially acquire and present apoptotic cell-associated antigen. J Exp Med. 2011;208:1789–97.
106. Aziz M, Matsuda A, Yang W-L, Jacob A, Wang P. Milk fat globule-epidermal growth factor-factor 8 attenuates neutrophil infiltration in acute lung injury via modulation of CXCR2. J Immunol. 2012;189:393–402.
107. Vandivier RW, Richens TR, Horstmann SA, Decathelineau AM, Ghosh M, Reynolds SD, Xiao Y-Q, Riches DW, Plumb J, Vachon E, Downey GP, Henson PM. Dysfunctional cystic fibrosis transmembrane conductance regulator inhibits phagocytosis of apoptotic cells with proinflammatory consequences. Am J Physiol Lung Cell Mol Physiol. 2009;297: L677–86.
108. Yoshida S, Minematsu N, Chubachi S, Nakamura H, Miyazaki M, Tsuduki K, Takahashi S, Miyasho T, Iwabuchi T, Takamiya R, Tateno H, Mouded M, Shapiro SD, Asano K, Betsuyaku T. Annexin V decreases PS-mediated macrophage efferocytosis and deteriorates elastase-induced pulmonary emphysema in mice. Am J Physiol Lung Cell Mol Physiol. 2012;303:L852–60.
109. Minemetsu N, Blumental-Perry A, Shapiro SD. Cigarette smoke inhibits engulfment of apoptotic cells by macrophages through inhibition of actin rearrangement. Am J Respir Cell Mol Biol. 2011;44:474–82.
110. Hodge S, Hodge G, Scicchitano R, Reynolds PN, Holmes M. Alveolar macrophages from subjects with chronic obstructive pulmonary disease are deficient in their ability to phagocytose apoptotic airway epithelial cells. Immunol Cell Biol. 2003;81:289–96.
111. Chogle A, Bu H-F, Wang X, Brown JB, Chou PM, Tan X-D. Milk fat globule-EGF factor 8 is a critical protein for healing of dextran sodium sulfate-induced acute colitis in mice. Mol Med. 2011;17:502–7.
112. Aziz MM, Ishihara S, Mishima Y, Oshima N, Moriyama I, Yuki T, Kadowaki Y, Rumi MAK, Amano Y, Kinoshita Y. MFG-E8 attenuates intestinal inflammation in murine experimental colitis by modulating osteopontin-dependent alphavbeta3 integrin signaling. J Immunol. 2009;182:7222–32.

113. Lee CS, Penberthy KK, Wheeler KM, Juncadella IJ, Vandenabeele P, Lysiak JJ, RavichandranKS. Boosting Apoptotic Cell Clearance by Colonic Epithelial Cells Attenuates Inflammation In Vivo. Immunity. 2016 Apr 19;44(4):807–20. doi: 10.1016/j.immuni.2016.02.005.
114. Kaur B, Brat DJ, Devi NS, van Meir EG. Vasculostatin, a proteolytic fragment of brain angiogenesis inhibitor 1, is an antiangiogenic and antitumorigenic factor. Oncogene. 2005;24:3632–42.
115. Nishimori H, Shiratsuchi T, Urano T, Kimura Y, Kiyono K, Tatsumi K, Yoshida S, Ono M, Kuwano M, Nakamura Y, Tokino T. A novel brain-specific p53-target gene, BAI1, containing thrombospondin type 1 repeats inhibits experimental angiogenesis. Oncogene. 1997;15:2145–50.
116. Cork SM, VAN Meir EG. Emerging roles for the BAI1 protein family in the regulation of phagocytosis, synaptogenesis, neurovasculature, and tumor development. J Mol Med (Berl). 2011;89:743–52.
117. Kaur B, Brat DJ, Calkins CC, van Meir EG. Brain angiogenesis inhibitor 1 is differentially expressed in normal brain and glioblastoma independently of p53 expression. Am J Pathol. 2003;162:19–27.
118. Stephenson JR, Paavola KJ, Schaefer SA, Kaur B, van Meir EG, Hall RA. Brain-specific angiogenesis inhibitor-1 signaling, regulation, and enrichment in the postsynaptic density. J Biol Chem. 2013;288:22248–56.
119. Zhu D, Hunter SB, Vertino PM, van Meir EG. Overexpression of MBD2 in glioblastoma maintains epigenetic silencing and inhibits the antiangiogenic function of the tumor suppressor gene BAI1. Cancer Res. 2011;71:5859–70.
120. Dulak AM, Stojanov P, Peng S, Lawrence MS, Fox C, Stewart C, Bandla S, Imamura Y, Schumacher SE, Shefler E, McKenna A, Carter SL, Cibulskis K, Sivachenko A, Saksena G, Voet D, Ramos AH, Auclair D, Thompson K, Sougnez C, Onofrio RC, Guiducci C, Beroukhim R, Zhou Z, Lin L, Lin J, Reddy R, Chang A, Landrenau R, Pennathur A, Ogino S, Luketich JD, Golub TR, Gabriel SB, Lander ES, Beer DG, Godfrey TE, Getz G, Bass AJ. Exome and whole-genome sequencing of esophageal adenocarcinoma identifies recurrent driver events and mutational complexity. Nat Genet. 2013;45:478–86.
121. Grimsley CM, Kinchen JM, Tosello-Trampont AC, Brugnera E, Haney LB, Lu M, Chen Q, Klingele D, Hengartner MO, Ravichandran KS. Dock180 and ELMO1 proteins cooperate to promote evolutionarily conserved Rac-dependent cell migration. J Biol Chem. 2004;279:6087–97.
122. Li H, Yang L, Fu H, Yan J, Wang Y, Guo H, Hao X, Xu X, Jin T, Zhang N. Association between Galphai2 and ELMO1/Dock180 connects chemokine signalling with Rac activation and metastasis. Nat Commun. 2013;4:1706.
123. Shi L, Zhang B, Sun X, Zhang X, Lv S, Li H, Wang X, Zhao C, Zhang H, Xie X, Wang Y, Zhang P. CC chemokine ligand 18(CCL18) promotes migration and invasion of lung cancer cells by binding to Nir1 through. Nir1-ELMO1/DOC180 signaling pathway. Mol Carcinog. 2016 Jan 12. doi: 10.1002/mc.22450.
124. Zhang B, Shi L, Lu S, Sun X, Liu Y, Li H, Wang X, Zhao C, Zhang H, Wang Y. Autocrine IL-8 promotes F-actin polymerization and mediate mesenchymal transition via ELMO1-NF-kappaB-Snail signaling in glioma. Cancer Biol Ther. 2015;16:898–911.
125. Furlow PW, Zhang S, Soong TD, Halberg N, Goodarzi H, Mangrum C, Wu YG, Elemento O, Tavazoie SF. Mechanosensitive pannexin-1 channels mediate microvascular metastatic cell survival. Nat Cell Biol. 2015;17:943–52.
126. Poon IK, Chiu YH, Armstrong AJ, Kinchen JM, Juncadella IJ, Bayliss DA, Ravichandran KS. Unexpected link between an antibiotic, pannexin channels and apoptosis. Nature. 2014;507:329–34.
127. Lemke G, Burstyn-Cohen T. TAM receptors and the clearance of apoptotic cells. Ann N Y Acad Sci. 2010;1209:23–9.

128. Lemke G, Rothlin CV. Immunobiology of the TAM receptors. Nat Rev Immunol. 2008;8:327–36.
129. Bosurgi L, Bernink JH, Delgado Cuevas V, Gagliani N, Joannas L, Schmid ET, Booth CJ, Ghosh S, Rothlin CV. Paradoxical role of the proto-oncogene Axl and Mer receptor tyrosine kinases in colon cancer. Proc Natl Acad Sci U S A. 2013;110:13091–6.
130. Rothlin CV, Carrera-Silva EA, Bosurgi L, Ghosh S. TAM receptor signaling in immune homeostasis. Annu Rev Immunol. 2015;33:355–91.

Chapter 3
Microenvironmental Effects of Cell Death in Malignant Disease

Christopher D. Gregory, Catriona A. Ford, and Jorine J.L.P. Voss

Abstract Although apoptosis is well recognized as a cell death program with clear anticancer roles, accumulating evidence linking apoptosis with tissue repair and regeneration indicates that its relationship with malignant disease is more complex than previously thought. Here we review how the responses of neighboring cells in the microenvironment of apoptotic tumor cells may contribute to the cell birth/cell death disequilibrium that provides the basis for cancerous tissue emergence and growth. We describe the bioactive properties of apoptotic cells and consider, in particular, how apoptosis of tumor cells can engender a range of responses including pro-oncogenic signals having proliferative, angiogenic, reparatory, and immunosuppressive features. Drawing on the parallels between wound healing, tissue regeneration and cancer, we propose the concept of the "onco-regenerative niche," a cell death-driven generic network of tissue repair and regenerative mechanisms that are hijacked in cancer. Finally, we consider how the responses to cell death in tumors can be targeted to provide more effective and long-lasting therapies.

Keywords Cell death • Apoptosis • Tumor microenvironment • Macrophage • Extracellular vesicle • Burkitt lymphoma • Starry-sky • Angiogenesis • Onco-regenerative niche • Anticancer therapy

3.1 Introduction: Imbalances of Cell Birth and Cell Death in Cancer

Cell gain is finely balanced ultimately by cell loss in normal tissue homeostasis. In cancer, by contrast, this balance is lost and cell gain (cell birth) outweighs cell loss (Fig. 3.1). Of possible cell loss mechanisms, which include cell migration, differentiation, and cell death, the latter plays crucial roles in regulating the sizes of

C.D. Gregory (✉) • C.A. Ford • J.J.L.P. Voss
MRC Centre for Inflammation Research, University of Edinburgh Queen's Medical Research Institute, Edinburgh EH16 4TJ, UK
e-mail: chris.gregory@ed.ac.uk

C.D. Gregory (ed.), *Apoptosis in Cancer Pathogenesis and Anti-cancer Therapy*, Advances in Experimental Medicine and Biology 930,
DOI 10.1007/978-3-319-39406-0_3

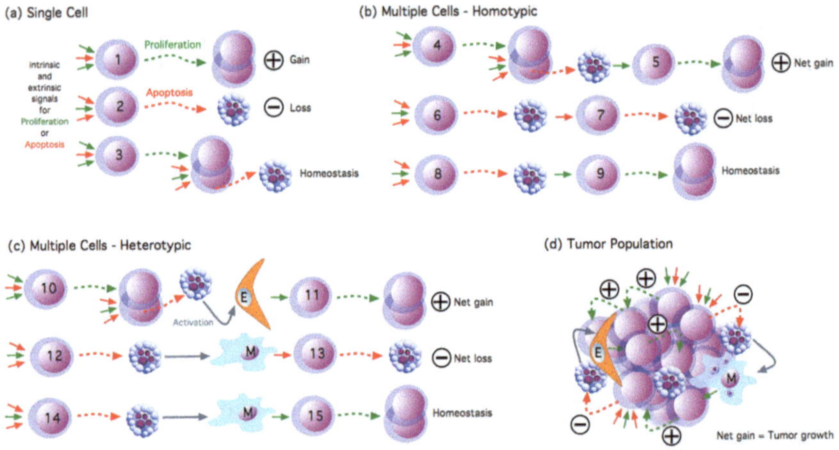

Fig. 3.1 Cell birth/cell death balances and imbalances and their relationships to the tumor microenvironment. Conceptual examples illustrating the impact of dying cells on tissue microenvironments. (**a**) Effects of proliferation or apoptosis of single cells. In the case of cell 1, the integration of external and internal mitogenic (*green arrow*) and apoptotic (*red arrow*) signals favors proliferation (*green dotted arrow*) and leads to cell gain. For cell 2, apoptosis signaling predominates, the cell dies (*red dotted arrow*), and the endpoint is cell loss. Cell 3 initially proliferates and one of its progeny subsequently undergoes apoptosis. In this case cell birth is balanced by cell death, illustrating homeostasis. (**b**) Examples of homotypic effects of dying cells in multicellular populations. In the *upper* scenario, cell 4 initially proliferates and one of its progeny undergoes apoptosis. Subsequently, this apoptotic cell provides mitogenic signal(s) to cell 5, a member of the same lineage as cell 4. Cell 5 proliferates, the result being net gain of cells. This illustrates compensatory proliferation. Cell 6 interacts with its relative, cell 7 as illustrated, which results in net loss of cells ("apoptosis-induced apoptosis") whereas cell 8 interacts with cell 9 in homeostatic mode via compensatory proliferation. (**c**) Conceptual illustrations of apoptotic cells influencing multicellular populations through heterotypic interactions. Cell 10 initially proliferates and one of its progeny subsequently dies. The latter cell then activates a member of another cell lineage within the tissue—for example, an endothelial cell (E). This cell then leads to a relative of cell 10, cell 11, receiving mitogenic signals (e.g., through improved vascularization) resulting in net gain of cells. In the middle example, death of cell 12 leads to activation of a different lineage (for example, an immune cell) which causes further death of cell 12's relatives such as cell 13, leading to net cell loss. In the lower example, the death of cell 14 stimulates compensatory proliferation in its relative, cell 15 via activation of a cell of another lineage (for example, a macrophage, M). (**d**) Application of the mechanisms illustrated in (**a–c**) to the tumor microenvironment, indicating how death of tumor cells can lead to positive as well as negative signals influencing tumor growth, ultimately with net growth being the outcome

normal and malignant cell populations. Although physiological cell death mechanisms have been vigorously pursued over recent years, our knowledge of the molecular mechanisms that regulate regression of cell populations through cell death lags behind that of population expansion through proliferation. Expansion of a rogue tissue through dysregulated imbalance of normal homeostatic mechanisms is the fundamental principle at the root of malignant disease which stems from loss of the key controlling elements of both cell proliferation and cell survival/death.

Genetic mutations that result in gain of function of oncoproteins like Bcl-2 and loss of function of tumor suppressors like p53 are classical examples of canonical pathways that lead to inappropriate survival of premalignant and malignant cells, promoting the cell birth/death imbalance in cancer. The resulting imbalance is dynamic: in other words, although cell death mechanisms may well be required to be inhibited in individual cells acquiring cancerous characteristics [1, 2], such mechanisms are often prominent in the populations of transformed cells in established tumors. Thus, it is not the case that regulated cell death mechanisms, notably apoptosis, are switched off in malignant disease. It follows that cell death mechanisms of ultimate advantage to the success of the malignant populations will be retained. Indeed, 'constitutive' apoptosis is often highly prominent in aggressive cancers. The critical point for successful establishment and progression of malignant disease is that the cell birth/cell death equation favors cell birth, the *net* effect being population expansion (Fig. 3.1). It is becoming clear that the dying component in this dynamic disequilibrium is not a passive entity but rather can feed positively into the cell population expansion and malignant evolutionary processes. This unexpected oncogenic property of dying cells is the main subject of this chapter.

Historically, in contrast to cell gain, cell loss has not been widely studied in the context of cancer. In cases where it has, there is a clear message: substantial cell loss either (a) logically virtually balances out cell gain in slow growing tumors; or (b) nonintuitively is associated with aggressive, rapidly growing tumors. Cell loss means that even rapidly growing tumors are actually substantially constrained in their volumetric growth rates, relative to the doubling time of the tumor cells themselves [3].

An example of a relatively 'balanced' tumor is basal cell carcinoma (BCC). One of the most common human tumors, BCC is a slow growing, locally invasive tumor which takes many months to double in size. However, average cell doubling times in BCC are only around 9 days. The relatively slow net expansion of the tumor cell population is likely to be due to cell death providing a significant balancing effect. Notably, histologically overt evidence of apoptosis—pyknotic nuclei—is common in these tumors [4]. Similar balancing effects have been reported for micrometastases [5]. A logical consequence of aggressive tumors outgrowing local anabolic resources is that, in rapidly growing tumors, cell death becomes more prominent as the tumors get progressively bigger, tumor population growth slowing in parallel [6]. Such death has commonly been described as necrotic—as this is easier to discern histologically—but apoptosis (again reported by the hallmark pyknotic nuclei of apoptotic cells) is also a conspicuous feature of tumors that have outgrown nutrient and oxygen supplies [7, 8].

Like mitosis, apoptosis is prominent in a wide variety of established, aggressive tumor types, including non-Hodgkin's lymphoma (NHL) [9], squamous cell carcinoma [10, 11], transitional cell carcinoma [12], hepatocellular carcinoma [13], and undoubtedly many others. The preferential association of constitutive apoptosis with aggressiveness of malignant disease is reflected in recent work indicating that cleaved caspase-3, the activated form of the apoptosis effector caspase-3, is a risk factor in gastric, ovarian, and cervical cancers [14]. The significance of a relatively high constitutive apoptosis index should not be underestimated since it not only represents a very high level of cell death but also links strongly to poor prognosis. Just as mitotic figures represent only a very brief visualization window on proliferating cells, so apoptosis in histological sections is only fleetingly evident because of the rapidity with which apoptotic cells disappear from microscopic view. Since they are efficiently phagocytosed and digested, apoptotic cells are only identifiable in tissues for around 1–3 hours [15]. Therefore, just as small numbers of mitotic figures are indicative of very rapid proliferation, so small numbers of apoptotic bodies are likely to indicate massive cell death in situ. In certain tumors such as "starry-sky" lymphomas (to be discussed in further detail later in this chapter), mitosis, apoptosis, and phagocytic clearance of apoptotic cells by macrophages can all be routinely observed in standard histological fields (Fig. 3.2). These lymphomas take their description from the infiltrating macrophages, which appear as bright "stars" in a darkly staining "sky" of tumor cells in standard histological sections.

Refsum and Berdal stated almost half a century ago: "Tumour growth must be looked upon as dependent on the process(es) of cell proliferation and cell loss, where changes in any of (these processes) can result in profound alterations in the clinical rate of tumour growth. The cancer problem is not only why cancer cells proliferate uncontrolled, but also why so many of these cells die…" [16]. In the ensuing decades, much has been learned about modes of cell loss and their consequences. As we discuss in the following, although it has greater renown for its tumor-suppressive properties, cell loss by apoptosis also has sinister oncogenic attributes.

3.2 A Poisoned Chalice: Tumor-Promoting as well as Tumor-Suppressing Roles of Apoptosis

It has been appreciated for well over a century since Paget's original 'seed and soil' suggestion in 1889 [17] that the tumor microenvironment plays critical roles in the aggressiveness of malignant disease. Tumors are rogue tissues, often with features resembling chronic wounds, described by Dvorak as "wounds that do not heal" [18]. It is beyond doubt that stromal cellular elements, the nontransformed cells of tumors, including fibroblasts, endothelial cells, fat cells, and inflammatory cells, participate in a two-way signaling conversation with the transformed cells in order to produce a deregulated, aggressive, invasive tissue. Less well appreciated are the roles tumor cell death can play in conditioning the tumor microenvironment. Apoptosis appears to act as a bipolar modulator of tumor growth and progression: on the one hand, it is well established that apoptosis functions to prevent oncogenesis, while on the other, as we detail as follows, apoptosis has tumor-promoting properties. But how can apoptosis function both as tumor suppressor and tumor promoter? A comparison between the role of apoptosis in single cells versus its roles in cell populations helps to unify this apparent paradox (Fig. 3.1).

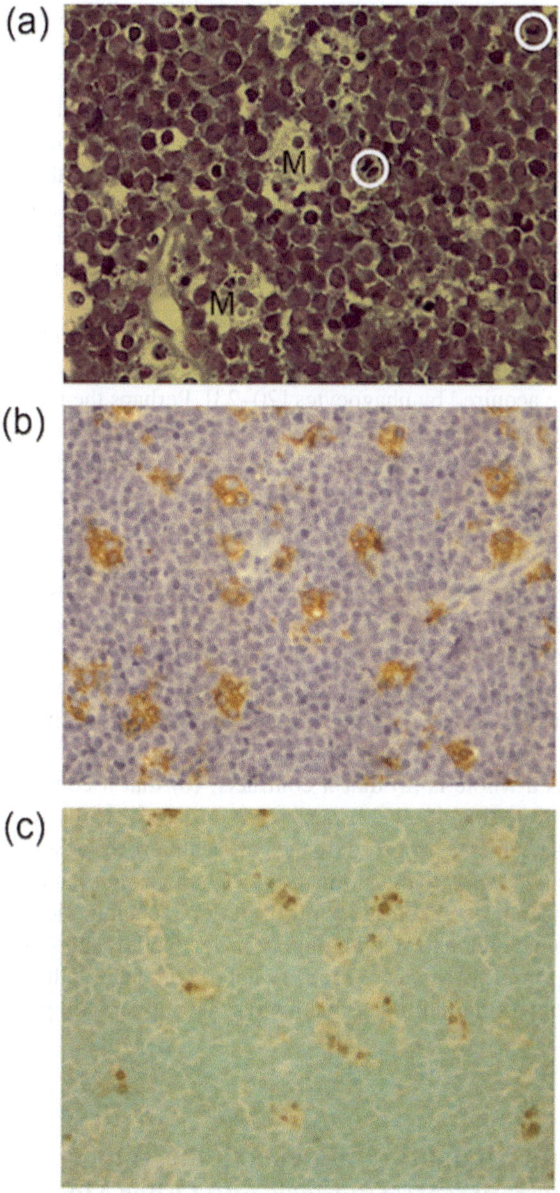

Fig. 3.2 Aggressive "starry-sky" non-Hodgkin's B-cell lymphoma. Standard hematoxylin and eosin (**a**) and immunohistochemical preparations (**b**, **c**) of the classical starry-sky lymphoma, Burkitt's lymphoma. (**a**) The starry-sky macrophages (M) are apparent as areas of brightness against a dark background of tumor cells. The macrophages contain apoptotic tumor cells in various states of degradation. Mitotic figures are also present (circled). (**b**) Macrophages labeled according to their expression of CD68. (**c**) Apoptotic cells revealed by in situ end labeling of cleaved DNA. Most are associated with the starry-sky macrophages

In a single cell, the cell-autonomous consequence of apoptosis is, of course, that the cell is deleted and removed from the population. This cell fate decision is based upon external signals, which interact with the internal context of the cell. This means that apparently identical external signals (for example, a TNF-family member or hypoxic stress) can have different consequences depending on the internal composition of the cell as well as its microenvironment [19]. In the early stages of oncogenesis, the decision to undergo apoptosis is likely to be an important mechanism for removal of individual premalignant cells that have acquired potentially dangerous genetic mutations. However, apoptosis—even of a single cell—cannot be regarded as an isolated event because there are consequences of such an event for the tissue in which that cell resides: for example, phagocytosis and other responses in neighboring cells (vide infra). Furthermore, genetic material from apoptotic tumor cells can be acquired by phagocytes [20–23]. Perhaps the most efficient way of getting rid of a potentially dangerous cell is by jettisoning it to the exterior environment as is seen in the sloughing off of apoptotic, terminally differentiated cells in epithelia. Even here, however, cells in lower layers have potential to respond to their dying upper neighbors. This is illustrated by the retrieval in some mammalian species of dying, epithelial cells by phagocytes, for example, in the guinea pig, dying apical enterocytes are engulfed by responsive macrophages from the lamina propria [24]. Therefore, an individual dying cell can impact profoundly upon its near and distant neighbors.

In a population of tumor cells, while the core apoptosis program is cell autonomous (whether it is triggered through intrinsic or extrinsic mechanisms), and the consequence for individual cells is loss of those cells, the net effect for the tumor cell population as a whole is (a) that it contracts; (b) that it continues to expand, though at a lower rate than if all cells survived; or paradoxically (c), that it grows with the aid of mechanisms triggered by the cell death process itself (Fig. 3.1). For example, as discussed elsewhere in this book (Chap. 4) and later in this chapter, cell death can engender compensatory proliferation, accumulation, and protumor activation of stromal elements and angiogenesis, as well as suppression of antitumor immune responses. In this way, the very act of cell deletion has real and sinister potential to activate a multitude of pathways that support cell gain in outpacing cell loss [25–28].

3.3 Cell Death Modalities Influencing the Tumor Microenvironment: Apoptosis and Other Mechanisms

Multiple cell death routes including apoptosis, necrosis, and autophagy-associated cell death are prominent in malignant disease, even in individual tumors. Indeed, the available evidence suggests that multiple forms of cell death are likely to occur in individual tumors, either constitutively via stress responses, or as a consequence of therapy. As well as apoptosis, necrosis, and autophagy-associated cell death, additional modalities such as anoikis, mitotic catastrophe, entosis, necroptosis, and others are all likely to contribute to the tumor microenvironment to a greater or lesser extent in concert with other cell fate processes including differentiation, proliferation, and senescence.

Caspase-dependent apoptosis is the most widely studied form of programmed cell death in cancer and is our main focus here. The seminal definition of apoptosis by Kerr, Wyllie, and Currie in their classic work was based on morphological features of dying cells in multiple contexts, not least malignant diseases [29]. These authors and others noted the peculiar hallmarks of dying tumor cells: the cell shrinkage, chromatin condensation, and the formation of vesicles and apoptotic bodies containing organelles apparently free from degradation in many different tumor types. Later, biochemical features of apoptosis subdivided this form of programmed cell death into caspase-dependent and caspase-independent categories (see [30]). The most renowned microenvironmental effects of apoptosis are anti-inflammatory and tolerogenic phagocytic clearance responses, which may be hijacked in tumors to provide them with immunological escape mechanisms [31]. However, certain cell death stimuli, notably anthracycline anticancer drugs, induce immunogenic apoptosis [32]. Currently, the microenvironmental effects of apoptosis and other forms of cell death in cancer are largely underestimated and underinvestigated.

3.3.1 Triggering of Apoptosis in Growing Tumors

A generic principle at the root of cell death in malignant disease is environmental stress: the nascent tumor must be located in and/or acquire the appropriate generative niche in which to evolve and grow and it is important that, at the population level, the niche fosters the imbalance of cell birth over cell death that is required for successful tumor growth. In fact it seems likely that many tumors or tumor cell clones fail to achieve persistent net growth and never become clinically problematic. Alternatively, death of such clones may 'feed' potentially more aggressive neoplastic or preneoplastic cells. Once established, an aggressively growing tumor rapidly outgrows its environment and it is logical that metabolic cellular stress stimuli—limitations in essential nutrient, growth factor, and oxygen supplies—feature prominently in triggering multiple cell death pathways. Cell death induced by therapy adds an additional dimension as does death due to the relatively inefficient seeding of malignant clones to metastatic sites [33] (Fig. 3.3).

Thus, multiple factors are likely to contribute to the cell fate decision-making processes that culminate in apoptosis, as well as other forms of cell death, in tumors [19, 35, 36]. Apoptotic signaling pathways may be initiated at multiple cellular locations including the plasma membrane (e.g., death receptor signaling; absence of growth factors), nucleus (irreparable DNA damage response), endoplasmic reticulum (e.g., unfolded protein response), and mitochondria. The latter organelles are central to the intrinsic apoptosis program, notably its initiation and also contribute to execution and amplification of extrinsic apoptosis pathways. Conditions of oxygen stress, glucose, and amino acid deprivation caused by rapid tumor growth

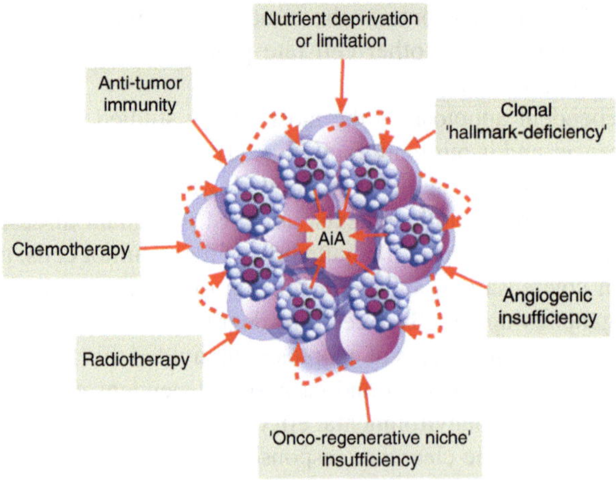

Fig. 3.3 Triggers for tumor cell apoptosis of the tumor microenvironment. Illustrative sources of proapoptotic stimuli in tumors. In addition to apoptosis induced by therapy or antitumor immunity, limited amounts of nutrients and oxygen as a consequence of insufficient angiogenesis are likely to be important in generating proapoptotic signals. Clonal 'hallmark-deficiency' refers to clones of tumor cells carrying genetic mutations but fail to acquire hallmark characteristics of cancer cells (for example, as defined by Hanahan and Weinberg [2]) necessary for survival in primary or alternatively metastatic tissue locations. The 'Onco-regenerative niche' refers to a putative collection of conserved tissue repair and regeneration mechanisms driven by cell damage and hijacked in cancer (see text for details). AiA: propagation of apoptosis triggering through the process of 'apoptosis-induced apoptosis' [34]

are therefore likely to have fundamental effects on tumor cell populations by eliciting proapoptotic signals at mitochondria and the endoplasmic reticulum [35, 36]. In order to maintain the cell birth/cell death imbalance in favor of tumor growth, it follows that the tumor microenvironment needs to adapt in a dynamic way. As we will discuss, responses to apoptosis play key roles in such adaptation.

3.3.2 Necrosis, Autophagy, and Other Mechanisms

In addition to our main focus on apoptosis, brief discussion of other forms of cell death is also warranted here since multiple death processes can contribute to the tumor microenvironment (see [37] for a recent review of accidental versus regulated forms of cell death). Importantly, different forms of cell death may condition the tumor microenvironment in contrasting ways such as anti-inflammatory versus pro-inflammatory, tolerogenic versus immunogenic, death promoting versus death inhibiting. The 'balance' of these effects has not yet been studied in detail but will undoubtedly impinge significantly on tumor evolution and patient outcome and the underlying mechanisms have potential to provide information on novel therapeutic targeting.

Among cell death processes, the classical antithesis of apoptosis is necrosis. Because of the relative ease by which necrotic death can be visualized histologically, this passive, catastrophic form of cell death was appreciated long before apoptosis or other forms of programmed cell death because it affects contiguous tracts of cells, rendering the process microscopically obvious [38]. In rapidly growing tumors, necrotic lesions are prominent due to the growing tumor population outpacing effective nutrient supplies and gaseous exchange. Such a stressful scenario also elicits other forms of cell death (vide infra). In stark contrast to apoptosis, necrosis is typically portrayed as proinflammatory and immunogenic (although this generalized principle has been challenged, see [39]) and it is by no means clear how necrosis contributes to tumor cell biology. Given the generality of necrosis in aggressive tumors and its association with poor prognosis, it seems likely that it fails to impart a negative influence beyond that of the cells immediately affected. The same may prove to be true of regulated necrosis (necroptosis), which has the same morphological features as necrosis [40].

Based on its origin from cell damage, it has been suggested that the main role of necrosis is to induce tissue repair responses [41] and indeed it has been proposed as a critical inducer of tumorigenesis [42]. Therefore in the context of tumor biology, necrosis, like apoptosis, may also prove to feed into the cell birth/cell death equation in complex ways. Furthermore, necrosis can effectively suppress antitumor T cell immunity [43]. In many, perhaps all, cases, necrotic lesions in tumors are accompanied or presaged by apoptotic or other death modalities. Furthermore, the microenvironmental effects of necrosis may, in principle, be dominantly subverted by the effects of alternative modes of cell death or vice versa. Although the relative dominance of cell death responses in various tissue damage or disease scenarios has not been studied in detail, it is of interest to note that, in terms of activating macrophage migratory responses, at least in the case of the Drosophila hemocyte, apoptosis dominates over acute wounding (laser-induced cell ablation) and developmental growth factor signals [44].

Autophagic cell death is also likely to play important roles in tumor microenvironmental conditioning alongside apoptosis and other forms of cell death [45]. Although autophagy is a cell survival mechanism that is triggered under conditions of stress, including the challenging tissue environment in cancer, its deregulation or failure can lead to cell death. In cancer its divergent roles are context dependent [46]. A most interesting case in point comes from recent metabolic studies in prostate cancer in which it was found that arginine starvation (by arginine deiminase) caused autophagic death of prostate cancer cells deficient in arginosuccinate synthetase. The cell death was characterized by mitochondrial depolarization and breakdown of chromatin in large autophagosomes [47]. Autophagy has been reported to limit inflammation associated with necrosis [48] and downregulate proinflammatory cytokine responses at the level of transcription [49]. In this anti-inflammatory context, autophagy broadly resembles apoptosis in its microenvironmental effects. Although its role in suppressing antitumor immunity (as is the case with apoptosis) has yet to be defined in detail, a recent study suggests that autophagic degradation of nuclear components, notably lamin B1 has tumor-suppressive properties through its capacity to induce oncogene-mediated senescence [50].

3.4 Biological Activities of Apoptotic Tumor Cells

The anti-inflammatory and tolerogenic effects of apoptosis are well established and have been reviewed extensively elsewhere (see Chaps. 2 and 7). By contrast, although it has been known for well over half a century that dying cells can promote tumor growth, this property has been underappreciated. Early studies showed that tumor cells that were destined to die (lethally irradiated or histoincompatible cells) exhibited substantial tumor growth-promoting activities when admixed with small numbers of compatible viable tumor cells in murine transplantable carcinoma, sarcoma, and lymphoma models [51–53]. Host systemic effects were ruled out by the key observation that the admixed populations must be present at the same anatomical site in order to achieve enhanced tumor growth [52]. Initial clarification of underlying mechanisms was provided by the results of investigations of ascites tumors transplanted in diffusion chambers, which separated the host microenvironment from the growing tumor cells (together with admixed damaged cells) by a 0.45 μm membrane. Critically, these studies showed that direct, contact-mediated effects of host cells in the immediate tumor microenvironment were not required in order for the growth-promoting effects of lethally irradiated tumor cells to be revealed. Thus, Ehrlich ascites and L1210 lymphoma cell growth were increased by orders of magnitude in the presence of lethally irradiated cells which would be presumed to have been committed to undergo apoptosis. Heat-killed, necrotic, cells induced only a small, statistically insignificant growth-promoting effect in comparison [54]. Although contributions of soluble factors or extracellular vesicles from host inflammatory or other cells in these models cannot be excluded, these studies suggested a 'feeder' principle in the tumor microenvironment emanating from dying or dead cells. In this respect, the dying/dead cells were thought of as a source of nutrients in much the same way as irradiated cells could be used as a source of feeder cells to clone tumor lines like HeLa [55].

These early studies serve to provide a firm grounding for the establishment of the principle that cell death can impart potentially significant oncogenic effects in the microenvironment of tumors. In our opinion, it seems likely, however, that there will prove to be exceptions to this principle and/or alternative, context-dependent effects, which are growth inhibitory and therefore tumor suppressive. In addition, it seems logical that dying cells will prove to have different properties from dead cells. One possibility is that, since dying cells are cleared so efficiently from tissues under normal circumstances (free dead cells never being observed in normal physiology), dead cells engender unwanted effects such as proinflammatory properties. It is noteworthy in the context of cancer that, at least in vitro in the case of hybridoma cells, cell death constrains growth and antibody productivity in the surviving cells [56]. These observations suggest that the positive and negative effects of cell death driven by multiple factors, including type of death, type of dying cell, phase of death, and type of responding cell are likely to be integrated in the tumor microenvironment to determine overall biological outcome.

3.4.1 Mechanisms

Accumulating evidence indicates that the biological activities of apoptotic cells, including effects of relevance to the tumor microenvironment, are mediated through three mechanisms: (1) intercellular contact, (2) production of extracellular vesicles, and (3) release of soluble factors.

Although the specific molecular details are far from clear (see [15, 39] for reviews), the recognition and binding mechanisms that are a prerequisite for the clearance of apoptotic cells through phagocytosis, together with the production of secreted factors directly by apoptotic cells, are associated with additional, nonphagocytic responses of interacting cells (including both phagocytes and non-phagocytes). These range from anti-inflammatory mediators and migratory responses to cell fate decisions such as cell survival, growth, differentiation, and death. Current knowledge is summarized in Table 3.1 and Fig. 3.4.

The exposure of phosphatidylserine (PS) is the most renowned plasma membrane moiety in the cell–cell interactions that contribute to these processes. Other alterations in the plasma membranes of apoptotic cells, involving additional lipids as well as carbohydrates, proteins, and nucleic acids, undoubtedly play key roles too. For example, apoptotic cells display increased levels of heat shock proteins, HSP25, HSP60, HSP70, and HSP90 at their surfaces [68] and translocation of HSP60 to the surface of apoptotic cells can promote phagocytosis [71]. Additional plasma membrane changes involve loss of 'don't-eat-me' signals such as the plasma membrane glycoproteins CD31 [63] and CD47 [64], along with exposure of the 'eat-me' signal PS. These changes lead to the engulfment of apoptotic cells and are linked to the production of anti-inflammatory mediators by phagocytes. As yet, it is unclear to what extent these changes are linked to other, nonphagocytic, responses to apoptotic cells, either in normal or malignant contexts. It is notable that loss of functional CD47 leads to phagocytosis of cells regardless of their commitment to apoptosis and anti-CD47 antibodies are providing promising biological therapeutics for non-Hodgkin's lymphoma and acute lymphoblastic leukemia [101, 102].

Although the extracellular vesicles (EVs) produced during apoptosis are often simply referred to as 'apoptotic bodies,' the latter term is not well defined and it seems likely that cell stress and apoptosis lead to active release of various categories of EVs ranging in size from small, exosome-like vesicles (30–100 nm in diameter) through microvesicles (~100–1000 nm) to large subcellular vesicles (>1000 nm, perhaps most generally described by the term 'apoptotic bodies'). Given the well-described capacity for extracellular vesicles to carry or display at their membrane surfaces multiple categories of biological molecules, including transmembrane protein receptors, translational and enzyme systems, DNA, and micro-RNAs [103, 104], it seems probable that EVs produced actively during cell stress and cell death will prove to have multiple biological functions, especially since EVs released in association with apoptosis harbor hundreds of proteins [97, 100]. It is notable and relevant to the concept of apoptosis leading to the production of functional vesicles that EVs can mediate the transfer of biologically active molecules between

Table 3.1 Examples of bioactive (or potentially bioactive) components of mammalian dying and dead cells of potential relevance to the tumor microenvironment

Factor	Category	Known properties	Comments	References
Phosphatidylserine (PS)	Membrane lipid[a]	Prophagocytic; anti-inflammatory; lipid homeostasis	Prototypic 'eat-me' signal; stimulates cholesterol export by phagocytes	[57, 59]
Phosphatidylethanolamine	Membrane lipid	Not yet clear		[60]
Sphingosine 1 phosphate	Released lipid	Promigratory		[61]
LPC	Released lipid	Promigratory		[62]
CD31 (loss)	Membrane protein	Prophagocytic	CD31 acts as "don't eat me" signal	[63]
CD47 (loss)	Membrane protein	Prophagocytic	CD47 acts as "don't eat me" signal	[64]
CX_3CL1 (fractalkine)	Membrane protein[a]; released protein[a]	Promigratory; prophagocytic		[65–67]
Heat shock proteins (HSPgp96,HSP25, HSP60, HSP70 and HSP90	Membrane protein; released protein	Prophagocytic; pro-inflammatory; immunogenic	Plasma membrane expression of some HSPs by apoptotic cells; proinflammatory release of HSPs is by necrotic cells	[68–71]
IL10	Released protein	Anti-inflammatory		[72]
TGF-β1	Released protein	Anti-inflammatory		[73]
Lactoferrin	Released protein	Antimigratory; anti-inflammatory	Acts as "keep out" signal	[74, 75]
Alpha-defensins	Released peptide	Anti-inflammatory	Released both by apoptotic and necrotic neutrophils	[76]
ATP, UTP	Released nucleotide	Pro-migratory		[77]
S19	Released protein	Pro-migratory		[78]
EMAP II	Released protein	Pro-migratory		[79]
Histones; MHC II	Released protein[a]; Membrane protein[a]	Not yet clear		[80, 81]
Annexin I	Membrane protein	Pro-phagocytic; anti-inflammatory	Externalized by late apoptotic cells with leaky membranes	[82]
Thrombospondin 1, heparin binding domain fragment	Released protein	Anti-inflammatory, tolerogenic		[83]
Epidermal growth factor	Released protein	Pro-survival	Promotes vascular repair by mesenchymal stem cells	[84]

Perlecan C-terminal fragment, LG3	Released protein	Pro-migratory	Perlecan fragment LG3 promotes vascular remodeling by vascular smooth muscle cells	[85]
Matrix metalloproteinases 2 and 12	Released protein[a]	Not yet clear		[27]
ICAM-3	Membrane protein[a]; released protein[a]	Pro-phagocytic; pro-migratory		[86, 87]
MCP-1	Released protein	Pro-migratory		[88]
Cytochrome c	Released protein	Toxic to neighboring cells; pro-inflammatory	Increases in circulation of patients with haematological malignancies	[89, 90]
Nucleosomes	Membrane DNA; Membrane protein	Autoantigens	Exposed on apoptotic cell blebs	[91, 92]
Ribonucleoproteins	Membrane protein	Autoantigens	Exposed on apoptotic cell blebs	[91]
Carbohydrates	Membrane moieties[a]	Pro-phagocytic		[93]
Negatively charged cell surface molecules	Membrane moieties	Angiogenic	Electrostatic signaling mechanism	[94]
Multiple proteins	Membrane moieties[a]; Released moieties[a]	Pro-survival and undefined activities	Pro-survival activity of TCTP on endothelial cells	[95]
"LPS-like" ACAMPs	Membrane moieties[a]; Released moieties[a]	Not yet clear	Example is laminin binding protein	[96]
Multiple proteins (>200)	Released proteins[a]	Not yet clear	Especially enriched in nuclear proteins	[97]
RNAs	Released RNA[a]	Not yet clear		[98]
Multiple cellular components	Membrane moieties[a]; Released moieties[a]	Not yet clear	Exclusion of nuclear components from certain vesicles	[99]
Multiple proteins (>500)	Membrane moieties[a]; Released moieties[a]	Immunogenic and undefined activities	Differences between relatively large (apoptotic bodies) and relatively small (exosome-like) vesicles. Immunogenic activity of 20S proteasome core in small vesicles	[100]

[a]Shown to be associated with extracellular vesicles released by apoptotic cells

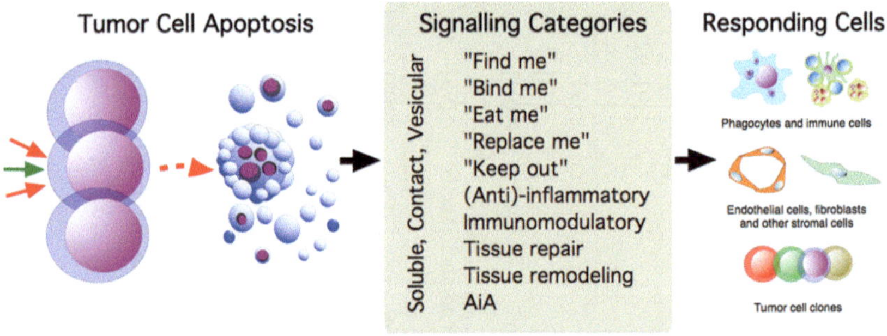

Fig. 3.4 Responses to apoptosis in the tumor microenvironment. Through direct contact or via release of extracellular vesicles and soluble factors, apoptotic cells communicate with a range of neighboring cells to elicit a multitude of biological responses that have implications for the growth and spread of malignant disease. Responding cells include phagocytes, other cells of the immune system, stromal cells and tumor cells themselves, including cancer stem cells. AiA apoptosis-induced apoptosis

cells. Furthermore, as a consequence of exposure of PS, which appears to be a general characteristic of EVs, these particles may elicit responses in nonapoptotic cells similar to the effects of apoptotic cells themselves. In this sense, apoptotic cell-derived EVs may act as complex intercellular signaling structures that inform the microenvironment of the apoptotic cell, fostering specific responses. For example, at early stages of activation, neutrophils release EVs which, like apoptotic cells, can inhibit inflammatory responses of macrophages and dendritic cells via the Mer receptor tyrosine kinase [105]. Furthermore, it is known that apoptotic cells can transfer biologically active genetic material, notably oncogenes, into phagocytes [21] and it seems likely that stressed or apoptotic cell-derived EVs have similar properties, especially since some are known to carry cargoes of DNA and/or microRNAs. The consequences of tumor cell-derived EVs for cancer pathogenesis and their potential as targets in following disease progression and therapeutic responses are well rehearsed [106]. However, the biological roles of apoptotic cell-derived EVs are ill defined, not least in cancer. To date, our knowledge of the function of apoptotic tumor-cell derived EVs is limited to their ability to activate chemotactic responses in mononuclear phagocytes via the chemokine CX_3CL1 (fractalkine) [65] and also by an ICAM-3-dependent mechanism [87]. They may also be involved in stimulating proliferation of rat insulinoma cells [107] and in expressing and processing certain matrix metalloproteinases [27]. It is important to note, however, that in none of these studies have apoptotic EVs been specifically discriminated from EVs produced as a consequence of other processes such as preapoptotic stress. We suggest that future work will uncover numerous additional cargoes and functions of apoptotic tumor cell-derived EVs of fundamental importance to human cancer pathogenesis, diagnosis, prognosis, and therapy. Future work will also uncover the mechanisms through which apoptotic EVs exert their effects, such as by receptor-mediated endocytic or phagocytic events, via membrane fusion or by secretion or diffusion of soluble factors from the vesicles. In the latter context, it has been shown that EVs produced prior to loss of membrane permeability of the originator apoptotic cell body have highly permeable membranes, permitting release of large macromolecules [97].

The release of biologically active soluble factors from apoptotic cells is well established (Table 3.1), especially in the context of inflammation control with factors such as IL-10, transforming growth factor-β1 (TGF-β1), α-defensins and lactoferrin being produced by various types of apoptotic cell [72–74, 76]. Chemotactic 'find-me' signals including lysophosphatidylcholine (LPC), CX_3CL1, nucleotides, and sphingosine-1-phosphate (S1P) are released by apoptotic cells in order to attract mononuclear phagocytes to sites of apoptosis [39, 108, 109]. Most, if not all, of these factors have pleiotropic properties, endowing apoptotic cells with the propensity to foster multiple responses of relevance to tumor establishment and growth, not least survival and proliferation of viable neighboring tumor cells as well as pro-oncogenic activation of stromal cellular elements of the tumor microenvironment. There seems little doubt that the capacity for release of biologically active moieties by apoptotic cells is currently underestimated. We return to this subject later when we consider compensatory proliferation and mitogen production such as prostaglandin E2 (PGE_2) [110] in response to apoptosis.

Finally, in this section, it is worth mentioning an intriguing function of apoptotic cells which appears to be the sequestration of chemokines. Specifically, apoptotic leukocytes can express increased functional CCR5 leading to sequestration of CCL3, CCL4, and CCL5 and this can be modulated positively by proresolution mediators such as lipoxin A_4, resolvin E1, and protectin D1, and negatively by pro-inflammatory mediators such as TNF-α [111]. On this basis it might be speculated that apoptotic cells in the tumor microenvironment could regulate the activity of chemokines through modulation of chemokine receptor expression, which in turn may be controlled on the apoptotic cells by the cytokine milieu of the tumor.

3.4.2 *Activities of Stressed Cells Versus Dying Cells*

The wide-ranging changes in phenotypes and activities of apoptotic tumor cells together with the responses they consequently elicit demonstrate that apoptosis is far from the silent process it was initially thought to be. It is important, however, to consider the extent to which the microenvironmental conditioning by dying cells in tumors originates from biological mediators produced (a) as a consequence of pre-apoptotic stress responses or alternatively (b) specifically as a result of activation of the apoptotic program and the cell moving past the 'point of no return' in this pathway, widely regarded as mitochondrial outer membrane permeabilization (MOMP). Prototypic examples of stress-response mediators, heat-shock proteins, notably HSP27 and HSP70, are known to inhibit apoptosis and may modulate oncogenesis. For example, HSP70 can inhibit apoptosis signaling through interference with apoptosis protease-activating factor-1 (Apaf-1) and apoptosis-inducing factor

(AIF) [112]. Furthermore, modulation by stress signals can induce a state of antitumor immunogenicity in previously tolerogenic apoptotic tumor cells [70]. Importantly, it should be appreciated that cellular stress responses are restorative, being aimed at cell survival rather than cell death [113]. Studies aimed at dissecting the molecular mechanisms responsible for the biological effects of apoptotic cells must carefully uncouple the underlying principles from those resulting from preapoptotic stress.

The associations between cell stress and cell death open up possibilities that common principles underlie the capabilities of stressed and dying cells to influence the pro-oncogenic tumor microenvironment. A tantalizing example here is cellular senescence, an overtly tumor-suppressive mechanism, which appears to be a state of chronic stress, designed to lock cells out of cycle and thereby prevent propagation of premalignant, damaged cells [114]. Potential parallels may exist between apoptotic tumor cells and senescent tumor cells participating in paracrine pro-oncogenic signaling events.

3.5 Responses to Cell Death in the Tumor Microenvironment

The most renowned responses to apoptotic cells emanate from the phagocytes that engulf them, the molecular mechanisms underlying phagocytosis, and anti-inflammatory signaling being the most widely studied, especially in macrophages (reviewed in Chaps. 2 and 9). It has been known for many years that apoptotic cells not only elicit engulfment responses and the production of anti-inflammatory mediators by phagocytes but also angiogenic and growth factors such as vascular endothelial growth factor (VEGF) and hepatocyte growth factor (HGF), respectively. We have previously proposed the '3Rs' of apoptotic cell clearance as *recognition, response,* and *removal* [15, 39] and, just as unconventional phagocytes such as epithelial cells, endothelial cells, and fibroblasts respond to their apoptotic cellular neighbors by engulfing them, it follows that a wide variety of different cell types are able to respond in other ways to apoptotic cells in their vicinity through direct intercellular contact and via soluble factors or EVs produced by the dying cells. Apart from phagocytic and anti-inflammatory responses, relatively little is known either about the variety of cells that can respond to their apoptotic neighbors or about the range of such responses (cf. [115–118]). Available evidence, however, indicates that multiple cell types can respond to apoptotic cells and that the nature of such responses can vary widely. While such responses are undoubtedly fundamental to normal physiological systems and normal tissue homeostasis, here, we consider these responses in the context of the microenvironment of malignant tissue.

The macrophage represents a focal point of cell death and cancer. Macrophages are almost invariably found in either normal or diseased tissues at sites of apoptosis and can accumulate in tumors in large numbers: for example, up to 50 % of breast cancer tissue can be comprised of macrophages [119]. The functional plasticity of macrophages makes them highly responsive to their microenvironments. Tumor-associated macrophages (TAMs) have the potential to exert innate antitumor immunity as appears to be the case in colorectal and gastric cancer. This stems from their adopting a proinflammatory phenotype in the tumor microenvironment that inhibits tumor cell proliferation and promotes T cell-mediated antitumor immunity [120]. In most cancers including breast cancer, lymphoma, cervical cancer, squamous carcinomas, and melanoma, however, TAMs are associated with poor prognosis [121]. Information supporting the view that tumor cell apoptosis is capable of driving protumor properties of TAMs has begun to emerge. Thus, in postpartum breast cancer, the phagocytic and anti-inflammatory receptor tyrosine kinase Mer that interacts with apoptotic cells via its ligands Gas-6 and Protein S (which bind to PS exposed on apoptotic cells) promotes cancer progression as a consequence of tumor cell apoptosis [122]. Furthermore, in prostate cancer, the engulfment of apoptotic tumor cells by macrophages induces the upregulation of milk fat globule epidermal growth factor (MFG-E8)—a PS-binding ligand that bridges apoptotic cells with $\alpha_v\beta_3$ and $\alpha_v\beta_5$ integrins on phagocytes—and enhancement of signal transducer and activator of transcription 3 (STAT3)-dependent transcription with resultant pro-oncogenic/anti-inflammatory polarization ("M2-like") of the macrophages [123]. (With respect to macrophage polarization, it is worth noting that 'canonical' phenotypes centered around M1 and M2 are unlikely to represent the true picture of the range of activation states of macrophages either in cancer or in tissue repair [27, 124, 125]; here, for convenience, we will use the terms M1-like and M2-like to indicate positioning of macrophages toward a particular activation state.) The results of these studies of prostate and breast cancer demonstrate that multiple receptors of macrophages are involved in apoptotic cell clearance and in anti-inflammatory responsiveness to apoptosis. It will be important in future studies to determine the relative importance of different phagocytic and anti-inflammatory receptors of macrophages in a wide range of cancers in which apoptosis is significant. It is tempting to speculate, for example, that Liver X receptor (LXR) is important in this context especially since LXR signaling has been shown to play key roles in apoptotic cell clearance by mouse macrophages and its associated anti-inflammatory responses and immunological tolerance [126].

We have recently carried out detailed studies of the role of apoptosis in the pathogenesis of aggressive starry-sky B-cell non-Hodgkin's lymphoma (NHL), prototypically Burkitt's lymphoma (BL) [27]. In these tumors frequent apoptosis can readily be observed histologically and the majority of the apoptotic tumor cells are found in association with TAMs, creating the distinctive starry-sky morphology (Fig. 3.2). Inhibition of apoptosis in a murine xenograft model of BL led to impaired tumor growth and substantially reduced tumor cell proliferation. Notably, when tumor cell apoptosis was suppressed, the hypoxic microenvironment of the tumor was enhanced due to impaired blood vessel formation, suggesting a role for tumor cell apoptosis in driving not only proliferation but also angiogenesis. In both patient samples and mouse models of Burkitt's lymphoma, a strong correlation between tumor cell apoptosis and macrophage numbers was observed indicating that death of tumor cells induces TAM accumulation. A proportion of these TAMs also express proliferative markers suggesting that apoptotic cell conditioning may stimulate the proliferation of starry-sky TAMs, helping to populate the growing tumor with macrophages. In situ gene expression profiling of starry-sky TAMs revealed a protumorigenic activation state characterized by transcriptional activation of anti-inflammatory, proangiogenic, and tissue remodeling pathways. For example, starry-sky TAMs were found to preferentially express high levels of *TGF-β1*, growth factors including *IGF-1* and *PDGF-CC*, and several genes associated with angiogenesis and metastasis including *MMP2*, *MMP3*, *MMP12*, *TIMP2*, *ANPEP*, *LGALS3*, *HMOX1*, and *GPNMB*. In addition, CD91, Mer, Axl, and Gas6 were found among apoptotic cell interaction molecules that were upregulated in starry-sky TAMs providing further evidence for the involvement of the Tyro-Axl-Mer receptor tyrosine kinase axis in TAM activation and extending the variety of phagocytic receptors of apoptotic cells of potential importance for tumor pathogenesis. Apoptotic tumor cells thus direct the activities of TAMs toward the acquisition of multiple and diverse protumorigenic properties that support the growth of tumor cells and subsequent oncogenic progression.

The accumulating evidence therefore indicates that tumor cell apoptosis can elicit specific tumor-promoting responses in breast cancer, prostate cancer, and NHL. A similar argument may also be made for melanoma [27] and recent results also indicate that apoptosis can drive hepatocarcinogenesis [127]. The current state of knowledge suggests that apoptosis of tumor cells may represent a generic, pro-oncogenic stimulus which acts through multiple molecular cell biological mechanisms affecting antitumor immunity, tumor population growth, and angiogenesis. While responses have thus far been focused mainly on the role of macrophages, additional nonmacrophage mechanisms are likely to prove to be important for tumor progression, especially since apoptotic cells are known to have direct effects on angiogenesis [94] and also have the potential to produce immunomodulatory factors, chemokines, cytokines, and mitogens directly (Table 3.1).

The mitogenic capacity of apoptosis is well illustrated by the phenomenon of compensatory proliferation of neighboring cells, first described in Drosophila. Developmental and wounding studies in the fly indicated that the pathways underlying the proliferative responses to apoptosis involve p53, WNT, TGF-β/BMP, JNK, and Hedgehog (Hh) signaling (see [128] for recent review). The pathways are context related and may be dependent or independent of executioner caspase activation. Compensatory proliferation in response to injury has also been described during regenerative responses in other species, including Hydra, Planaria, newts, frogs, zebrafish, and mice. In the latter, PGE_2, a key inflammatory and regulatory niche player, is produced to drive compensatory proliferation as a result of caspase-3-dependent activation of calcium-independent phospholipase A_2 (iPLA2) [110]. It is conceivable that PGE_2 could mediate its regenerative power in this context through activation of β-catenin following binding to the G protein-coupled receptor EP2. Additional plausible molecular players in compensatory proliferation that may be produced downstream of caspase activation and apoptosis include Sdf1, FGF20, and IL11 [128].

Apoptosis-driven compensatory proliferation of hematopoietic stem/progenitor cells (HSCs) has been implicated in thymic lymphoma development in mice subjected to low-dose irradiation. This pathway requires p53, which activates leukocyte apoptosis via the BH3-only proapoptosis protein, PUMA [129, 130]. Thus, *PUMA–/–* HSCs are protected from cell death but show reduced compensatory proliferation and stress-associated DNA damage with consequent inhibition of thymic lymphoma formation in response to gamma-irradiation. In this way it seems that p53, acting via its apoptosis-promoting properties, appears to promote tumorigenesis induced by DNA damage, an activity of p53 which contradicts its widely accepted antitumor function in securing the demise and deletion of cells carrying potentially oncogenic mutations. A possible alternative explanation in this scenario is that, although PUMA is clearly a mediator of p53-induced cell death in response to radiation, it may also have additional specific roles in thymic lymphomagenesis, especially since *PUMA–/–* mice still produced many different kinds of tumors (high- and low-grade lymphomas, sarcomas, and carcinomas) in response to irradiation [130].

Recent evidence indicates that, in stark contrast to compensatory proliferation, apoptotic cells can also induce further apoptosis in neighboring cells, thereby propagating the apoptosis signal from a single cell to a population. Thus, genetic studies in Drosophila wing disc development show that apoptosis in one compartment in the tissue can induce apoptosis in a neighboring compartment. The authors appropriately termed this effect "apoptosis-induced-apoptosis (AiA)" and identified the Drosophila TNF homolog Eiger as a key factor, triggered by JNK signaling, in mediating this response [34]. They also demonstrated a similar, TNF-α-mediated orchestration of hair follicle cell death in mice. These observations have obvious implications for the roles of apoptosis both in constraining cancer emergence and in delivering effective therapy.

Although it is well accepted that apoptosis represents a significant control mechanism in the pathogenesis of malignant disease through inhibitory mechanisms that limit cancer outgrowth, it is clear that responses to the process are likely to play important roles in modulating tumor growth and progression in positive as well as negative ways. In considering these effects of apoptotic cells, it is essential also to take into account the nonapoptotic roles of the caspases that execute the apoptosis program [131, 132] as well as features of apoptotic cells such as PS exposure that may have tumor modulatory effects independently of apoptosis. Examples here include the caspase-3-mediated activation of iPLA2-induced activation of migration of ovarian cancer cells [133] and of PS exposure on intratumoral endothelial cells [134] that occur in the absence of apoptosis. As previously proposed, we support the argument that responses to such effects may occur in parallel in the tumor microenvironment whether they are "apoptosis-like" effects or whether they are *bona fide* effects of the apoptosis program produced by dying cells.

3.6 Apoptosis and Angiogenesis: The Effects of Hypoxia

The microenvironments of malignant tumors are heterogeneous with respect to oxygen tension and hypoxia is a well-known stress factor in tumors. Typically, lethally low partial pressures of oxygen (including anoxia) in rapidly growing cancers lead to necrotic zones of dead tumor cells with cells at the borders of these ischemic regions undergoing hypoxia-induced apoptosis. In tumors in which apoptosis is constitutively prominent such as starry-sky NHL, it seems that the hypoxia caused by the rapidly proliferating tumor cells creates a dynamic cycle of malignant tissue expansion in which the hypoxia induces apoptosis that stimulates angiogenesis, and that this in turn permits further population growth with further hypoxia-induced apoptosis as the tumor cells outgrow their oxygen supply [27]. In this cyclical cause-and-effect feed-forward system, apoptosis of tumor cells may exert a key oncogenic response in endothelial cells which results in neovascularization.

In order for such a system to work, there also needs to be adaptation to hypoxia. Both transformed and nontransformed cells adapt to the various levels of hypoxia found in progressing tumors via the induction of hypoxia-inducible factors (HIF) and NF-κB [135]. These transcription factors are both activated in response to hypoxia following the inactivation of prolyl hydroxylases. HIF proteins function as a heterodimer of alpha and beta subunits but under normoxic conditions the alpha subunits are constitutively marked for degradation as a result of hydroxylation by prolyl hydroxylases; inhibition of these enzymes under hypoxic conditions enables the alpha subunits to translocate to the nucleus, dimerize with beta subunits, and regulate transcription [136]. Similarly, NF-κB is activated under hypoxic conditions via the inhibition of prolyl hydroxylases, leading to a suppression of hydroxylation of IκB kinase-β, which can then phosphorylate the NF-κB inhibitor, IκBα, leading to its degradation and a cessation of NF-κB inhibition [137]. As a consequence of HIF and NF-κB activation, hypoxia regulates a broad range of oncogenic cellular functions at the transcriptional level, including metabolism, proliferation, apoptosis, inflammation, angiogenesis, and tissue remodeling.

Available evidence indicates that, depending on cellular context (as well as oxygen tension level) hypoxia may either trigger or suppress apoptosis, for example, through differential effects of HIF on pro- or antiapoptotic Bcl-2-family protein expression. Furthermore, glycolytic enzymes can be induced by HIF 1α and, at least in certain tumor cell types, this is associated with suppression of apoptosis. It has been suggested that this may be important in the adaptation of tumor cells to aerobic glycolysis, otherwise known as the Warburg effect [138].

Hypoxic regions of tumors are frequently populated with TAMs that generally have protumorigenic properties. For example, in breast cancer, TAMs are most abundantly found in poorly vascularized regions of the tumor [119]. Indeed, hypoxia can regulate the expression of chemokines and their receptors including HIF1-regulated CXCL12/CXCR4 and CXCL8/CXCR8 to help navigate macrophages to regions of low oxygen tension [139, 140]. The exposure of TAMs to hypoxia also modulates their function. For example, in a BALB/c mammary adenocarcinoma model, tumors contained MHC II^{hi} and MHC II^{low} TAMs; the MHC II^{hi} TAMs expressed higher levels of proinflammatory genes including *NOS2*, *PTGS2*, *IL1b*, *IL6*, and *IL12b* whereas the MHC II^{low} TAMs expressed more anti-inflammatory genes such as *ARG1*, *CD163*, *STAB1*, and *MRC1* [141]. Hypoxic regions of these tumors were predominantly populated with MHC II^{low} TAMs, i.e., anti-inflammatory macrophages, which were poor at processing antigen and stimulating T cell-mediated responses. Hypoxic TAMs have also been reported to produce IL-10, which again contributes to the anti-inflammatory microenvironment, promoting evasion of antitumor T cell-based immunity [142]. In addition to promoting TAM enrichment and helping to reduce inflammation in the tumor, hypoxia-conditioned TAMs further promote tumor growth by stimulating neoangiogenesis. This is mediated by the HIF-driven upregulation and release of VEGF and also via the release of various cytokines including CXCL12, CCL2, CXCL8, CXCL1, CXCL13, and CCL5 [119, 142, 143]. Moreover, hypoxia-conditioned macrophages support metastasis by induction of the HIF1α-regulated gene, *MIF*, leading to enhanced release of MMP-9 [142].

The role of cell death in the hypoxia-associated accumulation and conditioning of TAMs is not yet understood, but given the colocalization of cell death and hypoxia, it seems probable that cell death will prove to be important in helping to drive cellular adaptation in the tumor microenvironment. Future studies investigating the links between (1) hypoxia, (2) regulation of tumor cell apoptosis, (3) TAM accumulation and activation, and (4) endothelial cell activation will substantially improve our knowledge of the fundamental molecular cell biology of cell death-driven oncogenic progression. As we discuss in the next section, we can draw close parallels between these associations and basic tissue repair mechanisms that may be driven by cell death in tumors.

3.7 Apoptosis and Tissue Repair Mechanisms in Cancer: The "Onco-Regenerative Niche"

Given the close parallels between wound healing mechanisms—i.e., normal responses to tissue damage—and inflammatory host responses in cancer, it seems reasonable to expect that cell damage, including stress responses leading to cell death, plays significant roles in both normal tissue repair and in cancer pathogenesis. Indeed as we have discussed earlier, while cancers have been described as "wounds that do not heal," it may be more accurate to describe them as "wounds that do not stop repairing" (Savill, personal communication). Sites of injury—and consequently sites of repair responses—have been found to be favorable for tumor formation and metastasis including the gastrointestinal tract [144], lung [145], and liver [146]. Since tissue regenerative processes are closely linked to tissue repair, with organs such as the gut, skin, and liver having especially remarkable powers of turnover and regeneration, it seems reasonable to propose the concept of the "onco-regenerative niche" (ORN), a complex, interconnected confraternity of cells and extracellular factors that contribute to malignant disease via conserved, fundamental tissue repair mechanisms. The linkage between cell death and tissue regeneration is well illustrated in mouse models of liver damage and regeneration [147, 148] providing further rationale for consideration of cell death as a key driver mechanism in the ORN.

An important cellular player in the ORN is likely to be the macrophage, since TAMs have widely accepted pro-oncogenic functions—including playing a central role in a cancer stem cell niche [149]—and it has long been recognized that the mononuclear phagocytic arm of the inflammatory response plays critically important roles in wound healing [150, 151]. In the initial stages of the wound healing response, macrophages are believed to help remove damaged tissue and cells and release proinflammatory cytokines, chemokines, and factors such as VEGF that drive subsequent angiogenesis and granulation tissue formation. During the resolution phase, apoptosis is the principal mechanism through which healing wounds are depleted of inflammatory cells, macrophages mediating their safe disposal [150]. Just as phagocytosis of apoptotic tumor cells can alter the phenotype and function of TAMs, macrophages engulfing apoptotic cells in wounds are driven toward an anti-inflammatory phenotype with expression of CD206 and arginase I [150]. The clearance of apoptotic cells by macrophages thus changes the wound microenvironment from proinflammatory to anti-inflammatory thereby limiting excessive tissue damage.

The final stage of wound repair is tissue remodeling. This involves the induction of angiogenesis to restore a normal blood supply to the region and fibroblast-driven extracellular matrix deposition. In tumors, the extracellular matrix is constantly remodeling and new blood vessels must form to sustain additional growth [152, 153]. Myofibroblasts are highly contractible fibroblasts that aid tissue remodeling during wound healing but die by apoptosis when the extracellular matrix is able to support the wound. In tumors they are often found to persist, perhaps due to the ongoing matrix remodeling of tumors and the presence of tumor and stromal-derived growth factors such as TGF-β1, FGF, and PDGF. Myofibroblasts in tumors can be significant sources of growth-promoting and metastatic factors [153] but nothing is yet known of the implications of their apoptosis or their responses to apoptosis in tumors.

Although details of the activation status and effector functions of macrophages in wound repair are awaited, the available evidence suggests strong similarities between activation states that predominate early and later in wound repair and the macrophage polarization that characterizes early and established tumors. Thus, both in cancer and in wound healing, early macrophages may predominantly display proinflammatory phenotypes ("M1-like") whereas at later times during wound healing and in established tumors, the dominant macrophage activation state is anti-inflammatory/reparatory ("M2-like") [154, 155] (see also Chap. 9). Notably, studies of diabetic mice in which wound healing is impaired, as is commonly the case in human diabetic patients, indicated that effective clearance of apoptotic cells, including apoptotic neutrophils and endothelial cells, by wound macrophages was closely coupled to effective wound healing. In particular, in line with the aforementioned "M1-like" to "M2-like" transition, proinflammatory wound macrophages appear to be driven to an anti-inflammatory activation state by apoptotic cell clearance mechanisms [156]. Similarly, age-related impairment of wound healing has been suggested to be associated with defective clearance of apoptotic cells by macrophages [157]. These observations highlight the importance of macrophage-dependent clearance of apoptotic cells in the formation of an anti-inflammatory microenvironment, which under normal circumstances is necessary for healing and repair of wounds but in the more sinister setting of cancer, apoptosis may promote tumor cell survival, proliferation, tissue remodeling, neovascularization, and metastasis.

We suggest that cell death plays a driving role in the establishment of the ORN and that responding normal host cells, especially in the environs of cell demise (or as a result of recruitment incited by cell death), perform critical oncogenic functions as a consequence. In this way multiple cellular players in the ORN, such as—in addition to macrophages—endothelial cells and surviving or emerging clones of transformed cells, could interact in tissue repair modes in order to foster tumor establishment, stemness of cancer clones, metastatic spread, and post-therapy relapse. Of direct relevance to this argument is the wound-healing response to cell death via the Mer receptor tyrosine kinase that drives metastasis in postpartum breast cancer [122]. It will be important to establish the critical tissue repair molecules that are central to the ORN. Given their role in wound healing and liver regeneration, notably in the latter their production by apoptotic hepatocytes [147], we speculate that Hedgehog ligands produced by dying tumor cells could provide important molecular components of the ORN. As we discuss in the next section, PGE_2 signaling may also be crucial, especially in the ORN that may characterize post-therapeutic disease relapse.

3.8 Opportunities for Improving Cancer Treatment

The pro-oncogenic, as well as anti-oncogenic, properties of tumor cell apoptosis suggest opportunities for novel anticancer therapies. Mainly, such opportunities remain undeveloped, both from the standpoint of, for example, either promoting AiA or inhibiting compensatory proliferation. As detailed elsewhere in this book, both chemotherapeutic and radiotherapeutic treatments induce various modes of cell death including apoptosis. The key requirement for successful therapy is sustained imbalance of the tumor cell birth/cell death equation in favor of cell death. Immunogenic cell death is one way in which this can be achieved [158]. Since TAMs play prominent roles in responding to apoptotic tumor cells in pro-oncogenic modes of accumulation and activation it follows that therapeutic strategies aimed at (1) reducing TAM accumulation, and (2) repolarizing TAMs to an antitumor phenotype to mediate the elimination of tumor cells could also be valuable. Both strategies have potential to limit protumor and anti-inflammatory signaling by TAMs, and to lead to enhanced antitumor immunity, resulting in decreased tumor growth and reduced metastasis.

Therapeutic strategies aimed at limiting the accumulation of TAMs in the tumor microenvironment have focused either on reducing recruitment of monocytes or suppressing macrophage survival in the tumor microenvironment. CCL2 and CSF-1 are abundantly expressed in many tumors and are believed to be among the most important factors contributing to macrophage infiltration [159]. Additionally, many tumors upregulate CSF-1 as a result of cytotoxic or ionizing radiation therapies, which can lead to further accumulation of TAMs [160, 161]. Preclinical murine models targeting CCL2, CSF-1, or their receptors (CCR2 and CSF-1R, respectively) either alone, or in combination with chemotherapy are showing potential at reducing TAM accumulation, and enhanced antitumor immunity leading to decreased tumor growth and reduced metastases [160–164]. A Phase 1B clinical trial using an anti-CCR2 antibody is currently underway combined with standard chemotherapy in pancreatic cancer patients (http://clinicaltrials.gov/show/NCT01413022). Recent work also highlights the importance of the CCL2/CCR2 axis in orchestrating macrophage-driven pulmonary metastasis in murine breast cancer. Notably, a key downstream signaling arm, CCL3/CCR1 was identified, which the authors suggest could prove to be a useful, low-toxicity therapeutic target for treating metastatic disease [165].

Several therapies have been shown to suppress macrophage survival in tumors. Trabectedin, a chemotherapeutic agent, induces apoptosis exclusively in mononuclear phagocytes via the extrinsic apoptotic pathway mediated by tumor necrosis factor-related apoptosis-inducing ligand (TRAIL) receptors, and can deplete macrophages in tumors, thereby contributing to inhibition of tumor growth and reduction in angiogenesis [166]. Liposomal bisphosphonates, including clodronate and zoledronic acid, are also capable of macrophage depletion and have shown to reduce tumor growth and tumor vascularization [167, 168], as well as suppression of bone and muscle metastasis [169] in murine cancer models. Additionally, immunotoxin-conjugated mAbs targeting TAM surface proteins, including for example scavenger receptor-A, CD52, and folate receptor β, have also shown potential at reducing TAM infiltrate and restricting tumor growth in murine models of ovarian cancer and glioblastoma [170–172].

Alternative therapeutic approaches have targeted the plasticity of macrophages and are directed at preventing TAM phenotypic changes that promote tumor growth and/or re-educate TAMs in favor of antitumor and proinflammatory signaling. These strategies are in line with findings that patients with high ratios of M1-like/M2-like TAMs more frequently experience a complete tumor resolution and higher overall survival than patients with low M1-like/M2-like ratios [173, 174]. Common strategies have sought to target TAMs directly to achieve repolarization. NF-κB activation can redirect TAMs to a tumoricidal M1-like phenotype [175] and several agents capable of activating NF-κB have been reported, including Toll-like receptor (TLR) agonists, anti-CD40 mAbs, and anti-IL-10R mAbs. CpG oligodeoxynucleotide (CpG-ODN), a TLR9 ligand can upregulate NF-κB activation in TAMs, leading to the production of IL-12 and TNF [176] and increased expression of MHC Class II,

CD86, CD80, CD40, and IFN-γ, while decreasing expression of IL-4Rα, IL-4, and IL-10 [177]. Combining CpG-ODN treatment with anti-IL-10R mAb and monocyte chemoattractant CCL16 was shown to rapidly change the phenotype of infiltrating macrophages and induced the rejection of various preexisting tumors in murine models [176]. CpG-ODN-therapies currently in clinical trials show modest activity but may be improved by identifying molecular characteristics of subgroups of patients that could potentially benefit from this treatment [178–180]. CD40 activation can also reverse immune suppression of macrophages and drive antitumor T cell responses. Addition of anti-CD40 mAb to patients with chemotherapy-naïve, surgically incurable pancreatic ductal adenocarcinoma has been shown to induce an influx of CD40-activated, tumoricidal macrophages and promote the depletion of the tumor stroma [181, 182]. Blockade of the CSF1/CSF1R axis has also been shown to change the TAM phenotype by reducing expression of TAM genes such as *MRC1*, *HMOX1,* and *ARG1*, while upregulating *IL1b* and led to an increase in survival when given early in disease or increased tumor regression when given at advanced stages of a xenograft model of proneural glioblastoma multiforme [183].

Other strategies to reprogram macrophages that have shown success in preclinical cancer models are intratumoral delivery of IL-21 [184], and low-dose irradiation of macrophages [185], both of which were found to switch macrophage activation to an antitumor, proinflammatory phenotype, which was accompanied by increased infiltration of T cells. Furthermore, two microRNAs, miR-155 [186] and miR-125b [187], have been identified that are associated with proinflammatory activation of macrophages and should be further investigated as therapies for redirecting TAMs.

An approach to alter TAM phenotype of particular relevance to this chapter is the targeting of apoptotic cell-mediated activation of macrophages. Such therapies are generally targeted against PS expressed on the membrane of apoptotic cells. Masking of PS by a mutant form of MFG-E8 has been shown to inhibit phagocytosis of apoptotic cells by macrophages and could also inhibit the enhanced production of IL-10 [188]. Furthermore, addition of annexin V, which binds to PS on apoptotic cells, has been shown to target irradiated lymphoma cells to CD8+ dendritic cells for in vivo clearance, leading to release of proinflammatory cytokines and regression of tumors [189]. Murine and rat models of prostate cancer have also shown success when chemotherapy or irradiation was combined with an anti-PS antibody, 2aG4, a variant of the human antibody bavituximab, compared to either therapy alone [190, 191]. Gene expression analysis showed that TAMs cultured in the presence of 2aG4 increased iNOS, inflammatory cytokines IL-12 and TNFα, and T cell costimulatory molecules (CD80, CD86, and MHC class II), and decreased expression of arginase I, immunosuppressive cytokines IL-10 and TGF-β, and VEGF-A [191]. Bavituximab, a human anti-PS antibody, is currently being tested in a clinical trial [192]. Additionally, chemotherapies that induce immunogenic cell death can limit apoptotic cell activation of macrophages, and instead signal via dendritic cells leading to stimulation of T cells, which improves therapeutic responses [158].

The notion that switching macrophages from M2-like procancer to M1-like anticancer activation states is undoubtedly simplistic and seems likely to be tumor specific or even tumor region specific, depending on the functional subtleties of the activation status of the particular TAMs and their biological activities in modulating tumor growth and progression. A relevant salutary observation is that macrophages activated in vitro with the classical M1, proinflammatory stimuli, IFNγ and TNFα have been reported to promote tissue (skeletal muscle) regeneration [124]. Furthermore, the inflammatory drive of the tumor pathogenesis will be important in approaches that target the polarization state of the macrophage. For example, although the receptor tyrosine kinases (RTKs) Mer and Axl are widely held as having strong oncogenic properties [193], which has led to the development of inhibitors of these RTKs for treatment of solid tumors [194], the oncogenic role of the Tyro, Axl, Mer-RTK axis may be tumor and/or cell context specific. Thus, in inflammation-driven colitis-linked cancer, loss of Mer and Axl actually promotes the underlying oncogenic processes through failure to dampen inflammation [195]. Furthermore, suppression of apoptotic cell clearance via the Tyro, Axl, Mer-RTK axis could also act to promote oncogenesis in an inflammatory cancer context. By contrast, the Tyro, Axl, Mer-RTK axis may activate protumor mechanisms through suppression of antitumor immunity or through driving oncogenic properties in tumor-associated macrophages.

An additional TAM therapeutic approach is to enlist them to target tumor cells, most likely through using their ability to perform antibody-dependent cellular phagocytosis (ADCP). Recent work by Montalvao et al. has suggested that the effectiveness of rituximab, an anti-CD20 therapy that is proven successful for treating B cell malignancies, is dependent on ADCP by Kupffer cells, the macrophages of the liver [196]. Similarly, antitumor mAb therapy in various murine carcinoma models resulted in rapid phagocytosis of tumor cells by Kupffer cells and inhibition of liver metastasis [197]. As we have already noted earlier, the antagonism of "don't-eat-me" signals such as CD47 is a potentially highly effective response in cancer therapy that harnesses the power of the phagocytes to engulf and degrade otherwise viable tumor cells.

Cell death induced by anticancer therapeutics may activate host responses that promote relapse. Recent work illustrates this effect especially in murine models of breast cancer and melanoma in which postradiotherapy or postchemotherapy, tumor repopulation was shown to be driven by caspase-3-dependent PGE_2 production generated by therapy-induced tumor cell apoptosis [25, 198]. Human studies of several cancer types indeed support the argument that caspase-3 could be an important therapeutic target [14]. In our view, the presumption that inducing cell death in tumors (by whatever means) should be the fundamental aim of cancer therapy is fatally flawed. Given the burgeoning evidence that host responses to tumor cell death—including therapy-induced death—may be firmly rooted in tissue repair and regeneration, it seems timely to reassess the therapeutic "death hit" in terms of the host response, especially since post-therapeutic relapse is such a common event. In addition to sustaining the anticancer effect through stimulating antitumor immunity as is the case with immunogenic cell death, combination therapies to induce tumor cell death and simultaneously to block death-induced development of the 'relapse niche' should be considered. Identification of the key molecular targets in this niche will be crucial to the development of effective and lasting novel anticancer therapies.

3.9 Conclusions and Future Perspectives

Far from being passive biological entities that simply signify reduction in malignant cell population sizes, apoptotic cells actively modulate the tumor microenvironment, eliciting multiple host responses, many having pro-oncogenic properties. Therefore, although evasion from apoptosis has accepted roles as an acquired characteristic of a cancer cell, its oncogenic features should be considered as part of the tumor microenvironmental contribution to the development of malignant disease. The surprising effects of apoptosis in promoting as well as inhibiting oncogenesis are reflected by recent evidence that p53, the prototypical tumor suppressor, also has tumor-promoting properties [199].

The notion that apoptosis may provide a generic driver mechanism that elicits tissue repair-like processes unifies the features of wound healing, tissue regeneration, and developmental remodeling that characterize malignant tumors. In this respect, the concept of the ORN constitutes a focus for the identification of future therapeutic targets. The critical question here is what molecular mechanisms determine the nature of the responses that emanate from apoptosis? What features of an apoptotic cell lead to compensatory proliferation, angiogenesis, antitumor immunity, or further cell death? As we have discussed, we are beginning to understand this important question and elucidate the answers. Undoubtedly, through thorough understanding of the molecular cell biology of the systemic as well as microenvironmental responses to apoptosis—and indeed other forms of cell death—in tumors, the future holds great promise for improved diagnosis, patient stratification, and long-lasting anticancer therapies.

References

1. Hanahan D, Weinberg RA. The hallmarks of cancer. Cell. 2000;100:57–70.
2. Hanahan D, Weinberg RA. Hallmarks of cancer: the next generation. Cell. 2011;144:646–74.
3. Wyllie AH. The biology of cell death in tumours. Anticancer Res. 1985;5:131–6.
4. Weinstein GD, Frost P. Cell proliferation in human basal cell carcinoma. Cancer Res. 1970;30:724–8.
5. Holmgren L, O'Reilly MS, Folkman J. Dormancy of micrometastases: balanced proliferation and apoptosis in the presence of angiogenesis suppression. Nat Med. 1995;1:149–53.
6. Clifton KH, Yatvin MB. Cell population growth and cell loss in the MTG-B mouse mammary carcinoma. Cancer Res. 1970;30:658–64.
7. Revesz L, Klein G. Quantitative studies on the multiplication of neoplastic cells in vivo. II. Growth curves of three ascites lymphomas. J Natl Cancer Inst. 1954;15:253–73.
8. Lala PK. Evaluation of the mode of cell death in Ehrlich ascites tumor. Cancer. 1972;29:261–6.
9. Leoncini L, Del Vecchio MT, Megha T, Barbini P, Galieni P, Pileri S, Sabattini E, Gherlinzoni F, Tosi P, Kraft R, et al. Correlations between apoptotic and proliferative indices in malignant non-Hodgkin's lymphomas. Am J Pathol. 1993;142:755–63.
10. Ohbu M, Saegusa M, Okayasu I. Apoptosis and cellular proliferation in oesophageal squamous cell carcinomas: differences between keratinizing and nonkeratinizing types. Virchows Archiv. 1995;427:271–6.

11. Naresh KN, Lakshminarayanan K, Pai SA, Borges AM. Apoptosis index is a predictor of metastatic phenotype in patients with early stage squamous carcinoma of the tongue: a hypothesis to support this paradoxical association. Cancer. 2001;91:578–84.
12. Jalalinadoushan M, Peivareh H, Azizzadeh Delshad A. Correlation between apoptosis and histological grade of transitional cell carcinoma of urinary bladder. Urol J. 2004;1:177–9.
13. Sun BH, Zhang J, Wang BJ, Zhao XP, Wang YK, Yu ZQ, Yang DL, Hao LJ. Analysis of in vivo patterns of caspase 3 gene expression in primary hepatocellular carcinoma and its relationship to p21(WAF1) expression and hepatic apoptosis. World J Gastroenterol. 2000;6:356–60.
14. Hu Q, Peng J, Liu W, He X, Cui L, Chen X, Yang M, Liu H, Liu S, Wang H. Elevated cleaved caspase-3 is associated with shortened overall survival in several cancer types. Int J Clin Exp Pathol. 2014;7:5057–70.
15. Gregory CD, Pound JD. Cell death in the neighbourhood: direct microenvironmental effects of apoptosis in normal and neoplastic tissues. J Pathol. 2011;223:177–94.
16. Refsum SB, Berdal P. Cell loss in malignant tumours in man. Eur J Cancer. 1967;3:235–6.
17. Langley RR, Fidler IJ. The seed and soil hypothesis revisited—the role of tumor-stroma interactions in metastasis to different organs. Int J Cancer. 2011;128:2527–35.
18. Dvorak HF. Tumors: wounds that do not heal. Similarities between tumor stroma generation and wound healing. N Engl J Med. 1986;315:1650–9.
19. Flusberg DA, Sorger PK. Surviving apoptosis: life-death signaling in single cells. Trends Cell Biol. 2015;25(8):446–58.
20. Bergsmedh A, Ehnfors J, Kawane K, Motoyama N, Nagata S, Holmgren L. DNase II and the Chk2 DNA damage pathway form a genetic barrier blocking replication of horizontally transferred DNA. Mol Cancer Res. 2006;4:187–95.
21. Bergsmedh A, Szeles A, Henriksson M, Bratt A, Folkman MJ, Spetz AL, Holmgren L. Horizontal transfer of oncogenes by uptake of apoptotic bodies. Proc Natl Acad Sci U S A. 2001;98:6407–11.
22. Ehnfors J, Kost-Alimova M, Persson NL, Bergsmedh A, Castro J, Levchenko-Tegnebratt T, Yang L, Panaretakis T, Holmgren L. Horizontal transfer of tumor DNA to endothelial cells in vivo. Cell Death Differ. 2009;16:749–57.
23. Holmgren L, Szeles A, Rajnavolgyi E, Folkman J, Klein G, Ernberg I, Falk KI. Horizontal transfer of DNA by the uptake of apoptotic bodies. Blood. 1999;93:3956–63.
24. Han H, Iwanaga T, Uchiyama Y, Fujita T. Aggregation of macrophages in the tips of intestinal villi in guinea pigs: their possible role in the phagocytosis of effete epithelial cells. Cell Tissue Res. 1993;271:407–16.
25. Huang Q, Li F, Liu X, Li W, Shi W, Liu FF, O'Sullivan B, He Z, Peng Y, Tan AC, Zhou L, Shen J, Han G, Wang XJ, Thorburn J, Thorburn A, Jimeno A, Raben D, Bedford JS, Li CY. Caspase 3-mediated stimulation of tumor cell repopulation during cancer radiotherapy. Nat Med. 2011;17:860–6.
26. Lauber K, Munoz LE, Berens C, Jendrossek V, Belka C, Herrmann M. Apoptosis induction and tumor cell repopulation: the yin and yang of radiotherapy. Radiat Oncol. 2011;6:176.
27. Ford CA, Petrova S, Pound JD, Voss JJ, Melville L, Paterson M, Farnworth SL, Gallimore AM, Cuff S, Wheadon H, Dobbin E, Ogden CA, Dumitriu IE, Dunbar DR, Murray PG, Ruckerl D, Allen JE, Hume DA, van Rooijen N, Goodlad JR, Freeman TC, Gregory CD. Oncogenic properties of apoptotic tumor cells in aggressive B cell lymphoma. Curr Biol. 2015;25:577–88.
28. Lauber K, Herrmann M. Tumor biology: with a little help from my dying friends. Curr Biol. 2015;25:R198–201.
29. Kerr JF, Wyllie AH, Currie AR. Apoptosis: a basic biological phenomenon with wide-ranging implications in tissue kinetics. Br J Cancer. 1972;26:239–57.
30. Green DR. Means to an end. Apoptosis and other cell death mechanisms. New York: Cold Spring Harbor Laboratory Press; 2011.
31. Savill J, Dransfield I, Gregory C, Haslett C. A blast from the past: clearance of apoptotic cells regulates immune responses. Nat Rev Immunol. 2002;2:965–75.

32. Zitvogel L, Kepp O, Kroemer G. Decoding cell death signals in inflammation and immunity. Cell. 2010;140:798–804.
33. Obenauf AC, Massague J. Surviving at a distance: organ specific metastasis. Trends Cancer. 2015;1:76–91.
34. Perez-Garijo A, Fuchs Y, Steller H. Apoptotic cells can induce non-autonomous apoptosis through the TNF pathway. eLife. 2013;2, e01004.
35. Galluzzi L, Bravo-San Pedro JM, Kroemer G. Organelle-specific initiation of cell death. Nat Cell Biol. 2014;16:728–36.
36. Green DR, Galluzzi L, Kroemer G. Cell biology. Metabolic control of cell death. Science. 2014;345:1250256.
37. Galluzzi L, Bravo-San Pedro JM, Vitale I, Aaronson SA, Abrams JM, Adam D, Alnemri ES, Altucci L, Andrews D, Annicchiarico-Petruzzelli M, Baehrecke EH, Bazan NG, Bertrand MJ, Bianchi K, Blagosklonny MV, Blomgren K, Borner C, Bredesen DE, Brenner C, Campanella M, Candi E, Cecconi F, Chan FK, Chandel NS, Cheng EH, Chipuk JE, Cidlowski JA, Ciechanover A, Dawson TM, Dawson VL, De Laurenzi V, De Maria R, Debatin KM, Di Daniele N, Dixit VM, Dynlacht BD, El-Deiry WS, Fimia GM, Flavell RA, Fulda S, Garrido C, Gougeon ML, Green DR, Gronemeyer H, Hajnoczky G, Hardwick JM, Hengartner MO, Ichijo H, Joseph B, Jost PJ, Kaufmann T, Kepp O, Klionsky DJ, Knight RA, Kumar S, Lemasters JJ, Levine B, Linkermann A, Lipton SA, Lockshin RA, Lopez-Otin C, Lugli E, Madeo F, Malorni W, Marine JC, Martin SJ, Martinou JC, Medema JP, Meier P, Melino S, Mizushima N, Moll U, Munoz-Pinedo C, Nunez G, Oberst A, Panaretakis T, Penninger JM, Peter ME, Piacentini M, Pinton P, Prehn JH, Puthalakath H, Rabinovich GA, Ravichandran KS, Rizzuto R, Rodrigues CM, Rubinsztein DC, Rudel T, Shi Y, Simon HU, Stockwell BR, Szabadkai G, Tait SW, Tang HL, Tavernarakis N, Tsujimoto Y, Vanden Berghe T, Vandenabeele P, Villunger A, Wagner EF, Walczak H, White E, Wood WG, Yuan J, Zakeri Z, Zhivotovsky B, Melino G, Kroemer G. Essential versus accessory aspects of cell death: recommendations of the NCCD 2015. Cell Death Differ. 2015;22:58–73.
38. Wyllie AH, Kerr JFR, Currie AR. Cell death: the significance of apoptosis. Int Rev Cytol. 1980;68:251–305.
39. Gregory CD, Pound JD. Microenvironmental influences of apoptosis in vivo and in vitro. Apoptosis. 2010;15:1029–49.
40. Degterev A, Hitomi J, Germscheid M, Ch'en IL, Korkina O, Teng X, Abbott D, Cuny GD, Yuan C, Wagner G, Hedrick SM, Gerber SA, Lugovskoy A, Yuan J. Identification of RIP1 kinase as a specific cellular target of necrostatins. Nat Chem Biol. 2008;4:313–21.
41. Medzhitov R, Janeway Jr CA. Decoding the patterns of self and nonself by the innate immune system. Science. 2002;296:298–300.
42. Vakkila J, Lotze MT. Inflammation and necrosis promote tumour growth. Nat Rev Immunol. 2004;4:641–8.
43. Gamrekelashvili J, Kruger C, von Wasielewski R, Hoffmann M, Huster KM, Busch DH, Manns MP, Korangy F, Greten TF. Necrotic tumor cell death in vivo impairs tumor-specific immune responses. J Immunol. 2007;178:1573–80.
44. Moreira S, Stramer B, Evans I, Wood W, Martin P. Prioritization of competing damage and developmental signals by migrating macrophages in the Drosophila embryo. Curr Biol. 2010;20:464–70.
45. Marino G, Niso-Santano M, Baehrecke EH, Kroemer G. Self-consumption: the interplay of autophagy and apoptosis. Nat Rev Mol Cell Biol. 2014;15(2):81–94.
46. White E. Deconvoluting the context-dependent role for autophagy in cancer. Nat Rev Cancer. 2012;12:401–10.
47. Changou CA, Chen YR, Xing L, Yen Y, Chuang FY, Cheng RH, Bold RJ, Ann DK, Kung HJ. Arginine starvation-associated atypical cellular death involves mitochondrial dysfunction, nuclear DNA leakage, and chromatin autophagy. Proc Natl Acad Sci U S A. 2014;111:14147–52.
48. Degenhardt K, Mathew R, Beaudoin B, Bray K, Anderson D, Chen G, Mukherjee C, Shi Y, Gelinas C, Fan Y, Nelson DA, Jin S, White E. Autophagy promotes tumor cell survival and restricts necrosis, inflammation, and tumorigenesis. Cancer Cell. 2006;10:51–64.

49. Wen Z, Fan L, Li Y, Zou Z, Scott MJ, Xiao G, Li S, Billiar TR, Wilson MA, Shi X, Fan J. Neutrophils counteract autophagy-mediated anti-inflammatory mechanisms in alveolar macrophage: role in posthemorrhagic shock acute lung inflammation. J Immunol. 2014;193:4623–33.
50. Dou Z, Xu C, Donahue G, Shimi T, Pan JA, Zhu J, Ivanov A, Capell BC, Drake AM, Shah PP, Catanzaro JM, Daniel Ricketts M, Lamark T, Adam SA, Marmorstein R, Zong WX, Johansen T, Goldman RD, Adams PD, Berger SL. Autophagy mediates degradation of nuclear lamina. Nature. 2015;527:105–9.
51. Revesz L. Effect of tumour cells killed by x-rays upon the growth of admixed viable cells. Nature. 1956;178:1391–2.
52. Revesz L. Effect of lethally damaged tumor cells upon the development of admixed viable cells. J Natl Cancer Inst. 1958;20:1157–86.
53. Ringertz N, Klein E, Revesz L. Growth of small compatible tumor implants in presence of admixed radiation-killed or incompatible tumor cells. Cancer. 1959;12:697–707.
54. Seelig KJ, Revesz L. Effect of lethally damaged tumour cells upon the growth of admixed viable cells in diffusion chambers. Br J Cancer. 1960;14:126–38.
55. Puck TT, Marcus PI, Cieciura SJ. Clonal growth of mammalian cells in vitro; growth characteristics of colonies from single HeLa cells with and without a feeder layer. J Exp Med. 1956;103:273–83.
56. Gregory CD, Pound JD, Devitt A, Wilson-Jones M, Ray P, Murray RJ. Inhibitory effects of persistent apoptotic cells on monoclonal antibody production in vitro: simple removal of non-viable cells improves antibody productivity by hybridoma cells in culture. MAbs. 2009;1:370–6.
57. Fadok VA, Voelker DR, Campbell PA, Cohen JJ, Bratton DL, Henson PM. Exposure of phosphatidylserine on the surface of apoptotic lymphocytes triggers specific recognition and removal by macrophages. J Immunol. 1992;148:2207–16.
58. Fadok VA, Bratton DL, Konowal A, Freed PW, Westcott JY, Henson PM. Macrophages that have ingested apoptotic cells in vitro inhibit proinflammatory cytokine production through autocrine/paracrine mechanisms involving TGF-β, PGE2, and PAF. J Clin Invest. 1998;101:890–8.
59. Fond AM, Lee CS, Schulman IG, Kiss RS, Ravichandran KS. Apoptotic cells trigger a membrane-initiated pathway to increase ABCA1. J Clin Invest. 2015;125:2748–58.
60. Emoto K, Toyama-Sorimachi N, Karasuyama H, Inoue K, Umeda M. Exposure of phosphatidylethanolamine on the surface of apoptotic cells. Exp Cell Res. 1997;232:430–4.
61. Gude DR, Alvarez SE, Paugh SW, Mitra P, Yu J, Griffiths R, Barbour SE, Milstien S, Spiegel S. Apoptosis induces expression of sphingosine kinase 1 to release sphingosine-1-phosphate as a "come-and-get-me" signal. FASEB J. 2008;22:2629–38.
62. Lauber K, Bohn E, Krober SM, Xiao YJ, Blumenthal SG, Lindemann RK, Marini P, Wiedig C, Zobywalski A, Baksh S, Xu Y, Autenrieth IB, Schulze-Osthoff K, Belka C, Stuhler G, Wesselborg S. Apoptotic cells induce migration of phagocytes via caspase-3- mediated release of a lipid attraction signal. Cell. 2003;113:717–30.
63. Brown S, Heinisch I, Ross E, Shaw K, Buckley CD, Savill J. Apoptosis disables CD31-mediated cell detachment from phagocytes promoting binding and engulfment. Nature. 2002;418:200–3.
64. Gardai SJ, McPhillips KA, Frasch SC, Janssen WJ, Starefeldt A, Murphy-Ullrich JE, Bratton DL, Oldenborg PA, Michalak M, Henson PM. Cell-surface calreticulin initiates clearance of viable or apoptotic cells through trans-activation of LRP on the phagocyte. Cell. 2005;123:321–34.
65. Truman LA, Ford CA, Pasikowska M, Pound JD, Wilkinson SJ, Dumitriu IE, Melville L, Melrose LA, Ogden CA, Nibbs R, Graham G, Combadiere C, Gregory CD. CX3CL1/fractalkine is released from apoptotic lymphocytes to stimulate macrophage chemotaxis. Blood. 2008;112:5026–36.
66. Leonardi-Essmann F, Emig M, Kitamura Y, Spanagel R, Gebicke-Haerter PJ. Fractalkine-upregulated milk-fat globule EGF factor-8 protein in cultured rat microglia. J Neuroimmunol. 2005;160:92–101.

67. Miksa M, Amin D, Wu R, Ravikumar TS, Wang P. Fractalkine-induced MFG-E8 leads to enhanced apoptotic cell clearance by macrophages. Mol Med. 2007;13:553–60.
68. Sapozhnikov AM, Ponomarev ED, Tarasenko TN, Telford WG. Spontaneous apoptosis and expression of cell surface heat-shock proteins in cultured EL-4 lymphoma cells. Cell Prolif. 1999;32:363–78.
69. Basu S, Binder RJ, Suto R, Anderson KM, Srivastava PK. Necrotic but not apoptotic cell death releases heat shock proteins, which deliver a partial maturation signal to dendritic cells and activate the NF-kappa B pathway. Int Immunol. 2000;12:1539–46.
70. Feng H, Zeng Y, Whitesell L, Katsanis E. Stressed apoptotic tumor cells express heat shock proteins and elicit tumor-specific immunity. Blood. 2001;97:3505–12.
71. Goh YC, Yap CT, Huang BH, Cronshaw AD, Leung BP, Lai PB, Hart SP, Dransfield I, Ross JA. Heat-shock protein 60 translocates to the surface of apoptotic cells and differentiated megakaryocytes and stimulates phagocytosis. Cell Mol Life Sci. 2011;68:1581–92.
72. Gao Y, Herndon JM, Zhang H, Griffith TS, Ferguson TA. Antiinflammatory effects of CD95 ligand (FasL)-induced apoptosis. J Exp Med. 1998;188:887–96.
73. Chen W, Frank ME, Jin W, Wahl SM. TGF-beta released by apoptotic T cells contributes to an immunosuppressive milieu. Immunity. 2001;14:715–25.
74. Bournazou I, Pound JD, Duffin R, Bournazos S, Melville LA, Brown SB, Rossi AG, Gregory CD. Apoptotic human cells inhibit migration of granulocytes via release of lactoferrin. J Clin Invest. 2009;119:20–32.
75. Bournazou I, Mackenzie KJ, Duffin R, Rossi AG, Gregory CD. Inhibition of eosinophil migration by lactoferrin. Immunol Cell Biol. 2010;88:220–3.
76. Miles K, Clarke DJ, Lu W, Sibinska Z, Beaumont PE, Davidson DJ, Barr TA, Campopiano DJ, Gray M. Dying and necrotic neutrophils are anti-inflammatory secondary to the release of alpha-defensins. J Immunol. 2009;183:2122–32.
77. Elliott MR, Chekeni FB, Trampont PC, Lazarowski ER, Kadl A, Walk SF, Park D, Woodson RI, Ostankovich M, Sharma P, Lysiak JJ, Harden TK, Leitinger N, Ravichandran KS. Nucleotides released by apoptotic cells act as a find-me signal to promote phagocytic clearance. Nature. 2009;461:282–6.
78. Horino K, Nishiura H, Ohsako T, Shibuya Y, Hiraoka T, Kitamura N, Yamamoto T. A monocyte chemotactic factor, S19 ribosomal protein dimer, in phagocytic clearance of apoptotic cells. Lab Invest. 1998;78:603–17.
79. Knies UE, Behrensdorf HA, Mitchell CA, Deutsch U, Risau W, Drexler HC, Clauss M. Regulation of endothelial monocyte-activating polypeptide II release by apoptosis. Proc Natl Acad Sci U S A. 1998;95:12322–7.
80. Thery C, Boussac M, Veron P, Ricciardi-Castagnoli P, Raposo G, Garin J, Amigorena S. Proteomic analysis of dendritic cell-derived exosomes: a secreted subcellular compartment distinct from apoptotic vesicles. J Immunol. 2001;166:7309–18.
81. Zirngibl M, Furnrohr BG, Janko C, Munoz LE, Voll RE, Gregory CD, Schett G, Herrmann M. Loading of nuclear autoantigens prototypically recognized by systemic lupus erythematosus sera into late apoptotic vesicles requires intact microtubules and myosin light chain kinase activity. Clin Exp Immunol. 2015;179:39–49.
82. Blume KE, Soeroes S, Waibel M, Keppeler H, Wesselborg S, Herrmann M, Schulze-Osthoff K, Lauber K. Cell surface externalization of annexin A1 as a failsafe mechanism preventing inflammatory responses during secondary necrosis. J Immunol. 2009;183:8138–47.
83. Krispin A, Bledi Y, Atallah M, Trahtemberg U, Verbovetski I, Nahari E, Zelig O, Linial M, Mevorach D. Apoptotic cell thrombospondin-1 and heparin-binding domain lead to dendritic-cell phagocytic and tolerizing states. Blood. 2006;108:3580–9.
84. Soulez M, Sirois I, Brassard N, Raymond MA, Nicodeme F, Noiseux N, Durocher Y, Pshezhetsky AV, Hebert MJ. Epidermal growth factor and perlecan fragments produced by apoptotic endothelial cells co-ordinately activate ERK1/2-dependent antiapoptotic pathways in mesenchymal stem cells. Stem Cells. 2010;28:810–20.
85. Soulez M, Pilon EA, Dieude M, Cardinal H, Brassard N, Qi S, Wu SJ, Durocher Y, Madore F, Perreault C, Hebert MJ. The perlecan fragment LG3 is a novel regulator of obliterative remodeling associated with allograft vascular rejection. Circ Res. 2012;110:94–104.

86. Moffatt OD, Devitt A, Bell ED, Simmons DL, Gregory CD. Macrophage recognition of ICAM-3 on apoptotic leukocytes. J Immunol. 1999;162:6800–10.
87. Torr EE, Gardner DH, Thomas L, Goodall DM, Bielemeier A, Willetts R, Griffiths HR, Marshall LJ, Devitt A. Apoptotic cell-derived ICAM-3 promotes both macrophage chemoattraction to and tethering of apoptotic cells. Cell Death Differ. 2012;19:671–9.
88. Kobara M, Sunagawa N, Abe M, Tanaka N, Toba H, Hayashi H, Keira N, Tatsumi T, Matsubara H, Nakata T. Apoptotic myocytes generate monocyte chemoattractant protein-1 and mediate macrophage recruitment. J Appl Physiol. 2008;104:601–9.
89. Renz A, Berdel WE, Kreuter M, Belka C, Schulze-Osthoff K, Los M. Rapid extracellular release of cytochrome c is specific for apoptosis and marks cell death in vivo. Blood. 2001;98:1542–8.
90. Codina R, Vanasse A, Kelekar A, Vezys V, Jemmerson R. Cytochrome c-induced lymphocyte death from the outside in: inhibition by serum leucine-rich alpha-2-glycoprotein-1. Apoptosis. 2010;15:139–52.
91. Casciola-Rosen LA, Anhalt G, Rosen A. Autoantigens targeted in systemic lupus erythematosus are clustered in two populations of surface structures on apoptotic keratinocytes. J Exp Med. 1994;179:1317–30.
92. Radic M, Marion T, Monestier M. Nucleosomes are exposed at the cell surface in apoptosis. J Immunol. 2004;172:6692–700.
93. Bilyy RO, Shkandina T, Tomin A, Munoz LE, Franz S, Antonyuk V, Kit YY, Zirngibl M, Furnrohr BG, Janko C, Lauber K, Schiller M, Schett G, Stoika RS, Herrmann M. Macrophages discriminate glycosylation patterns of apoptotic cell-derived microparticles. J Biol Chem. 2012;287:496–503.
94. Weihua Z, Tsan R, Schroit AJ, Fidler IJ. Apoptotic cells initiate endothelial cell sprouting via electrostatic signaling. Cancer Res. 2005;65:11529–35.
95. Sirois I, Raymond MA, Brassard N, Cailhier JF, Fedjaev M, Hamelin K, Londono I, Bendayan M, Pshezhetsky AV, Hebert MJ. Caspase-3-dependent export of TCTP: a novel pathway for antiapoptotic intercellular communication. Cell Death Differ. 2011;18:549–62.
96. Tennant I, Pound JD, Marr LA, Willems JJ, Petrova S, Ford CA, Paterson M, Devitt A, Gregory CD. Innate recognition of apoptotic cells: novel apoptotic cell-associated molecular patterns revealed by crossreactivity of anti-LPS antibodies. Cell Death Differ. 2013;20:698–708.
97. Wickman GR, Julian L, Mardilovich K, Schumacher S, Munro J, Rath N, Zander SA, Mleczak A, Sumpton D, Morrice N, Bienvenut WV, Olson MF. Blebs produced by actin-myosin contraction during apoptosis release damage-associated molecular pattern proteins before secondary necrosis occurs. Cell Death Differ. 2013;20:1293–305.
98. Crescitelli R, Lasser C, Szabo TG, Kittel A, Eldh M, Dianzani I, Buzas EI, Lotvall J. Distinct RNA profiles in subpopulations of extracellular vesicles: apoptotic bodies, microvesicles and exosomes. J extracellular vesicles. 2013;2.
99. Atkin-Smith GK, Tixeira R, Paone S, Mathivanan S, Collins C, Liem M, Goodall KJ, Ravichandran KS, Hulett MD, Poon IK. A novel mechanism of generating extracellular vesicles during apoptosis via a beads-on-a-string membrane structure. Nat Commun. 2015;6:7439.
100. Dieude M, Bell C, Turgeon J, Beillevaire D, Pomerleau L, Yang B, Hamelin K, Qi S, Pallet N, Beland C, Dhahri W, Cailhier JF, Rousseau M, Duchez AC, Levesque T, Lau A, Rondeau C, Gingras D, Muruve D, Rivard A, Cardinal H, Perreault C, Desjardins M, Boilard E, Thibault P, Hebert MJ. The 20S proteasome core, active within apoptotic exosome-like vesicles, induces autoantibody production and accelerates rejection. Sci Trans Med. 2015;7:318ra200.
101. Chao MP, Alizadeh AA, Tang C, Myklebust JH, Varghese B, Gill S, Jan M, Cha AC, Chan CK, Tan BT, Park CY, Zhao F, Kohrt HE, Malumbres R, Briones J, Gascoyne RD, Lossos IS, Levy R, Weissman IL, Majeti R. Anti-CD47 antibody synergizes with rituximab to promote phagocytosis and eradicate non-Hodgkin lymphoma. Cell. 2010;142:699–713.
102. Chao MP, Alizadeh AA, Tang C, Jan M, Weissman-Tsukamoto R, Zhao F, Park CY, Weissman IL, Majeti R. Therapeutic antibody targeting of CD47 eliminates human acute lymphoblastic leukemia. Cancer Res. 2011;71:1374–84.

103. Thery C, Ostrowski M, Segura E. Membrane vesicles as conveyors of immune responses. Nat Rev Immunol. 2009;9:581–93.
104. Buzas EI, Gyorgy B, Nagy G, Falus A, Gay S. Emerging role of extracellular vesicles in inflammatory diseases. Nat Rev Rheumatol. 2014;10:356–64.
105. Eken C, Martin PJ, Sadallah S, Treves S, Schaller M, Schifferli JA. Ectosomes released by polymorphonuclear neutrophils induce a MerTK-dependent anti-inflammatory pathway in macrophages. J Biol Chem. 2010;285:39914–21.
106. D'Souza-Schorey C, Clancy JW. Tumor-derived microvesicles: shedding light on novel microenvironment modulators and prospective cancer biomarkers. Genes Dev. 2012;26: 1287–99.
107. Bonner C, Bacon S, Concannon CG, Rizvi SR, Baquie M, Farrelly AM, Kilbride SM, Dussmann H, Ward MW, Boulanger CM, Wollheim CB, Graf R, Byrne MM, Prehn JH. INS-1 cells undergoing caspase-dependent apoptosis enhance the regenerative capacity of neighboring cells. Diabetes. 2010;59:2799–808.
108. Peter C, Wesselborg S, Herrmann M, Lauber K. Dangerous attraction: phagocyte recruitment and danger signals of apoptotic and necrotic cells. Apoptosis. 2010;15(9):1007–28.
109. Arandjelovic S, Ravichandran KS. Phagocytosis of apoptotic cells in homeostasis. Nat Immunol. 2015;16:907–17.
110. Li F, Huang Q, Chen J, Peng Y, Roop DR, Bedford JS, Li CY. Apoptotic cells activate the "phoenix rising" pathway to promote wound healing and tissue regeneration. Sci Signal. 2010;3:ra13–13.
111. Ariel A, Fredman G, Sun YP, Kantarci A, Van Dyke TE, Luster AD, Serhan CN. Apoptotic neutrophils and T cells sequester chemokines during immune response resolution through modulation of CCR5 expression. Nat Immunol. 2006;7:1209–16.
112. Garrido C, Gurbuxani S, Ravagnan L, Kroemer G. Heat shock proteins: endogenous modulators of apoptotic cell death. Biochem Biophys Res Commun. 2001;286:433–42.
113. Baixauli F, Lopez-Otin C, Mittelbrunn M. Exosomes and autophagy: coordinated mechanisms for the maintenance of cellular fitness. Front Immunol. 2014;5:403.
114. Coppe JP, Desprez PY, Krtolica A, Campisi J. The senescence-associated secretory phenotype: the dark side of tumor suppression. Annu Rev Pathol. 2010;5:99–118.
115. Birge RB, Ucker DS. Innate apoptotic immunity: the calming touch of death. Cell Death Differ. 2008;15:1096–102.
116. Cocco RE, Ucker DS. Distinct modes of macrophage recognition for apoptotic and necrotic cells are not specified exclusively by phosphatidylserine exposure. Mol Biol Cell. 2001;12:919–30.
117. Cvetanovic M, Mitchell JE, Patel V, Avner BS, Su Y, van der Saag PT, Witte PL, Fiore S, Levine JS, Ucker DS. Specific recognition of apoptotic cells reveals a ubiquitous and unconventional innate immunity. J Biol Chem. 2006;281:20055–67.
118. Cvetanovic M, Ucker DS. Innate immune discrimination of apoptotic cells: repression of proinflammatory macrophage transcription is coupled directly to specific recognition. J Immunol. 2004;172:880–9.
119. Lewis CE, Leek R, Harris A, McGee JO. Cytokine regulation of angiogenesis in breast cancer: the role of tumor-associated macrophages. J Leukoc Biol. 1995;57:747–51.
120. Ong SM, Tan YC, Beretta O, Jiang D, Yeap WH, Tai JJ, Wong WC, Yang H, Schwarz H, Lim KH, Koh PK, Ling KL, Wong SC. Macrophages in human colorectal cancer are pro-inflammatory and prime T cells towards an anti-tumour type-1 inflammatory response. Eur J Immunol. 2012;42:89–100.
121. Zhang QW, Liu L, Gong CY, Shi HS, Zeng YH, Wang XZ, Zhao YW, Wei YQ. Prognostic significance of tumor-associated macrophages in solid tumor: a meta-analysis of the literature. PLoS One. 2012;7, e50946.
122. Stanford JC, Young C, Hicks D, Owens P, Williams A, Vaught DB, Morrison MM, Lim J, Williams M, Brantley-Sieders DM, Balko JM, Tonetti D, Earp HS, Cook RS. Efferocytosis produces a prometastatic landscape during postpartum mammary gland involution. J Clin Invest. 2014;124(11):4737–52.

123. Soki FN, Koh AJ, Jones JD, Kim YW, Dai J, Keller ET, Pienta KJ, Atabai K, Roca H, McCauley LK. Polarization of prostate cancer-associated macrophages is induced by milk fat globule-EGF factor 8 (MFG-E8)-mediated efferocytosis. J Biol Chem. 2014;289:24560–72.
124. Novak ML, Weinheimer-Haus EM, Koh TJ. Macrophage activation and skeletal muscle healing following traumatic injury. J Pathol. 2014;232:344–55.
125. Murray PJ, Allen JE, Biswas SK, Fisher EA, Gilroy DW, Goerdt S, Gordon S, Hamilton JA, Ivashkiv LB, Lawrence T, Locati M, Mantovani A, Martinez FO, Mege JL, Mosser DM, Natoli G, Saeij JP, Schultze JL, Shirey KA, Sica A, Suttles J, Udalova I, van Ginderachter JA, Vogel SN, Wynn TA. Macrophage activation and polarization: nomenclature and experimental guidelines. Immunity. 2014;41:14–20.
126. A-Gonzalez N, Bensinger SJ, Hong C, Beceiro S, Bradley MN, Zelcer N, Deniz J, Ramirez C, Diaz M, Gallardo G, de Galarreta CR, Salazar J, Lopez F, Edwards P, Parks J, Andujar M, Tontonoz P, Castrillo A. Apoptotic cells promote their own clearance and immune tolerance through activation of the nuclear receptor LXR. Immunity. 2009;31:245–58.
127. Kondylis V, Polykratis A, Ehlken H, Ochoa-Callejero L, Straub BK, Krishna-Subramanian S, Van TM, Curth HM, Heise N, Weih F, Klein U, Schirmacher P, Kelliher M, Pasparakis M. NEMO prevents steatohepatitis and hepatocellular carcinoma by inhibiting RIPK1 kinase activity-mediated hepatocyte apoptosis. Cancer Cell. 2015;28:582–98.
128. Fuchs Y, Steller H. Live to die another way: modes of programmed cell death and the signals emanating from dying cells. Nat Rev Mol Cell Biol. 2015;16:329–44.
129. Michalak EM, Vandenberg CJ, Delbridge AR, Wu L, Scott CL, Adams JM, Strasser A. Apoptosis-promoted tumorigenesis: gamma-irradiation-induced thymic lymphomagenesis requires Puma-driven leukocyte death. Genes Dev. 2010;24:1608–13.
130. Labi V, Erlacher M, Krumschnabel G, Manzl C, Tzankov A, Pinon J, Egle A, Villunger A. Apoptosis of leukocytes triggered by acute DNA damage promotes lymphoma formation. Genes Dev. 2010;24:1602–7.
131. Miura M. Apoptotic and nonapoptotic caspase functions in animal development. Cold Spring Harb Perspect Biol. 2012;4.
132. Connolly PF, Jager R, Fearnhead HO. New roles for old enzymes: killer caspases as the engine of cell behavior changes. Front Physiol. 2014;5:149.
133. Zhao X, Wang D, Zhao Z, Xiao Y, Sengupta S, Xiao Y, Zhang R, Lauber K, Wesselborg S, Feng L, Rose TM, Shen Y, Zhang J, Prestwich G, Xu Y. Caspase-3-dependent activation of calcium-independent phospholipase A2 enhances cell migration in non-apoptotic ovarian cancer cells. J Biol Chem. 2006;281:29357–68.
134. Ran S, Thorpe PE. Phosphatidylserine is a marker of tumor vasculature and a potential target for cancer imaging and therapy. Int J Radiat Oncol Biol Phys. 2002;54:1479–84.
135. Riboldi E, Porta C, Morlacchi S, Viola A, Mantovani A, Sica A. Hypoxia-mediated regulation of macrophage functions in pathophysiology. Int Immunol. 2013;25:67–75.
136. Kaelin Jr WG, Ratcliffe PJ. Oxygen sensing by metazoans: the central role of the HIF hydroxylase pathway. Mol Cell. 2008;30:393–402.
137. Cummins EP, Berra E, Comerford KM, Ginouves A, Fitzgerald KT, Seeballuck F, Godson C, Nielsen JE, Moynagh P, Pouyssegur J, Taylor CT. Prolyl hydroxylase-1 negatively regulates IkappaB kinase-beta, giving insight into hypoxia-induced NFkappaB activity. Proc Natl Acad Sci U S A. 2006;103:18154–9.
138. Sendoel A, Hengartner MO. Apoptotic cell death under hypoxia. Physiology (Bethesda). 2014;29:168–76.
139. Schioppa T, Uranchimeg B, Saccani A, Biswas SK, Doni A, Rapisarda A, Bernasconi S, Saccani S, Nebuloni M, Vago L, Mantovani A, Melillo G, Sica A. Regulation of the chemokine receptor CXCR4 by hypoxia. J Exp Med. 2003;198:1391–402.
140. Biswas SK, Allavena P, Mantovani A. Tumor-associated macrophages: functional diversity, clinical significance, and open questions. Semin Immunopathol. 2013;35:585–600.
141. Movahedi K, Laoui D, Gysemans C, Baeten M, Stange G, Van den Bossche J, Mack M, Pipeleers D, In't Veld P, De Baetselier P, Van Ginderachter JA. Different tumor microenviron-

ments contain functionally distinct subsets of macrophages derived from Ly6C(high) monocytes. Cancer Res. 2010;70:5728–39.
142. Obeid E, Nanda R, Fu YX, Olopade OI. The role of tumor-associated macrophages in breast cancer progression (review). Int J Oncol. 2013;43:5–12.
143. Solinas G, Germano G, Mantovani A, Allavena P. Tumor-associated macrophages (TAM) as major players of the cancer-related inflammation. J Leukoc Biol. 2009;86:1065–73.
144. Hartnett L, Egan LJ. Inflammation, DNA methylation and colitis-associated cancer. Carcinogenesis. 2012;33:723–31.
145. Walter ND, Rice PL, Redente EF, Kauvar EF, Lemond L, Aly T, Wanebo K, Chan ED. Wound healing after trauma may predispose to lung cancer metastasis: review of potential mechanisms. Am J Respir Cell Mol Biol. 2011;44:591–6.
146. Wang XW, Hussain SP, Huo TI, Wu CG, Forgues M, Hofseth LJ, Brechot C, Harris CC. Molecular pathogenesis of human hepatocellular carcinoma. Toxicology. 2002;181–182: 43–7.
147. Jung Y, Witek RP, Syn WK, Choi SS, Omenetti A, Premont R, Guy CD, Diehl AM. Signals from dying hepatocytes trigger growth of liver progenitors. Gut. 2010;59:655–65.
148. Lu WY, Bird TG, Boulter L, Tsuchiya A, Cole AM, Hay T, Guest RV, Wojtacha D, Man TY, Mackinnon A, Ridgway RA, Kendall T, Williams MJ, Jamieson T, Raven A, Hay DC, Iredale JP, Clarke AR, Sansom OJ, Forbes SJ. Hepatic progenitor cells of biliary origin with liver repopulation capacity. Nat Cell Biol. 2015;17(8):971–83.
149. Lu H, Clauser KR, Tam WL, Frose J, Ye X, Eaton EN, Reinhardt F, Donnenberg VS, Bhargava R, Carr SA, Weinberg RA. A breast cancer stem cell niche supported by juxtacrine signalling from monocytes and macrophages. Nat Cell Biol. 2014;16:1105–17.
150. Koh TJ, DiPietro LA. Inflammation and wound healing: the role of the macrophage. Expert Rev Mol Med. 2011;13, e23.
151. Novak ML, Koh TJ. Phenotypic transitions of macrophages orchestrate tissue repair. Am J Pathol. 2013;183:1352–63.
152. Schafer M, Werner S. Cancer as an overhealing wound: an old hypothesis revisited. Nat Rev Mol Cell Biol. 2008;9:628–38.
153. Byun JS, Gardner K. Wounds that will not heal: pervasive cellular reprogramming in cancer. Am J Pathol. 2013;182:1055–64.
154. Mantovani A, Sica A. Macrophages, innate immunity and cancer: balance, tolerance, and diversity. Curr Opin Immunol. 2010;22:231–7.
155. Wynn TA, Chawla A, Pollard JW. Macrophage biology in development, homeostasis and disease. Nature. 2013;496:445–55.
156. Khanna S, Biswas S, Shang Y, Collard E, Azad A, Kauh C, Bhasker V, Gordillo GM, Sen CK, Roy S. Macrophage dysfunction impairs resolution of inflammation in the wounds of diabetic mice. PLoS One. 2010;5, e9539.
157. Swift ME, Burns AL, Gray KL, DiPietro LA. Age-related alterations in the inflammatory response to dermal injury. J Invest Dermatol. 2001;117:1027–35.
158. Kroemer G, Galluzzi L, Kepp O, Zitvogel L. Immunogenic cell death in cancer therapy. Annu Rev Immunol. 2013;31:51–72.
159. Murdoch C, Giannoudis A, Lewis CE. Mechanisms regulating the recruitment of macrophages into hypoxic areas of tumors and other ischemic tissues. Blood. 2004;104:2224–34.
160. DeNardo DG, Brennan DJ, Rexhepaj E, Ruffell B, Shiao SL, Madden SF, Gallagher WM, Wadhwani N, Keil SD, Junaid SA, Rugo HS, Hwang ES, Jirstrom K, West BL, Coussens LM. Leukocyte complexity predicts breast cancer survival and functionally regulates response to chemotherapy. Cancer Discov. 2011;1:54–67.
161. Xu J, Escamilla J, Mok S, David J, Priceman S, West B, Bollag G, McBride W, Wu L. CSF1R signaling blockade stanches tumor-infiltrating myeloid cells and improves the efficacy of radiotherapy in prostate cancer. Cancer Res. 2013;73:2782–94.
162. Loberg RD, Ying C, Craig M, Day LL, Sargent E, Neeley C, Wojno K, Snyder LA, Yan L, Pienta KJ. Targeting CCL2 with systemic delivery of neutralizing antibodies induces prostate cancer tumor regression in vivo. Cancer Res. 2007;67:9417–24.

163. Rozel S, Galban CJ, Nicolay K, Lee KC, Sud S, Neeley C, Snyder LA, Chenevert TL, Rehemtulla A, Ross BD, Pienta KJ. Synergy between anti-CCL2 and docetaxel as determined by DW-MRI in a metastatic bone cancer model. J Cell Biochem. 2009;107:58–64.
164. Sanford DE, Belt BA, Panni RZ, Mayer A, Deshpande AD, Carpenter D, Mitchem JB, Plambeck-Suess SM, Worley LA, Goetz BD, Wang-Gillam A, Eberlein TJ, Denardo DG, Goedegebuure SP, Linehan DC. Inflammatory monocyte mobilization decreases patient survival in pancreatic cancer: a role for targeting the CCL2/CCR2 axis. Clin Cancer Res. 2013;19:3404–15.
165. Kitamura T, Qian BZ, Soong D, Cassetta L, Noy R, Sugano G, Kato Y, Li J, Pollard JW. CCL2-induced chemokine cascade promotes breast cancer metastasis by enhancing retention of metastasis-associated macrophages. J Exp Med. 2015;212:1043–59.
166. Germano G, Frapolli R, Belgiovine C, Anselmo A, Pesce S, Liguori M, Erba E, Uboldi S, Zucchetti M, Pasqualini F, Nebuloni M, van Rooijen N, Mortarini R, Beltrame L, Marchini S, Fuso Nerini I, Sanfilippo R, Casali PG, Pilotti S, Galmarini CM, Anichini A, Mantovani A, D'Incalci M, Allavena P. Role of macrophage targeting in the antitumor activity of trabectedin. Cancer Cell. 2013;23:249–62.
167. Zeisberger SM, Odermatt B, Marty C, Zehnder-Fjallman AH, Ballmer-Hofer K, Schwendener RA. Clodronate-liposome-mediated depletion of tumour-associated macrophages: a new and highly effective antiangiogenic therapy approach. Br J Cancer. 2006;95:272–81.
168. Coscia M, Quaglino E, Iezzi M, Curcio C, Pantaleoni F, Riganti C, Holen I, Monkkonen H, Boccadoro M, Forni G, Musiani P, Bosia A, Cavallo F, Massaia M. Zoledronic acid repolarizes tumour-associated macrophages and inhibits mammary carcinogenesis by targeting the mevalonate pathway. J Cell Mol Med. 2010;14:2803–15.
169. Hiraoka K, Zenmyo M, Watari K, Iguchi H, Fotovati A, Kimura YN, Hosoi F, Shoda T, Nagata K, Osada H, Ono M, Kuwano M. Inhibition of bone and muscle metastases of lung cancer cells by a decrease in the number of monocytes/macrophages. Cancer Sci. 2008;99: 1595–602.
170. Pulaski HL, Spahlinger G, Silva IA, McLean K, Kueck AS, Reynolds RK, Coukos G, Conejo-Garcia JR, Buckanovich RJ. Identifying alemtuzumab as an anti-myeloid cell antiangiogenic therapy for the treatment of ovarian cancer. J Transl Med. 2009;7:49.
171. Bak SP, Walters JJ, Takeya M, Conejo-Garcia JR, Berwin BL. Scavenger receptor-A-targeted leukocyte depletion inhibits peritoneal ovarian tumor progression. Cancer Res. 2007;67:4783–9.
172. Nagai T, Tanaka M, Tsuneyoshi Y, Xu B, Michie SA, Hasui K, Hirano H, Arita K, Matsuyama T. Targeting tumor-associated macrophages in an experimental glioma model with a recombinant immunotoxin to folate receptor beta. Cancer Immunol Immunother. 2009;58:1577–86.
173. Petrillo M, Zannoni GF, Martinelli E, Pedone Anchora L, Ferrandina G, Tropeano G, Fagotti A, Scambia G. Polarisation of tumor-associated macrophages toward M2 phenotype correlates with poor response to chemoradiation and reduced survival in patients with locally advanced cervical cancer. PLoS One. 2015;10, e0136654.
174. Pantano F, Berti P, Guida FM, Perrone G, Vincenzi B, Amato MM, Righi D, Dell'aquila E, Graziano F, Catalano V, Caricato M, Rizzo S, Muda AO, Russo A, Tonini G, Santini D. The role of macrophages polarization in predicting prognosis of radically resected gastric cancer patients. J Cell Mol Med. 2013;17:1415–21.
175. Biswas SK, Lewis CE. NF-kappaB as a central regulator of macrophage function in tumors. J Leukoc Biol. 2010;88:877–84.
176. Guiducci C, Vicari AP, Sangaletti S, Trinchieri G, Colombo MP. Redirecting in vivo elicited tumor infiltrating macrophages and dendritic cells towards tumor rejection. Cancer Res. 2005;65:3437–46.
177. Buhtoiarov IN, Sondel PM, Wigginton JM, Buhtoiarova TN, Yanke EM, Mahvi DA, Rakhmilevich AL. Anti-tumour synergy of cytotoxic chemotherapy and anti-CD40 plus CpG-ODN immunotherapy through repolarization of tumour-associated macrophages. Immunology. 2011;132:226–39.

178. Weber JS, Zarour H, Redman B, Trefzer U, O'Day S, van den Eertwegh AJ, Marshall E, Wagner S. Randomized phase 2/3 trial of CpG oligodeoxynucleotide PF-3512676 alone or with dacarbazine for patients with unresectable stage III and IV melanoma. Cancer. 2009;115:3944–54.
179. Carpentier A, Metellus P, Ursu R, Zohar S, Lafitte F, Barrie M, Meng Y, Richard M, Parizot C, Laigle-Donadey F, Gorochov G, Psimaras D, Sanson M, Tibi A, Chinot O, Carpentier AF. Intracerebral administration of CpG oligonucleotide for patients with recurrent glioblastoma: a phase II study. Neuro Oncol. 2010;12:401–8.
180. Zent CS, Smith BJ, Ballas ZK, Wooldridge JE, Link BK, Call TG, Shanafelt TD, Bowen DA, Kay NE, Witzig TE, Weiner GJ. Phase I clinical trial of CpG oligonucleotide 7909 (PF-03512676) in patients with previously treated chronic lymphocytic leukemia. Leuk Lymphoma. 2012;53:211–7.
181. Beatty GL, Chiorean EG, Fishman MP, Saboury B, Teitelbaum UR, Sun W, Huhn RD, Song W, Li D, Sharp LL, Torigian DA, O'Dwyer PJ, Vonderheide RH. CD40 agonists alter tumor stroma and show efficacy against pancreatic carcinoma in mice and humans. Science. 2011;331:1612–6.
182. Beatty GL, Torigian DA, Chiorean EG, Saboury B, Brothers A, Alavi A, Troxel AB, Sun W, Teitelbaum UR, Vonderheide RH, O'Dwyer PJ. A phase I study of an agonist CD40 monoclonal antibody (CP-870,893) in combination with gemcitabine in patients with advanced pancreatic ductal adenocarcinoma. Clin Cancer Res. 2013;19:6286–95.
183. Pyonteck SM, Akkari L, Schuhmacher AJ, Bowman RL, Sevenich L, Quail DF, Olson OC, Quick ML, Huse JT, Teijeiro V, Setty M, Leslie CS, Oei Y, Pedraza A, Zhang J, Brennan CW, Sutton JC, Holland EC, Daniel D, Joyce JA. CSF-1R inhibition alters macrophage polarization and blocks glioma progression. Nat Med. 2013;19(10):1264–72.
184. Xu M, Liu M, Du X, Li S, Li H, Li X, Li Y, Wang Y, Qin Z, Fu YX, Wang S. Intratumoral delivery of IL-21 overcomes anti-Her2/Neu resistance through shifting tumor-associated macrophages from M2 to M1 phenotype. J Immunol. 2015;194:4997–5006.
185. Klug F, Prakash H, Huber PE, Seibel T, Bender N, Halama N, Pfirschke C, Voss RH, Timke C, Umansky L, Klapproth K, Schakel K, Garbi N, Jager D, Weitz J, Schmitz-Winnenthal H, Hammerling GJ, Beckhove P. Low-dose irradiation programs macrophage differentiation to an iNOS(+)/M1 phenotype that orchestrates effective T cell immunotherapy. Cancer Cell. 2013;24:589–602.
186. O'Connell RM, Taganov KD, Boldin MP, Cheng G, Baltimore D. MicroRNA-155 is induced during the macrophage inflammatory response. Proc Natl Acad Sci U S A. 2007;104:1604–9.
187. Chaudhuri AA, So AY, Sinha N, Gibson WS, Taganov KD, O'Connell RM, Baltimore D. MicroRNA-125b potentiates macrophage activation. J Immunol. 2011;187:5062–8.
188. Asano K, Miwa M, Miwa K, Hanayama R, Nagase H, Nagata S, Tanaka M. Masking of phosphatidylserine inhibits apoptotic cell engulfment and induces autoantibody production in mice. J Exp Med. 2004;200:459–67.
189. Bondanza A, Zimmermann VS, Rovere-Querini P, Turnay J, Dumitriu IE, Stach CM, Voll RE, Gaipl US, Bertling W, Poschl E, Kalden JR, Manfredi AA, Herrmann M. Inhibition of phosphatidylserine recognition heightens the immunogenicity of irradiated lymphoma cells in vivo. J Exp Med. 2004;200:1157–65.
190. He J, Yin Y, Luster TA, Watkins L, Thorpe PE. Antiphosphatidylserine antibody combined with irradiation damages tumor blood vessels and induces tumor immunity in a rat model of glioblastoma. Clin Cancer Res. 2009;15:6871–80.
191. Yin Y, Huang X, Lynn KD, Thorpe PE. Phosphatidylserine-targeting antibody induces M1 macrophage polarization and promotes myeloid-derived suppressor cell differentiation. Cancer Immunol Res. 2013;1:256–68.
192. Chalasani P, Marron M, Roe D, Clarke K, Iannone M, Livingston RB, Shan JS, Stopeck AT. A phase I clinical trial of bavituximab and paclitaxel in patients with HER2 negative metastatic breast cancer. Cancer Med. 2015;4:1051–9.

193. Verma A, Warner SL, Vankayalapati H, Bearss DJ, Sharma S. Targeting Axl and Mer kinases in cancer. Mol Cancer Ther. 2011;10:1763–73.
194. Sheridan C. First Axl inhibitor enters clinical trials. Nat Biotechnol. 2013;31:775–6.
195. Bosurgi L, Bernink JH, Delgado Cuevas V, Gagliani N, Joannas L, Schmid ET, Booth CJ, Ghosh S, Rothlin CV. Paradoxical role of the proto-oncogene Axl and Mer receptor tyrosine kinases in colon cancer. Proc Natl Acad Sci U S A. 2013;110(32):13091–6.
196. Montalvao F, Garcia Z, Celli S, Breart B, Deguine J, Van Rooijen N, Bousso P. The mechanism of anti-CD20-mediated B cell depletion revealed by intravital imaging. J Clin Invest. 2013;123:5098–103.
197. Gul N, Babes L, Siegmund K, Korthouwer R, Bogels M, Braster R, Vidarsson G, Ten Hagen TL, Kubes P, van Egmond M. Macrophages eliminate circulating tumor cells after monoclonal antibody therapy. J Clin Invest. 2014;124:812–23.
198. Donato AL, Huang Q, Liu X, Li F, Zimmerman MA, Li CY. Caspase 3 promotes surviving melanoma tumor cell growth after cytotoxic therapy. J Invest Dermatol. 2014;134:1686–92.
199. Kruiswijk F, Labuschagne CF, Vousden KH. p53 in survival, death and metabolic health: a lifeguard with a licence to kill. Nat Rev Mol Cell Biol. 2015;16:393–405.

Chapter 4
Apoptotic Caspases in Promoting Cancer: Implications from Their Roles in Development and Tissue Homeostasis

Catherine Dabrowska, Mingli Li, and Yun Fan

Abstract Apoptosis, a major form of programmed cell death, is an important mechanism to remove extra or unwanted cells during development. In tissue homeostasis apoptosis also acts as a monitoring machinery to eliminate damaged cells in response to environmental stresses. During these processes, caspases, a group of proteases, have been well defined as key drivers of cell death. However, a wealth of evidence is emerging which supports the existence of many other non-apoptotic functions of these caspases, which are essential not only in proper organism development but also in tissue homeostasis and post-injury recovery. In particular, apoptotic caspases in stress-induced dying cells can activate mitogenic signals leading to proliferation of neighbouring cells, a phenomenon termed apoptosis-induced proliferation. Apparently, such non-apoptotic functions of caspases need to be controlled and restrained in a context-dependent manner during development to prevent their detrimental effects. Intriguingly, accumulating studies suggest that cancer cells are able to utilise these functions of caspases to their advantage to enable their survival, proliferation and metastasis in order to grow and progress. This book chapter will review non-apoptotic functions of the caspases in development and tissue homeostasis with focus on how these cellular processes can be hijacked by cancer cells and contribute to tumourigenesis.

Keywords Apoptosis • Caspase • Non-apoptotic function • Apoptosis-induced proliferation • Development • Tissue homeostasis • Cancer

4.1 Introduction

Apoptosis was first identified as a form of cell death by its distinct morphological characteristics including cellular shrinkage, chromosome condensation, nuclear fragmentation and formation of apoptotic bodies [1, 2]. Studies in *C. elegans* then

C. Dabrowska • M. Li • Y. Fan (✉)
School of Biosciences, University of Birmingham, Edgbaston, Birmingham B15 2TT, UK
e-mail: y.fan@bham.ac.uk

C.D. Gregory (ed.), *Apoptosis in Cancer Pathogenesis and Anti-cancer Therapy*, Advances in Experimental Medicine and Biology 930,
DOI 10.1007/978-3-319-39406-0_4

uncovered that apoptosis is genetically controlled and plays critical roles during development to remove unwanted or unnecessary cells [3, 4]. Such function of apoptosis further extends to maintenance of tissue homeostasis by eliminating damaged or unfit cells [5, 6]. Apoptosis has therefore been viewed as a monitoring programme to identify and kill potentially harmful cells that may develop into cancer. Consistent with this idea, evading apoptosis has been considered as a hallmark of cancer [7, 8].

The key components of the apoptotic machinery are caspases, a family of cysteine proteases which cleave their substrates leading to cell death [9, 10]. Recently however, in addition to their functions in apoptosis, caspases are becoming better understood in their multifunctional nature with an increasing number of non-apoptotic functions discovered. We acknowledge the abundance of high quality reviews which have described with clarity the non-apoptotic functions of caspases in the context of development and tissue regeneration [11–14]. This chapter therefore focuses on the roles of caspases in sustaining cancers and promoting their spread which seems to contradict what we know about their roles in apoptosis. There is now certainly a great wealth of evidence to show that the apoptotic caspases actually have multiple functions other than executing cell death, and cancer cells can hijack these activities to directly promote their growth, metastasis and recurrence after therapy. Here we have synthesised the evidence present in the current literature supporting this claim, to highlight that the caspases do indeed have a role in progressing cancers. Issues that may exist in current cancer therapies for particular patient subsets are also discussed.

4.2 The Apoptotic Machinery: Functions of the Apoptotic Caspases

Apoptosis is an evolutionarily conserved mechanism in multicellular organisms, allowing correct pattern formation during development and the removal of cells which are detrimental to the health and survival of the organism [15–18]. The pathways leading to apoptosis have been elucidated in many organisms, including *C. elegans*, *Drosophila* and mammals, which are summarised in Fig. 4.1. A noticeable family of key components in these apoptosis pathways are the caspases. By definition, caspases are cysteine-aspartic acid proteases. They cleave their substrates after the aspartic acid residue which features at the end of short tetrapeptide motifs [9, 19]. In addition to their functions in apoptosis, caspases are also well known for their roles in inflammatory responses [20–22]. For example, there are 18 known mammalian caspases among which caspases-2, -3, -6, -7, -8, -9 and -10 function in apoptosis and thus have been classified as apoptotic caspases [14, 23]. This review focuses on these caspases, especially caspases-3, -7, -8 and -9 due to their reported multiple non-apoptotic functions.

Under normal cellular conditions all apoptotic caspases are present as inactive pro-caspases, called zymogens, which consist of a prodomain, a small subunit and

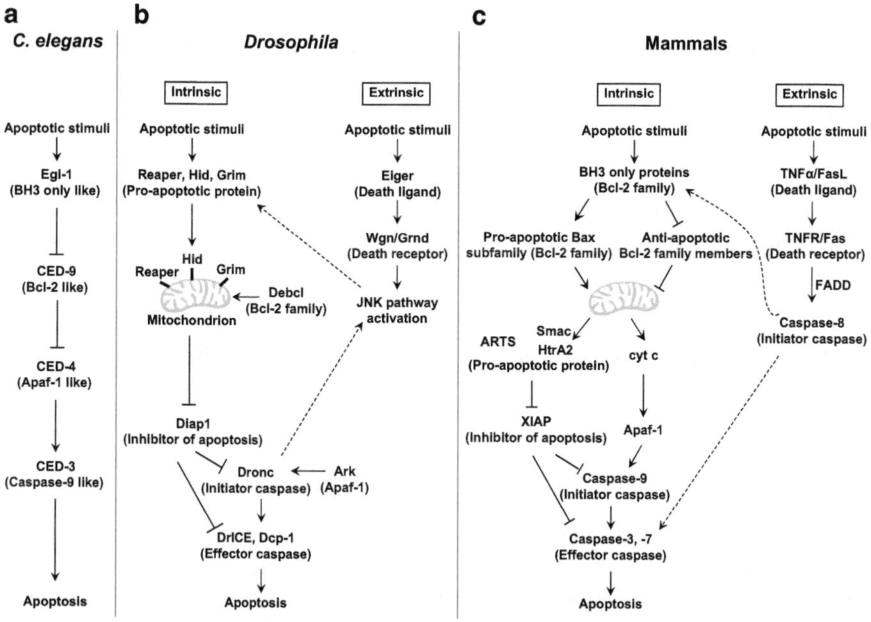

Fig. 4.1 Evolutionarily conserved apoptosis pathways. Schematic comparison of major components in the apoptosis pathways in *C. elegans* (**a**), *Drosophila* (**b**) and mammals (**c**). Protein homology is indicated in brackets. See text for more details

a large subunit [19, 24]. They require cleavage in apoptotic cellular conditions to become activated. Based on the structure of N-terminal prodomains, apoptotic caspases can be subdivided into the initiator (or apical) and effector (or executioner) caspases. The initiator caspases have elongated prodomains which contain either the death effector domains (DED, e.g. for caspase 8) or the caspase-recruitment domains (CARD, e.g. for caspase 9). In contrast, the effector caspases have small prodomains. These caspases also have distinct functions and substrates during the process of apoptosis [25–27]. The initiator caspases cleave inactive pro-effector caspases and activate them. They are therefore also called apical caspases. In contrast, effector caspases, once activated by the initiator caspases, further cleave their broad range of cellular proteins leading to execution of cell death. They therefore have another name as executioner caspases. For simplicity, terms of initiator and effector caspases are used in this review.

4.2.1 Apoptosis in C. elegans

The caspases were first identified in *C. elegans* in which 131 cells undergo apoptosis during development by the action of a simple and linear pathway (Fig. 4.1a) [3, 4, 16]. Before an apoptotic stimulus is detected by a cell, CED-4, a homologue of the mammalian adaptor protein apoptosis activating factor 1 (Apaf-1), exists as a dimer, which is sequestered on the outer leaflet of the outer membrane of the mitochondria by contact with a Bcl-2 family member called CED-9 [28]. Upon apoptotic stimulus, Egl-1, a pro-apoptotic BH3-only protein (Bcl-2 homology 3), is expressed, binding CED-9, thus releasing CED-4. CED-4 is then free to form a tetramer. Once the CED-4 tetramer is assembled, it can cleave and activate the caspase CED-3, which in turn activates other downstream apoptotic effector proteins leading to cell death (Fig. 4.1a) [28, 29].

4.2.2 The Intrinsic Apoptosis Pathway in Drosophila *and Mammals*

Unlike the linear pathway in *C. elegans*, apoptotic pathways of extrinsic and intrinsic origin have been identified in both *Drosophila* and mammals (Fig. 4.1). The intrinsic pathway has been extensively studied in *Drosophila* (Fig. 4.1b). Initially, apoptotic stimuli cause the expression of the pro-apoptotic genes of the RHG family: mainly *reaper*, *hid* (*head involution defective*) and *grim* [30, 31]. These gene products act to relieve the repression exerted by the inhibitors of apoptosis (IAPs) [32–34], which, under normal cellular conditions, inhibit activities of the *Drosophila* initiator caspases such as Dronc [35, 36] and effector caspases such as DrICE and Dcp-1 [37, 38]. The major IAP in *Drosophila* is Diap1 which functions as an E3-ubiquitin ligase. Under no apoptotic stimuli, it binds to Dronc via its own BIR2 domain and causes ubiquitin to be tagged to Dronc [36]. Such ubiquitylation was believed to stimulate degradation of Dronc via the proteasome. However, a recent genetic analysis suggests that Diap1-mediated ubiquitylation blocks processing and activation of Dronc but does not lead to its protein degradation [39]. When RHG proteins antagonise Diap1 by competitively binding to its BIR domains, Diap1 can no longer perform its function on inhibiting Dronc [36, 40, 41]. From here released Dronc, although inactive, can induce formation of the apoptosome by the adapter protein Ark [42, 43]. Upon such interaction Dronc can autocleave and become activated. Activated Dronc further cleaves and activates its downstream effector caspases, mainly DrICE and Dcp-1, leading to apoptosis (Fig. 4.1b) [44–46]. Notably, pro-apoptotic proteins need to localise to the mitochondria and execute their apoptotic functions in *Drosophila* [47–54]. Two Bcl-2 family members, Debcl and Buffy, have been identified in *Drosophila* [55–59]. Debcl is localised to the mitochondria and has pro-apoptotic functions, while Buffy may localise to endoplasmic reticula to carry out its own anti-apoptotic roles [60].

In the mammalian intrinsic pathway, the mitochondrion plays a central and more decidedly important role (Fig. 4.1c). The Bcl-2 family proteins can be subdivided into three groups: the BH3-only proteins (such as Bid, Bad, Bik, Bim, Noxa and Puma), the pro-apoptotic Bax subfamily members (such as Bax, Bak and Bok) and the anti-apoptotic Bcl-2 family members (such as Bcl-2 and Bcl-XL) [61–64]. In response to apoptosis, BH3-only proteins either activate the Bax subfamily members or antagonise the anti-apoptotic Bcl-2 members to regulate mitochondrial outer membrane permeabilisation (MOMP) which then leads to release of cytochrome c (cyt c). Released cyt c binds to the adaptor protein Apaf-1, via the WD repeat domain at the carboxy terminus of Apaf-1, forming the apoptosome. Pro-caspase-9 can in turn interact with Apaf-1 in the apoptosome, via their mutual CARD domains [65]. Pro-caspase-9 then autocleaves and becomes active [66]. The activated caspase-9 further cleaves its downstream effector caspases, caspase-3 and -7, to trigger apoptotic cell death [63]. In addition to cyt c, pro-apoptotic proteins such as Smac (or Diablo) and HtrA2 (or Omi) are also released from mitochondria during the process of MOMP [67–69]. Similar to what happens in *Drosophila*, these pro-apoptotic proteins antagonise IAPs such as XIAP leading to activation of caspase-9, -3 and -7 and apoptosis. In addition to Smac and HtrA2, another mammalian IAP antagonist is ARTS which is not released from mitochondria [70, 71]. Similar to the RHG proteins in *Drosophila*, it is localised to the mitochondrial outer membrane and inhibits XIAP [72].

4.2.3 *The Extrinsic Apoptosis Pathway in* Drosophila *and Mammals*

In contrast to the intrinsic pathway, the extrinsic pathway is initiated by the binding of a death ligand to a death receptor in the cell (Fig. 4.1c). In mammals, examples of the death ligands are tumour necrosis factor (TNF) family members including Fas ligand (FasL) and TNF-related apoptosis-inducing ligand (TRAIL) [73–75]. These ligands bind to their specific receptors Fas and DR4/5, forming complexes. Once such a ligand–receptor complex is formed, the adaptor protein Fas-associated Death Domain (FADD) can bind the cytosolic region of Fas and DR4/5. There, FADD acts as a platform on to which pro-caspase-8 can bind, by interaction of the death effector domain (DED) of FADD with the DED at the extended N-terminus of pro-caspase-8, forming the death-inducing signalling complex (DISC) [76–78]. Due to receptor clustering in the plasma membrane, the pro-caspase-8 monomers are brought within close proximity of each other in DISC complexes, and once in this newly established close proximity they can autocleave and become activated [79, 80]. Upon activation, caspase-8 can then cause the cleavage and activation of effector caspases caspase-3 and -7 leading to cell death [9, 25]. Homologues of death ligands, receptors and their functions in apoptosis induction have also been found in *Drosophila* (Fig. 4.1b). There is only one TNF homologue, Eiger (Egr), identified in *Drosophila* so far [81, 82]. Two TNF receptors including Wengen (Wgn) and, more recently, Grindelwald (Grnd) have been reported [83, 84]. Activation of Egr triggers both apoptosis and a type of non-apoptotic cell death through the Jun N-terminal Kinase (JNK) pathway, a stress-response signalling pathway [81, 82, 85, 86]. For the aspect of apoptosis, JNK induces expression of pro-apoptotic genes and activation of the apoptotic machinery [82, 86]. Interestingly, in stress-induced apoptosis, the initiator caspase Dronc can activate not only effector caspases DrICE and Dcp-1 but also JNK which then feedback to the apoptosis pathway to further amplify it [87]. Notably, although different in their nature of inducing apoptosis, connections between the extrinsic and intrinsic pathways also exist in mammals. Caspase-8 can act on the pro-apoptotic BH3-only proteins such as BID leading to activation of the intrinsic pathway which further ensures a robust apoptotic response [88–90].

4.3 Apoptosis, Development and Non-apoptotic Functions of Caspases

Apoptosis and development are interconnected. On the one hand, apoptotic caspases were originally identified as key players in the developmental programme [3]. Their apoptotic functions are critical for removal of extra cells produced at the early stage of development and elimination of unwanted cells in tissue patterning and morphogenesis [16]. A recent study on Apaf-1 knock-out mice suggests that apoptosis is required to remove Fgf8 morphogen-producing cells and terminate Fgf8 production at the correct developmental time, thus ensuring proper development of the forebrain [91]. Apoptotic cells can even actively drive epithelial folding during morphogenesis [92] and cell extrusion during tissue repair [93]. These examples have certainly underlined the developmental role of apoptosis. On the other hand, it is also becoming clear that the developmental programme can modulate cellular apoptotic responses. Many key components in the apoptosis pathway can be targeted by the developmental programme to define distinct cellular susceptibilities to apoptosis. For example, in *Drosophila* third instar larvae, a pulse of hormone ecdysone increases the whole organismal sensitivity to apoptosis by upregulating the basal level of Ark, Dronc and DrICE [94]. Furthermore, in the developing *Drosophila* eye tissue, multiple mechanisms were employed to control cellular levels of IAPs as well as pro-apoptotic proteins [95, 96]. Similarly, in mouse embryos, primed stem cells are very sensitive to apoptosis due to their low levels of BIM regulated by microRNAs [97]. Therefore, cellular apoptosis susceptibility can be modulated by developmental programmes. However, the links between apoptotic caspases and development go far beyond death. Increasing evidence is now demonstrating the actual, true multifunctional nature of the caspases with somewhat surprising and fascinating roles in diverse cellular processes. These functions include regulating immune responses, promoting cell proliferation, and regulating cell differentiation and fate specification which have been extensively reviewed elsewhere [1, 11, 13, 14, 21, 25, 98, 99]. Here, we highlight some of these non-apoptotic functions, in particular roles of caspases in tissue homeostasis, in the context of cancer development.

4.4 Caspases in Tissue Homeostasis: Apoptosis-Induced Proliferation (AiP)

Organisms are constantly exposed to environmental stresses. Damaged cells are frequently removed by apoptosis. Meanwhile, new cells are generated by proliferation to compensate for the cell loss thus to maintain tissue homeostasis. For example, up to 60 % of cells in the developing *Drosophila* wing epithelial tissue can be lost in response to radiation without affecting final adult wing size and morphology [100]. A similar phenomenon has also been found in the processes of wound healing and liver regeneration in mammals [101]. Apparently, tissue homeostasis is important for tissue function to remain optimal and critical to organism survival. Evidence in multiple organisms including *Hydra* [102], *Drosophila* [103–105] and mouse [101] is now demonstrating that apoptotic caspases have non-apoptotic functions to trigger compensatory proliferation, a process therefore termed apoptosis-induced proliferation (AiP) or apoptosis-induced compensatory cell proliferation [106–109]. For simplicity we use the term apoptosis-induced proliferation (AiP) in this review. Recent studies in *Drosophila* have provided mechanistic insights into how AiP occurs (Fig. 4.2a, b) [103–105, 110–113]. Intriguingly, depending on the developmental state of the affected tissue, i.e. proliferating versus differentiating tissues, either initiator or effector caspases drive distinct mechanisms of AiP in *Drosophila* [103].

4.4.1 *The Initiator Caspase-Driven AiP in* Drosophila

The molecular mechanism of AiP was first addressed in *Drosophila* by taking advantage of caspase inhibitors [104, 111, 112]. P35, a baculovirus inhibitor of apoptosis, acts as a peudosubstrate of *Drosophila* effector caspases, e.g. DrICE and Dcp-1 [114]. Expression of P35 thus blocks activity of DrICE and Dcp-1 and execution of cell death. To determine how stress-induced apoptotic cells may contribute to compensatory proliferation, such cells were kept "undead" by P35 (i.e. the apoptotic machinery is activated but execution of cell death is blocked). Surprisingly, "undead" cells stimulate overgrowth of surrounding tissues despite the presence of P35 [104, 111, 112]. This suggests that dying cells release mitogenic signals to induce AiP independent of effector caspases. Further loss-of-function analyses revealed that the initiator caspase Dronc, which is not inhibited by P35, actually coordinates apoptosis and AiP (Fig. 4.2a). It appears that, at least in the "undead" model of AiP, Dronc activates JNK in dying cells leading to activation of several mitogenic signalling pathways including the Wingless (Wg, a homologue of the mammalian Wnt) and Decapentaplegic (Dpp, a TGF-β-like homologue of the mammalian BMP) signalling pathways which are required for AiP [112, 115]. *Drosophila* homologue of p53 is also required for AiP, probably through its role in a feedback regulatory loop including JNK, p53 and pro-apoptotic genes [87, 113]. However, one concern of the "undead" model of AiP is that it may not represent what happens in the physiological process of AiP [115–117]. For example, it has been suggested that Wg and Dpp are not required for AiP when there are no "undead" cells [115]. Nevertheless, a *Drosophila* model of regenerative growth without using P35 has identified Wg as an important factor which is induced in response to tissue damage and is required for tissue regeneration [118]. In addition to these, a recent genetic screen using both an "undead" model and a P35-independent regenerative model has discovered a role of EGFR signalling in AiP and tissue regenerative growth [110]. In this process, JNK transcriptionally induces Spi, one of EGF ligands in

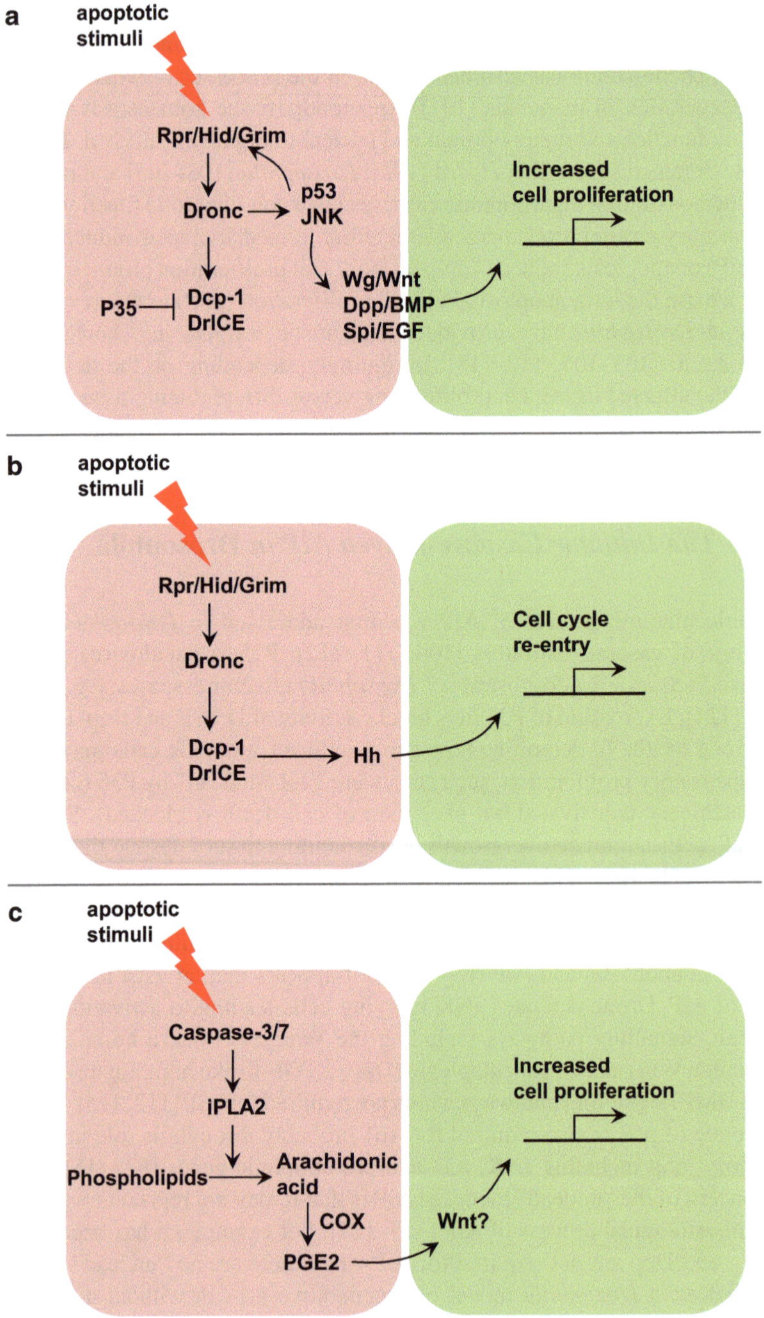

Fig. 4.2 Apoptosis-induced proliferation in *Drosophila* and mammals. Molecular mechanisms of apoptosis-induced proliferation (AiP) in proliferating (**a**) versus differentiating (cell cycle exited, **b**) tissues in *Drosophila* and in mammals (**c**). A stressed but "undead" cell (**a**, *left*, in *red*) or an apoptotic cell (**b** and **c**, *left*, in *red*) and a cell that is induced to undergo AiP (*right*, in *green*) are shown. See text for more details. *iPLA2* calcium-independent phospholipase A2, *COX* cyclooxygenases, *PGE2* prostaglandin E2

Drosophila, in dying cells which then activates proliferation of neighbouring cells via EGFR signalling. JNK can also activate the transcription factor Yorkie (Yki) in the Hippo signalling pathway to regulate AiP in developing *Drosophila* wing tissues [119, 120]. Interestingly, such a role of Yki in AiP seems to be tissue specific as it is not required for AiP in proliferating eye tissues [110].

4.4.2 The Effector Caspase-Driven AiP in Drosophila

A second form of AiP was identified in the differentiating *Drosophila* eye tissue which is a monolayer epithelium with differentiated photoreceptor cells at the apical side and cell cycle exited but unspecified cells at the basal side [103]. At the late third instar larval stage, both types of cells have relatively low susceptibility to apoptosis presumably due to their post-mitotic status and protection of survival signals such as high Diap1 and the EGFR signalling [95]. Therefore, under apoptotic stresses, e.g. expression of the pro-apoptotic gene *hid*, these cells do not die immediately. Instead, the stressed photoreceptor neurons release Hedgehog (Hh), another evolutionarily conserved growth signalling ligand, to trigger cell cycle re-entry of unspecified cells (Fig. 4.2b). Such an AiP event can be blocked by P35 or double mutants of DrICE and Dcp-1 suggesting an effector caspase-driven form of AiP is employed in the differentiating eye tissue [103]. Interestingly, mechanisms of AiP seem to be operated in a context-dependent manner. This is best shown in the developing *Drosophila* eye tissue. The late third larval eye tissue consists of an anterior proliferating portion where all cells are actively dividing and a posterior differentiating portion where most of the cells present have exited the cell cycle. The initiator caspase-driven AiP appears to be employed in proliferating tissues, while the effector caspase-driven AiP is employed in differentiating tissues [103]. However, what controls such distinction is not yet known.

4.4.3 AiP in Other Organisms Including Mammals

In addition to *Drosophila*, roles of AiP in regeneration have also been implicated in other multicellular organisms particularly in *Hydra*, *Xenopus* and mouse [101, 102, 106, 121]. In the freshwater *Hydra*, both head and foot can regenerate completely after bisection at the midgastric area. Massive localised apoptosis was observed for the head regenerating tip, but not the foot regenerating counterpart, preceding increase of cell proliferation [102]. Interestingly, ectopic activation of apoptosis at the foot regenerating tip resulted in regeneration of head instead of foot. In this process, caspases activate Wnt3, a homologue of *Drosophila* Wg, in dying cells leading to regenerative proliferation [102]. This study suggests that apoptosis can direct certain regenerative programmes. Similar requirements of caspases in regeneration were also reported for *Xenopus* tadpole tail regeneration which is abolished by inhibiting caspase-3 [121]. Notably, in other regeneration models such as planaria and newt, massive apoptosis at the amputation site has been observed [122–124]. However, it is not yet clear whether apoptotic caspases actually drive release of mitogenic signals such as Wnt, TGF-β and Hh in these processes. More recently, roles of AiP in mammals were reported in mouse models of wound healing and liver regeneration [101, 125]. The rate of skin wound healing and liver regrowth after partial hepatectomy was significantly reduced in caspase-3 or -7 deficient mice due to impaired post-injury cell proliferation. It was further revealed that activated caspase-3 and -7 cleave calcium-independent phospholipase A2 (iPLA2) to increase its catalytic ability and promote synthesis of prostaglandin E2 (PGE2). Release of PGE2 from the dying cell then induces compensatory proliferation (Fig. 4.2c) [101]. Although detailed mechanisms on how PGE2 triggers compensatory proliferation are not yet revealed, the link between PGE2 and the Wnt signalling cascade has been established in both zebrafish and mice [126]. PGE2 binds to EP2, a G-protein coupled receptor, leading to activation of β-catenin, a key intracellular transducer of Wnt signalling [127–129].

4.5 Caspases in Cancer Development: Non-apoptotic Functions

Current cancer therapies such as chemo- and radiotherapies frequently aim to activate apoptosis of cancer cells. Therefore, activating apoptosis has long been viewed as an "anti-cancer" process. However, increasing evidence is now suggesting that apoptotic caspases can play oncogenic roles through their non-apoptotic functions (Fig. 4.3). As discussed earlier the roles of the apoptotic caspases are essential in proper organism development and tissue homeostasis. Apparently different functional aspects of caspases needs to be tightly controlled and restrained by cellular contexts in order to prevent their detrimental effects. In the context of cancer, these non-apoptotic functions of caspases can be hijacked to ensure survival of cancer cells and promote their spread. Thus, the multifunctional nature of the apoptotic caspases is becoming clinically important.

4.5.1 Caspases Promote Cell Survival and Cell Proliferation

The crucial function of caspases in cell survival and proliferation has been reported during development. Targeted disruption of caspase-8 in mice causes embryonic lethality, a feature not shared by the other caspases [78, 130]. Caspase-$8^{-/-}$ mouse embryos exhibited abnormal phenotypes prior to death, namely hyperaemia, with the number of haematopoietic precursors significantly reduced [78]. This suggests that caspase-8 is required for either maintenance or proliferation of haematopoietic precursors. As further support for this view, depletion of caspase-8 in lymphoid tissues inhibits antigen-induced T and B lymphocyte proliferation [131–133].

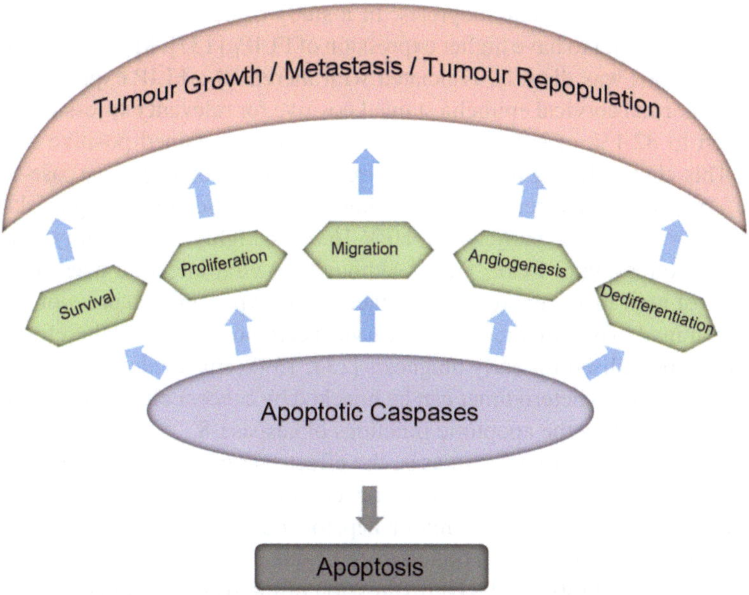

Fig. 4.3 Schematic diagram of non-apoptotic functions of caspases that may contribute to various aspects of cancer progression

Although it was originally thought that caspase-8 regulates cell proliferation in these cases, it is more likely that caspase-8 has pro-survival functions due to its inhibitory role on necroptosis, another form of programmed cell death [130, 134]. The key factors involved in such regulation are caspase-8, the long isoform of cellular FLICE-like inhibitory protein ($FLIP_L$), and two kinases, RIPK1 and RIPK3, which are required for activation of necroptosis. $FLIP_L$ is structurally similar to caspase-8 but without its catalytic activity [135]. It can bind to pro-caspase-8, forming a heterodimer which prevents caspase-8 from completing its apoptotic functions by occupying all binding sites of caspase-8 in the DISC. This consequently prevents caspase-8 homodimer formation. Therefore, when the $FLIP_L$ levels are low, homodimerisation of pro-caspase-8 occurs which activates caspase-8 for its apoptotic function. In contrast when $FLIP_L$ levels are high, e.g. triggered by survival signals mediated by a transcription factor NFκB, formation of the pro-caspase-8-$FLIP_L$ heterodimer does not trigger apoptosis. Instead, it can bind to the RIPK1-containing complex to suppress its activation of RIPK3 and necrotic cell death, although the underlying mechanism remains unclear [134]. Hence, the level of $FLIP_L$ is crucial for caspase-8-regulated cell survival. Interestingly, an increase in $FLIP_L$ expression has been detected in a variety of tumour types, including B-cell chronic lymphocytic leukaemia, pancreatic cancer and ovarian cancer, amongst many others [136, 137]. Down-regulating $FLIP_L$ levels in tumours sensitises the cells to apoptosis [136, 138]. This is most likely due to decreased ability for caspase-8-$FLIP_L$ heterodimers to form and increased ability of caspase-8 homodimerisation, which can then activate caspase-3 and apoptosis. In a study of cervical cancers, high-grade tumours were found to have higher expression of $FLIP_L$ [137]. Moreover, increasing grade of lesions was directly associated with increased c-FLIP expression, where 12.5 % of normal cervical epithelia stained positive for relevant expression of $FLIP_L$ compared to 82.1 % of squamous cervical carcinomas stained positive for $FLIP_L$ [137]. This shows the significance in correlation of uncontrolled caspase-8-$FLIP_L$ dimer formation and cancer progression. Interestingly, infection by high-risk human papillomavirus (HPV), particularly HPV-16, was highly significantly correlated with high expression of $FLIP_L$ [139]. Although the viral infection does not explain the cause of high $FLIP_L$ expression in other cancer types, high expression of $FLIP_L$ was determined to be a marker of early cervical carcinogenesis and therefore has the potential to be utilised for early diagnosis [137, 139]. This evidence highlights that the caspase-8-$FLIP_L$ heterodimer can be hijacked by cancer cells to promote tumour survival, by avoiding the apoptotic functions of caspase-8.

In addition to the initiator caspases, the effector caspases have also been implicated in promoting cell survival and cell cycle progression. In cultured cancer cell lines with their origin in leukaemia or hepatocellular or cervical carcinoma, caspase-3 and -7 are found to be required in cell cycle progression through the G1 and G2/M checkpoints [140, 141]. Overexpression of the BIR2 domain of XIAP inhibits caspase-3 and -7, and when added to cells also induced cell cycle arrest. In contrast, inhibition of caspase-9 by expression of the BIR3 domain of XIAP did not cause the same effect, which indicates that caspase-3 and -7 have functions independent of caspase-9 activity. Although it is not yet clear how caspase-3 and -7 may promote cell cycle progression without being cleaved by caspase-9, the anaphase-promoting complex/cyclosome (APC/C), which regulates degradation of various cell cycle regulators through ubiquitylation, failed to form when caspase-3 and -7 were inhibited [140]. This suggests that pro-caspase-3 and -7 may contribute to cell proliferation. Interestingly, direct substrates of caspases including cell cycle regulators can also promote cell survival or cell cycle progression at least in some circumstances. For example, the cyclin-dependent kinase inhibitor $P27^{Kip1}$ can be cleaved by caspase-3 which then becomes activated and anti-apoptotic to protect human leukemic cells from death [142]. In addition to this, a more recent study suggested that caspase-3 can act as a sensor to extracellular stresses, therefore determining whether the cells live or not [143]. In this study, caspase-3-knockout mice become more sensitive to UV radiation with increased number of cells undergoing necrosis compared to the control animals. In response to doxorubicin, an anticancer drug inducing apoptosis of cardiomyocytes, the caspase-3-deficient mice also show significantly increased number of apoptotic cardiomyocytes which die through caspase-7 instead [143]. Caspase-3, but not caspase-6 and -7, cleaves the p120 RasGAP protein in vitro to activate a kinase, Akt, leading to survival functions of PI3K signalling [144, 145]. Consistently, Akt activity, indicated by the level of phosphorylated Akt, increases in response to stresses such as UV radiation and doxorubicin injection. But such increase is strongly reduced in caspase-3-knockout mice. Knock-in mice with a RasGAP mutant resistant to caspase-3 cleavage can restore their apoptotic sensitivity [143]. Given these findings of caspases in cell survival and proliferation, could they contribute to tumourigenesis? As discussed later (see Sect. 5.3), the answer becomes clear by discovering roles of AiP in tumour reoccurrence following cytotoxic cancer therapies.

4.5.2 Caspases and Metastasis

Metastasis is a crucial process to understand in cancer progression as it is the cause of approximately 90 % of cancer-related deaths [8]. It is an incredibly complex process consisting of multiple key steps for a cancer cell, or a group of cancer cells, to progress through [146]. These steps include breaking away from a bulk tumour, disseminating in the blood or lymph, exiting the circulation, then establishing and repopulating at a new site, where a secondary tumour forms. Interestingly, caspases have been implicated in aiding some of these steps through their non-apoptotic functions in cell migration, angiogenesis and possibly cell dedifferentiation.

Caspases have been reported for their functions in controlling cell motility during development. In *Drosophila*, Dronc, the caspase-9 homologue, is required for migration of border cells in the ovary [147], a process critical for oocyte development. In mammals, caspase-$8^{-/-}$ mouse embryos die with a circulatory failure suggesting roles of caspase-8 in migration of endothelial cells [78, 148]. Similarly, in cancer-specific studies, caspase-3 and its downstream targets have been implicated in causing tumour cell migration, thus contributing to achieving metastasis. In ovarian cancer cells, caspase-3 has been shown to be involved in the process of initiating cell migration via activation of arachidonic acid, the precursor of PGE2, similarly as in the context of AiP described earlier (Fig. 4.2c) [128, 149, 150]. Ovarian cancer cells have strong migratory responses towards laminin-10/11, a protein component of the extracellular matrix. This is probably due to the high levels of β1 integrin in ovarian cancer cells, because binding of laminin-10/11 to β1 integrin leads to a moderate increase of caspase-3 activity [150]. Although the intermediate molecules determining caspase-3 activation from integrin–laminin binding are unknown, Zhao et al. [150] determined that moderate increase of caspase-3 activity does not lead to apoptosis, instead, it cleaves iPLA2 and activates its enzymatic activity to produce arachidonic acid and then PGE2. Consistently, pan caspase inhibitors, caspase-3-specific inhibitors or knockdown of iPLA2 inhibits migration of these cells. Interestingly, cleaved iPLA2 also activates Akt survival signalling to protect these cells from apoptosis [150]. This further enhances cancer cell migration. Not surprisingly arachidonic acid has also been implicated to be the driving factor of cell migration in other cancers including prostate cancers [151]. Further support for roles of caspase-3 in cell migration comes from a study on lung cancer metastasis [152]. In this study, however, a protease-independent function of caspase-3 was suggested to promote metastasis. The authors used A549 cells, derived from high malignancy lung adenocarcinoma cells with high levels of caspase-3, for their study. Knockdown of caspase-3 in A549 cells diminishes their metastatic activities in the lungs when these cells were injected into nude mice via the tail vein suggesting roles of caspase-3 in promoting metastasis. Consistently, ectopic expression of caspase-3 in MCF-7 cells, derived from caspase-3-deficient and low malignancy breast cancer cells, enhances metastatic ability of these cells. Following these findings, the authors then found that high levels of caspase-3 actually lead to high activity of the extracellular signal-regulated kinases (ERK) which are required for the observed lung metastasis. However intriguingly, such increased ERK activity and cell migration are not affected by the caspase inhibitor Z-DEVD-FMK. Furthermore, expression of protease-dead mutants of caspase-3 in MCF-7 cells still enhances their migration through increased ERK activities. Although it is not yet clear, the acid sphingomyelinase and its downstream signal molecule ceramide were suggested to be the molecules linking caspase-3 and ERK [152]. Interestingly, another mechanism of caspase-3-dependent cell migration has been reported for the "undead" cells in *Drosophila* models [153]. In this case, DrICE, a caspase-3 homologue in *Drosophila*, activates JNK leading to cell migration and tissue invasion. Therefore, cellular contexts may determine how caspase-3 promotes cell motility.

Following migration and invasion of cancer cells, angiogenesis is essential to further cancer progression, enabling tumour growth above a diameter of 1 mm and metastasis [146]. Knockdown of caspase-8 suppressed vascular endothelial growth factor (VEGF)-mediated angiogenic signalling [154]. Interestingly, such requirement of caspase-8 in promoting angiogenesis is not affected by Ac-IETD-cho, a caspase-8 inhibitor that maintains high levels of pro-caspase-8. In contrast, the same study also showed that caspase-8 is required in TRAIL signalling to antagonise angiogenesis which can be inhibited by Ac-IETD-cho [154]. Therefore, procaspase-8 and caspase-8 appear to have distinct functions during angiogenesis mediated by VEGF.

Another cellular process that can potentially impact on cancer metastasis is cell dedifferentiation. Although it is still a subject of debate, existence of "cancer stem cells", a small fraction of stem cell-like cancer progenitor cells, may facilitate or even establish the metastatic colonies for cancer progression [146]. If this is true, maintenance and reprogramming, thus dedifferentiation, of cancer cells may be crucial in the process of metastasis which, again, may involve caspases. Notably, both caspase-8 and caspase-3 are required for the dedifferentiation of murine fibroblasts to form induced pluripotent stem cells (iPSCs) in vitro [155]. Activation of caspase-8 and -3 is induced by expression of Oct-4, one of the four transcription factors used to programme iPSCs. By inhibiting caspase-8 the cells were completely unable to develop into iPSCs, whereas some could if only caspase-3 was inhibited suggesting potential roles of other effector caspases such as caspase-7 in induction of iPSCs. The authors further showed that the caspases act upon retinoblastoma susceptibility protein (Rb), but how from here the phenotype of a pluripotent stem cell is produced is unknown although p53 and its downstream cell cycle regulator p21 have been implicated in the process [155]. Interestingly, studies of human tumours in relation to their Oct-4 expression showed that tumours expressing high levels of Oct-4 resulted in increased metastases, shorter survival and furthered disease progression in comparison to tumours low in Oct-4 expression [156]. A recent study further sorted murine breast cancer 4T1 cells with either high or low Oct-4 expression and tested their tumourigenic potential in vivo by injecting sorted cells into the mouse mammary glands [157]. The results support that Oct-4 can enhance cancer stem cell properties. This fits in vitro data and hypotheses theorising on the capacity of cancer stem cells in disease progression, though more studies are required in a greater range of tumour types.

4.5.3 Caspases in Tumour Repopulation Following Cytotoxic Cancer Therapies

Cytotoxic therapies exert their anti-tumour properties by inducing apoptosis as a result of DNA damage [158]. As discussed earlier, AiP is a process utilised in non-cancerous tissue in order to maintain tissue homeostasis that allows tissue regeneration and recovery from damage. Consequently, this means that cytotoxic therapies can potentially induce not only cell death but also the AiP pathway which may in fact counteract cancer treatment. Tumours, to some extent, are comparable to standard developmental tissues [159], and, conceivably, when damaged they can respond in the same way to regenerate and to compensate for the inflicted damage, thus to repopulate and reoccur. Recent studies on AiP in cancer models have suggested this is the case. In one study, experiments were conducted to find out how caspase-3 is responsible for promoting accelerated tumour repopulation following cytotoxic therapy in 4T1 murine breast cancer cells [160]. It was found that the AiP pathway, activated in either cancer cells or stromal cells, could become hijacked by cancer cells following radiotherapy, causing accelerated tumour repopulation in vitro and in vivo, in nude mice. These were also confirmed with human breast cancer cell lines in nude mice [160]. The results gained in this study were further developed in studies on metastatic melanoma, showing that chemotherapy too can result in AiP and tumour repopulation [161]. As in the tissue regeneration mouse model, it is PGE2 which is secreted from apoptotic cells and stimulates recipient living cells to proliferate in the tumour repopulation model (Fig. 4.2c) [160, 161]. The authors also found that tumours with elevated caspase-3 were more resistant to radiotherapy than those with reduced caspase-3 [160]. This at first seems paradoxical; however, with regards to the AiP model, this observation is logical. Higher caspase-3 expression allows for greater production of PGE2, which in turn stimulates the increased growth rate of surviving cells, thus ensuring the maintenance of a larger tumour mass. Huang et al. found that the therapy sensitive cells were induced to undergo apoptosis, and the release of prostaglandins from the therapy sensitive cells caused the therapy-resistant cells to proliferate at an accelerated rate and repopulate the treated tumour [160].

Statistical studies have been conducted to give a measure of how higher expression of particular caspases in tumours can affect outcome and survival likelihood. In a study of breast cancer, 103 out of 137 tumours were deemed to have high levels of caspase-3, although some activity was noticed in all of the tumours [162]. Increased caspase-3 level significantly correlated with worsened survival of the patients sampled and, in the tumours sampled, caspase-3 was only found in the cytoplasm, not the nucleus where the apoptotic target of caspase-3 resides. This suggests a possible mechanistic block preventing the effector caspase-3 from reaching its target molecule, the inhibitor of caspase-activated DNase (iCAD), to free caspase-activated DNase (CAD) which can cause DNA fragmentation and subsequent cell death [162]. Another study assessing implications of caspase-3 in gastric, ovarian, cervical and colorectal cancers concluded that patients possessing tumours which expressed higher caspase-3 had shortened survival time and also found that caspase-3 expression was significantly associated with tumour stage [163]. Both studies concluded that higher caspase-3 expression resulted in worsened prognosis. Notably, participants of these studies had not undergone any form of therapy. However, these findings of statistical significance were further confirmed by Huang et al. [160], on patients who had undergone radiotherapy or chemotherapy.

Given these new insights of mechanisms causing tumour repopulation following cytotoxic therapy, if repopulation is to be prohibited in tumours, then the AiP pathway needs to be blocked while still allowing caspase-3 to carry out its apoptotic functions. As described for AiP (Fig. 4.2c), PGE2 is synthesised from arachidonic acid by cyclooxygenases (COX). Thus, COX inhibitors in theory should prevent the AiP pathway from progressing. This has been shown in practice, where administering a COX inhibitor in conjunction with cytotoxic therapy significantly decreases rate of tumour repopulation [160, 161]. Therefore, use of a COX inhibitor in conjunction with the cytotoxic therapy may benefit patients possessing tumours with high levels of caspase-3. Notably, caspase-3 may not be the only component in the apoptosis pathway that can promote cancer tumourigenesis as suggested by studies on lymphoma [164, 165]. Further mechanistic understanding of AiP in various cellular contexts will be the key to explore its clinical significance. Interestingly, in addition to AiP, engulfment of apoptotic cells by macrophages can create a tumour-promotive microenvironment by releasing signalling molecules [166–168] and regulating various aspects of tumour progression [169]. Again, caspases play key roles here. Activation of iPLA2 by caspase-3 leads to production of lysophosphatidylcholine (LPC), as well as PGE2, from dying cells [170]. LPCs, together with several other molecules such as sphingosine-1-phosphates (S1Ps) and the nucleotides ATP and UTP, recruit macrophages to engulf apoptotic cells [171]. Therefore, apoptotic caspases can promote tumourigenesis directly through AiP or indirectly through recruiting macrophages [128]. This is further discussed in Chap. 3.

4.6 Concluding Remarks

For many years, the apoptotic function of caspases has been considered, both in developmental settings and in a cancer setting, where activation of apoptotic proteins is considered to be essential in causing cell death and reducing tumour burden [8]. While these considerations of caspase function remain valid, increasing evidence suggests that non-apoptotic functions of the apoptotic caspases exist in a context-dependent manner. These functions are crucial in development and tissue homeostasis, where caspases have been implicated in stem cell pool maintenance by enhancing survival pathways and in AiP for tissue recovery upon cell loss, about which we have learned a lot from *Drosophila* models. Intriguingly, a wide range of non-apoptotic functions of caspases have been implicated in promoting tumour growth, metastasis and recurrence post-cytotoxic therapy (Fig. 4.3). It is therefore worthwhile to consider not only how to kill the tumour cells, but also how to prevent tumour spread and repopulation in cancer treatments. Further understanding of molecular mechanisms and cellular contexts leading to various non-apoptotic functions of caspases would certainly be beneficial.

Acknowledgments We apologise to our colleagues for omitting many relevant publications owing to space limitations. This work was supported by Marie Curie Career Integration Grant (CIG) 630846 from the European Union's Seventh Framework Programme (FP7) and the Birmingham Fellowship, University of Birmingham, UK.

References

1. Jacobson MD, Weil M, Raff MC. Programmed cell death in animal development. Cell. 1997;88:347–54.
2. Kerr JF, Wyllie AH, Currie AR. Apoptosis: a basic biological phenomenon with wide-ranging implications in tissue kinetics. Br J Cancer. 1972;26:239–57.
3. Ellis HM, Horvitz HR. Genetic control of programmed cell death in the nematode C. elegans. Cell. 1986;44:817–29.
4. Ellis RE, Yuan JY, Horvitz HR. Mechanisms and functions of cell death. Annu Rev Cell Biol. 1991;7:663–98.
5. Miura M. Active participation of cell death in development and organismal homeostasis. Dev Growth Differ. 2011;53:125–36.
6. Thompson CB. Apoptosis in the pathogenesis and treatment of disease. Science. 1995;267:1456–62.
7. Hanahan D, Weinberg RA. The hallmarks of cancer. Cell. 2000;100:57–70.
8. Hanahan D, Weinberg RA. Hallmarks of cancer: the next generation. Cell. 2011;144: 646–74.
9. Hengartner MO. The biochemistry of apoptosis. Nature. 2000;407:770–6.
10. Thornberry NA, Lazebnik Y. Caspases: enemies within. Science. 1998;281:1312–6.
11. Connolly PF, Jager R, Fearnhead HO. New roles for old enzymes: killer caspases as the engine of cell behavior changes. Front Physiol. 2014;5:149.
12. Kuranaga E. Caspase signaling in animal development. Dev Growth Differ. 2011;53: 137–48.
13. Miura M. Apoptotic and nonapoptotic caspase functions in animal development. Cold Spring Harb Perspect Biol. 2012;4:a008664.
14. Shalini S, Dorstyn L, Dawar S, Kumar S. Old, new and emerging functions of caspases. Cell Death Differ. 2015;22:526–39.
15. Degterev A, Yuan J. Expansion and evolution of cell death programmes. Nat Rev Mol Cell Biol. 2008;9:378–90.
16. Fuchs Y, Steller H. Programmed cell death in animal development and disease. Cell. 2011;147:742–58.
17. Green DR, Galluzzi L, Kroemer G. Cell biology. Metabolic control of cell death. Science. 2014;345:1250256.

18. Yamaguchi Y, Miura M. Programmed cell death in neurodevelopment. Dev Cell. 2015;32: 478–90.
19. Earnshaw WC, Martins LM, Kaufmann SH. Mammalian caspases: structure, activation, substrates, and functions during apoptosis. Annu Rev Biochem. 1999;68:383–424.
20. Lamkanfi M, Dixit VM. Mechanisms and functions of inflammasomes. Cell. 2014;157: 1013–22.
21. Lamkanfi M, Festjens N, Declercq W, Vanden Berghe T, Vandenabeele P. Caspases in cell survival, proliferation and differentiation. Cell Death Differ. 2007;14:44–55.
22. Vanaja SK, Rathinam VA, Fitzgerald KA. Mechanisms of inflammasome activation: recent advances and novel insights. Trends Cell Biol. 2015;25:308–15.
23. Eckhart L, Ballaun C, Hermann M, Vandeberg JL, Sipos W, Uthman A, Fischer H, Tschachler E. Identification of novel mammalian caspases reveals an important role of gene loss in shaping the human caspase repertoire. Mol Biol Evol. 2008;25:831–41.
24. Degterev A, Boyce M, Yuan J. A decade of caspases. Oncogene. 2003;22:8543–67.
25. Chowdhury I, Tharakan B, Bhat GK. Caspases—an update. Comp Biochem Physiol B Biochem Mol Biol. 2008;151:10–27.
26. Kumar S. Caspase function in programmed cell death. Cell Death Differ. 2007;14:32–43.
27. Yan N, Shi Y. Mechanisms of apoptosis through structural biology. Annu Rev Cell Dev Biol. 2005;21:35–56.
28. Lettre G, Hengartner MO. Developmental apoptosis in C. elegans: a complex CEDnario. Nat Rev Mol Cell Biol. 2006;7:97–108.
29. Conradt B, Xue D. Programmed cell death. WormBook; 2005. p. 1–13.
30. Kornbluth S, White K. Apoptosis in Drosophila: neither fish nor fowl (nor man, nor worm). J Cell Sci. 2005;118:1779–87.
31. Steller H. Regulation of apoptosis in Drosophila. Cell Death Differ. 2008;15:1132–8.
32. Goyal L, McCall K, Agapite J, Hartwieg E, Steller H. Induction of apoptosis by Drosophila reaper, hid and grim through inhibition of IAP function. EMBO J. 2000;19:589–97.
33. Lisi S, Mazzon I, White K. Diverse domains of THREAD/DIAP1 are required to inhibit apoptosis induced by REAPER and HID in Drosophila. Genetics. 2000;154:669–78.
34. Wang SL, Hawkins CJ, Yoo SJ, Muller HA, Hay BA. The Drosophila caspase inhibitor DIAP1 is essential for cell survival and is negatively regulated by HID. Cell. 1999;98:453–63.
35. Meier P, Silke J, Leevers SJ, Evan GI. The Drosophila caspase DRONC is regulated by DIAP1. EMBO J. 2000;19:598–611.
36. Wilson R, Goyal L, Ditzel M, Zachariou A, Baker DA, Agapite J, Steller H, Meier P. The DIAP1 RING finger mediates ubiquitination of Dronc and is indispensable for regulating apoptosis. Nat Cell Biol. 2002;4:445–50.
37. Hawkins CJ, Wang SL, Hay BA. A cloning method to identify caspases and their regulators in yeast: identification of Drosophila IAP1 as an inhibitor of the Drosophila caspase DCP-1. Proc Natl Acad Sci U S A. 1999;96:2885–90.
38. Yan N, Wu JW, Chai J, Li W, Shi Y. Molecular mechanisms of DrICE inhibition by DIAP1 and removal of inhibition by Reaper, Hid and Grim. Nat Struct Mol Biol. 2004;11:420–8.
39. Lee TV, Fan Y, Wang S, Srivastava M, Broemer M, Meier P, Bergmann A. Drosophila IAP1-mediated ubiquitylation controls activation of the initiator caspase DRONC independent of protein degradation. PLoS Genet. 2011;7, e1002261.
40. Ryoo HD, Bergmann A, Gonen H, Ciechanover A, Steller H. Regulation of Drosophila IAP1 degradation and apoptosis by reaper and ubcD1. Nat Cell Biol. 2002;4:432–8.
41. Yoo SJ, Huh JR, Muro I, Yu H, Wang L, Wang SL, Feldman RM, Clem RJ, Muller HA, Hay BA. Hid, Rpr and Grim negatively regulate DIAP1 levels through distinct mechanisms. Nat Cell Biol. 2002;4:416–24.
42. Chai J, Shi Y. Apoptosome and inflammasome: conserved machineries for caspase activation. Natl Sci Rev. 2014;1:101–18.
43. Pang Y, Bai XC, Yan C, Hao Q, Chen Z, Wang JW, Scheres SH, Shi Y. Structure of the apoptosome: mechanistic insights into activation of an initiator caspase from Drosophila. Genes Dev. 2015;29:277–87.

44. Dorstyn L, Kumar S. A biochemical analysis of the activation of the Drosophila caspase DRONC. Cell Death Differ. 2008;15:461–70.
45. Hawkins CJ, Yoo SJ, Peterson EP, Wang SL, Vernooy SY, Hay BA. The Drosophila caspase DRONC cleaves following glutamate or aspartate and is regulated by DIAP1, HID, and GRIM. J Biol Chem. 2000;275:27084–93.
46. Snipas SJ, Drag M, Stennicke HR, Salvesen GS. Activation mechanism and substrate specificity of the Drosophila initiator caspase DRONC. Cell Death Differ. 2008;15:938–45.
47. Abdelwahid E, Yokokura T, Krieser RJ, Balasundaram S, Fowle WH, White K. Mitochondrial disruption in Drosophila apoptosis. Dev Cell. 2007;12:793–806.
48. Claveria C, Albar JP, Serrano A, Buesa JM, Barbero JL, Martinez AC, Torres M. Drosophila grim induces apoptosis in mammalian cells. EMBO J. 1998;17:7199–208.
49. Claveria C, Caminero E, Martinez AC, Campuzano S, Torres M. GH3, a novel proapoptotic domain in Drosophila Grim, promotes a mitochondrial death pathway. EMBO J. 2002;21:3327–36.
50. Freel CD, Richardson DA, Thomenius MJ, Gan EC, Horn SR, Olson MR, Kornbluth S. Mitochondrial localization of Reaper to promote inhibitors of apoptosis protein degradation conferred by GH3 domain-lipid interactions. J Biol Chem. 2008;283:367–79.
51. Haining WN, Carboy-Newcomb C, Wei CL, Steller H. The proapoptotic function of Drosophila Hid is conserved in mammalian cells. Proc Natl Acad Sci U S A. 1999;96:4936–41.
52. Morishita J, Kang MJ, Fidelin K, Ryoo HD. CDK7 regulates the mitochondrial localization of a tail-anchored proapoptotic protein, Hid. Cell Rep. 2013;5:1481–8.
53. Olson MR, Holley CL, Gan EC, Colon-Ramos DA, Kaplan B, Kornbluth S. A GH3-like domain in reaper is required for mitochondrial localization and induction of IAP degradation. J Biol Chem. 2003;278:44758–68.
54. Thomenius M, Freel CD, Horn S, Krieser R, Abdelwahid E, Cannon R, Balasundaram S, White K, Kornbluth S. Mitochondrial fusion is regulated by Reaper to modulate Drosophila programmed cell death. Cell Death Differ. 2011;18:1640–50.
55. Brachmann CB, Jassim OW, Wachsmuth BD, Cagan RL. The Drosophila bcl-2 family member dBorg-1 functions in the apoptotic response to UV-irradiation. Curr Biol. 2000;10: 547–50.
56. Colussi PA, Quinn LM, Huang DC, Coombe M, Read SH, Richardson H, Kumar S. Debcl, a proapoptotic Bcl-2 homologue, is a component of the Drosophila melanogaster cell death machinery. J Cell Biol. 2000;148:703–14.
57. Igaki T, Kanuka H, Inohara N, Sawamoto K, Nunez G, Okano H, Miura M. Drob-1, a Drosophila member of the Bcl-2/CED-9 family that promotes cell death. Proc Natl Acad Sci U S A. 2000;97:662–7.
58. Quinn L, Coombe M, Mills K, Daish T, Colussi P, Kumar S, Richardson H. Buffy, a Drosophila Bcl-2 protein, has anti-apoptotic and cell cycle inhibitory functions. EMBO J. 2003;22:3568–79.
59. Zhang H, Huang Q, Ke N, Matsuyama S, Hammock B, Godzik A, Reed JC. Drosophila proapoptotic Bcl-2/Bax homologue reveals evolutionary conservation of cell death mechanisms. J Biol Chem. 2000;275:27303–6.
60. Doumanis J, Dorstyn L, Kumar S. Molecular determinants of the subcellular localization of the Drosophila Bcl-2 homologues DEBCL and BUFFY. Cell Death Differ. 2007;14:907–15.
61. Gabriel B, Sureau F, Casselyn M, Teissie J, Petit PX. Retroactive pathway involving mitochondria in electroloaded cytochrome c-induced apoptosis. Protective properties of Bcl-2 and Bcl-XL. Exp Cell Res. 2003;289:195–210.
62. Vander Heiden MG, Chandel NS, Williamson EK, Schumacker PT, Thompson CB. Bcl-xL regulates the membrane potential and volume homeostasis of mitochondria. Cell. 1997;91:627–37.
63. Xiong S, Mu T, Wang G, Jiang X. Mitochondria-mediated apoptosis in mammals. Protein Cell. 2014;5:737–49.

64. Yang J, Liu X, Bhalla K, Kim CN, Ibrado AM, Cai J, Peng TI, Jones DP, Wang X. Prevention of apoptosis by Bcl-2: release of cytochrome c from mitochondria blocked. Science. 1997;275:1129–32.
65. Srinivasula SM, Ahmad M, Fernandes-Alnemri T, Alnemri ES. Autoactivation of procaspase-9 by Apaf-1-mediated oligomerization. Mol Cell. 1998;1:949–57.
66. Kumar S, Colussi PA. Prodomains—adaptors—oligomerization: the pursuit of caspase activation in apoptosis. Trends Biochem Sci. 1999;24:1–4.
67. Green DR, Kroemer G. The pathophysiology of mitochondrial cell death. Science. 2004;305:626–9.
68. Silke J, Verhagen AM, Ekert PG, Vaux DL. Sequence as well as functional similarity for DIABLO/Smac and Grim, Reaper and Hid? Cell Death Differ. 2000;7:1275.
69. Suzuki Y, Imai Y, Nakayama H, Takahashi K, Takio K, Takahashi R. A serine protease, HtrA2, is released from the mitochondria and interacts with XIAP, inducing cell death. Mol Cell. 2001;8:613–21.
70. Gottfried Y, Rotem A, Lotan R, Steller H, Larisch S. The mitochondrial ARTS protein promotes apoptosis through targeting XIAP. EMBO J. 2004;23:1627–35.
71. Larisch S, Yi Y, Lotan R, Kerner H, Eimerl S, Tony Parks W, Gottfried Y, Birkey Reffey S, De Caestecker MP, Danielpour D, Book-Melamed N, Timberg R, Duckett CS, Lechleider RJ, Steller H, Orly J, Kim SJ, Roberts AB. A novel mitochondrial septin-like protein, ARTS, mediates apoptosis dependent on its P-loop motif. Nat Cell Biol. 2000;2:915–21.
72. Edison N, Zuri D, Maniv I, Bornstein B, Lev T, Gottfried Y, Kemeny S, Garcia-Fernandez M, Kagan J, Larisch S. The IAP-antagonist ARTS initiates caspase activation upstream of cytochrome C and SMAC/Diablo. Cell Death Differ. 2012;19:356–68.
73. Ashkenazi A, Dixit VM. Death receptors: signaling and modulation. Science. 1998;281:1305–8.
74. Lavrik I, Golks A, Krammer PH. Death receptor signaling. J Cell Sci. 2005;118:265–7.
75. Peter ME, Krammer PH. Mechanisms of CD95 (APO-1/Fas)-mediated apoptosis. Curr Opin Immunol. 1998;10:545–51.
76. Juo P, Kuo CJ, Yuan J, Blenis J. Essential requirement for caspase-8/FLICE in the initiation of the Fas-induced apoptotic cascade. Curr Biol. 1998;8:1001–8.
77. Medema JP, Scaffidi C, Kischkel FC, Shevchenko A, Mann M, Krammer PH, Peter ME. FLICE is activated by association with the CD95 death-inducing signaling complex (DISC). EMBO J. 1997;16:2794–804.
78. Varfolomeev EE, Schuchmann M, Luria V, Chiannilkulchai N, Beckmann JS, Mett IL, Rebrikov D, Brodianski VM, Kemper OC, Kollet O, Lapidot T, Soffer D, Sobe T, Avraham KB, Goncharov T, Holtmann H, Lonai P, Wallach D. Targeted disruption of the mouse Caspase 8 gene ablates cell death induction by the TNF receptors, Fas/Apo1, and DR3 and is lethal prenatally. Immunity. 1998;9:267–76.
79. Chang DW, Xing Z, Capacio VL, Peter ME, Yang X. Interdimer processing mechanism of procaspase-8 activation. EMBO J. 2003;22:4132–42.
80. Guicciardi ME, Gores GJ. Life and death by death receptors. FASEB J. 2009;23:1625–37.
81. Igaki T, Kanda H, Yamamoto-Goto Y, Kanuka H, Kuranaga E, Aigaki T, Miura M. Eiger, a TNF superfamily ligand that triggers the Drosophila JNK pathway. EMBO J. 2002;21:3009–18.
82. Moreno E, Yan M, Basler K. Evolution of TNF signaling mechanisms: JNK-dependent apoptosis triggered by Eiger, the Drosophila homolog of the TNF superfamily. Curr Biol. 2002;12:1263–8.
83. Andersen DS, Colombani J, Palmerini V, Chakrabandhu K, Boone E, Rothlisberger M, Toggweiler J, Basler K, Mapelli M, Hueber AO, Leopold P. The Drosophila TNF receptor Grindelwald couples loss of cell polarity and neoplastic growth. Nature. 2015;522:482–6.
84. Kanda H, Igaki T, Kanuka H, Yagi T, Miura M. Wengen, a member of the Drosophila tumor necrosis factor receptor superfamily, is required for Eiger signaling. J Biol Chem. 2002;277:28372–5.
85. Kanda H, Igaki T, Okano H, Miura M. Conserved metabolic energy production pathways govern Eiger/TNF-induced nonapoptotic cell death. Proc Natl Acad Sci U S A. 2011;108:18977–82.

86. Ma X, Huang J, Yang L, Yang Y, Li W, Xue L. NOPO modulates Egr-induced JNK-independent cell death in Drosophila. Cell Res. 2012;22:425–31.
87. Shlevkov E, Morata G. A dp53/JNK-dependant feedback amplification loop is essential for the apoptotic response to stress in Drosophila. Cell Death Differ. 2012;19:451–60.
88. Anto RJ, Mukhopadhyay A, Denning K, Aggarwal BB. Curcumin (diferuloylmethane) induces apoptosis through activation of caspase-8, BID cleavage and cytochrome c release: its suppression by ectopic expression of Bcl-2 and Bcl-xl. Carcinogenesis. 2002;23:143–50.
89. Kantari C, Walczak H. Caspase-8 and bid: caught in the act between death receptors and mitochondria. Biochim Biophys Acta. 2011;1813:558–63.
90. Tang D, Lahti JM, Kidd VJ. Caspase-8 activation and bid cleavage contribute to MCF7 cellular execution in a caspase-3-dependent manner during staurosporine-mediated apoptosis. J Biol Chem. 2000;275:9303–7.
91. Nonomura K, Yamaguchi Y, Hamachi M, Koike M, Uchiyama Y, Nakazato K, Mochizuki A, Sakaue-Sawano A, Miyawaki A, Yoshida H, Kuida K, Miura M. Local apoptosis modulates early mammalian brain development through the elimination of morphogen-producing cells. Dev Cell. 2013;27:621–34.
92. Monier B, Gettings M, Gay G, Mangeat T, Schott S, Guarner A, Suzanne M. Apico-basal forces exerted by apoptotic cells drive epithelium folding. Nature. 2015;518:245–8.
93. Kuipers D, Mehonic A, Kajita M, Peter L, Fujita Y, Duke T, Charras G, Gale JE. Epithelial repair is a two-stage process driven first by dying cells and then by their neighbours. J Cell Sci. 2014;127:1229–41.
94. Kang Y, Bashirullah A. A steroid-controlled global switch in sensitivity to apoptosis during Drosophila development. Dev Biol. 2014;386:34–41.
95. Fan Y, Bergmann A. Multiple mechanisms modulate distinct cellular susceptibilities toward apoptosis in the developing Drosophila eye. Dev Cell. 2014;30:48–60.
96. Hilgers V, Bushati N, Cohen SM. Drosophila microRNAs 263a/b confer robustness during development by protecting nascent sense organs from apoptosis. PLoS Biol. 2010;8, e1000396.
97. Pernaute B, Spruce T, Smith KM, Sanchez-Nieto JM, Manzanares M, Cobb B, Rodriguez TA. MicroRNAs control the apoptotic threshold in primed pluripotent stem cells through regulation of BIM. Genes Dev. 2014;28:1873–8.
98. Kuranaga E. Beyond apoptosis: caspase regulatory mechanisms and functions in vivo. Genes Cells. 2012;17:83–97.
99. Kuranaga E, Miura M. Nonapoptotic functions of caspases: caspases as regulatory molecules for immunity and cell-fate determination. Trends Cell Biol. 2007;17:135–44.
100. Haynie J, Bryant P. The effects of X-rays on the proliferation dynamics of cells in the imaginal wing disc of Drosophila melanogaster. Wilhelm Rouxs Arch Dev Biol. 1977;183:85–100.
101. Li F, Huang Q, Chen J, Peng Y, Roop DR, Bedford JS, Li CY. Apoptotic cells activate the "phoenix rising" pathway to promote wound healing and tissue regeneration. Sci Signal. 2010;3:ra13.
102. Chera S, Ghila L, Dobretz K, Wenger Y, Bauer C, Buzgariu W, Martinou JC, Galliot B. Apoptotic cells provide an unexpected source of Wnt3 signaling to drive hydra head regeneration. Dev Cell. 2009;17:279–89.
103. Fan Y, Bergmann A. Distinct mechanisms of apoptosis-induced compensatory proliferation in proliferating and differentiating tissues in the Drosophila eye. Dev Cell. 2008;14:399–410.
104. Huh JR, Guo M, Hay BA. Compensatory proliferation induced by cell death in the Drosophila wing disc requires activity of the apical cell death caspase Dronc in a nonapoptotic role. Curr Biol. 2004;14:1262–6.
105. Kondo S, Senoo-Matsuda N, Hiromi Y, Miura M. DRONC coordinates cell death and compensatory proliferation. Mol Cell Biol. 2006;26:7258–68.
106. Bergmann A, Steller H. Apoptosis, stem cells, and tissue regeneration. Sci Signal. 2010;3:re8.
107. Fan Y, Bergmann A. Apoptosis-induced compensatory proliferation. The cell is dead. Long live the cell! Trends Cell Biol. 2008;18:467–73.
108. Mollereau B, Perez-Garijo A, Bergmann A, Miura M, Gerlitz O, Ryoo HD, Steller H, Morata G. Compensatory proliferation and apoptosis-induced proliferation: a need for clarification. Cell Death Differ. 2013;20:181.

109. Ryoo HD, Bergmann A. The role of apoptosis-induced proliferation for regeneration and cancer. Cold Spring Harb Perspect Biol. 2012;4:a008797.
110. Fan Y, Wang S, Hernandez J, Yenigun VB, Hertlein G, Fogarty CE, Lindblad JL, Bergmann A. Genetic models of apoptosis-induced proliferation decipher activation of JNK and identify a requirement of EGFR signaling for tissue regenerative responses in Drosophila. PLoS Genet. 2014;10, e1004131.
111. Perez-Garijo A, Martin FA, Morata G. Caspase inhibition during apoptosis causes abnormal signalling and developmental aberrations in Drosophila. Development. 2004;131:5591–8.
112. Ryoo HD, Gorenc T, Steller H. Apoptotic cells can induce compensatory cell proliferation through the JNK and the Wingless signaling pathways. Dev Cell. 2004;7:491–501.
113. Wells BS, Yoshida E, Johnston LA. Compensatory proliferation in Drosophila imaginal discs requires Dronc-dependent p53 activity. Curr Biol. 2006;16:1606–15.
114. Hay BA, Wolff T, Rubin GM. Expression of baculovirus P35 prevents cell death in Drosophila. Development. 1994;120:2121–9.
115. Perez-Garijo A, Shlevkov E, Morata G. The role of Dpp and Wg in compensatory proliferation and in the formation of hyperplastic overgrowths caused by apoptotic cells in the Drosophila wing disc. Development. 2009;136:1169–77.
116. Bergantinos C, Corominas M, Serras F. Cell death-induced regeneration in wing imaginal discs requires JNK signalling. Development. 2010;137:1169–79.
117. Martin FA, Perez-Garijo A, Morata G. Apoptosis in Drosophila: compensatory proliferation and undead cells. Int J Dev Biol. 2009;53:1341–7.
118. Smith-Bolton RK, Worley MI, Kanda H, Hariharan IK. Regenerative growth in Drosophila imaginal discs is regulated by Wingless and Myc. Dev Cell. 2009;16:797–809.
119. Sun G, Irvine KD. Regulation of Hippo signaling by Jun kinase signaling during compensatory cell proliferation and regeneration, and in neoplastic tumors. Dev Biol. 2011;350:139–51.
120. Sun G, Irvine KD. Ajuba family proteins link JNK to Hippo signaling. Sci Signal. 2013;6:81.
121. Tseng AS, Adams DS, Qiu D, Koustubhan P, Levin M. Apoptosis is required during early stages of tail regeneration in Xenopus laevis. Dev Biol. 2007;301:62–9.
122. Pellettieri J, Fitzgerald P, Watanabe S, Mancuso J, Green DR, Sanchez Alvarado A. Cell death and tissue remodeling in planarian regeneration. Dev Biol. 2010;338:76–85.
123. Pellettieri J, Sanchez Alvarado A. Cell turnover and adult tissue homeostasis: from humans to planarians. Annu Rev Genet. 2007;41:83–105.
124. Vlaskalin T, Wong CJ, Tsilfidis C. Growth and apoptosis during larval forelimb development and adult forelimb regeneration in the newt (Notophthalmus viridescens). Dev Genes Evol. 2004;214:423–31.
125. Jung Y, Witek RP, Syn WK, Choi SS, Omenetti A, Premont R, Guy CD, Diehl AM. Signals from dying hepatocytes trigger growth of liver progenitors. Gut. 2010;59:655–65.
126. Goessling W, North TE, Loewer S, Lord AM, Lee S, Stoick-Cooper CL, Weidinger G, Puder M, Daley GQ, Moon RT, Zon LI. Genetic interaction of PGE2 and Wnt signaling regulates developmental specification of stem cells and regeneration. Cell. 2009;136:1136–47.
127. Castellone MD, Teramoto H, Williams BO, Druey KM, Gutkind JS. Prostaglandin E2 promotes colon cancer cell growth through a Gs-axin-beta-catenin signaling axis. Science. 2005;310:1504–10.
128. Jager R, Fearnhead HO. "Dead cells talking": the silent form of cell death is not so quiet. Biochem Res Int. 2012;2012:453838.
129. Moon RT, Kohn AD, de Ferrari GV, Kaykas A. WNT and beta-catenin signalling: diseases and therapies. Nat Rev Genet. 2004;5:691–701.
130. van Raam BJ, Salvesen GS. Proliferative versus apoptotic functions of caspase-8 Hetero or homo: the caspase-8 dimer controls cell fate. Biochim Biophys Acta. 2012;1824:113–22.
131. Beisner DR, Chen IL, Kolla RV, Hoffmann A, Hedrick SM. Cutting edge: innate immunity conferred by B cells is regulated by caspase-8. J Immunol. 2005;175:3469–73.
132. Salmena L, Lemmers B, Hakem A, Matysiak-Zablocki E, Murakami K, Au PY, Berry DM, Tamblyn L, Shehabeldin A, Migon E, Wakeham A, Bouchard D, Yeh WC, Mcglade JC,

Ohashi PS, Hakem R. Essential role for caspase 8 in T-cell homeostasis and T-cell-mediated immunity. Genes Dev. 2003;17:883–95.
133. Su H, Bidere N, Zheng L, Cubre A, Sakai K, Dale J, Salmena L, Hakem R, Straus S, Lenardo M. Requirement for caspase-8 in NF-kappaB activation by antigen receptor. Science. 2005;307:1465–8.
134. Oberst A, Green DR. It cuts both ways: reconciling the dual roles of caspase 8 in cell death and survival. Nat Rev Mol Cell Biol. 2011;12:757–63.
135. Zhou XD, Yu JP, Liu J, Luo HS, Chen HX, Yu HG. Overexpression of cellular FLICE-inhibitory protein (FLIP) in gastric adenocarcinoma. Clin Sci (Lond). 2004;106:397–405.
136. Ghavami S, Hashemi M, Ande SR, Yeganeh B, Xiao W, Eshraghi M, Bus CJ, Kadkhoda K, Wiechec E, Halayko AJ, Los M. Apoptosis and cancer: mutations within caspase genes. J Med Genet. 2009;46:497–510.
137. Ili CG, Brebi P, Tapia O, Sandoval A, Lopez J, Garcia P, Leal P, Sidransky D, Guerrero-Preston R, Roa JC. Cellular FLICE-like inhibitory protein long form (c-FLIPL) overexpression is related to cervical cancer progression. Int J Gynecol Pathol. 2013;32:316–22.
138. Sharp DA, Lawrence DA, Ashkenazi A. Selective knockdown of the long variant of cellular FLICE inhibitory protein augments death receptor-mediated caspase-8 activation and apoptosis. J Biol Chem. 2005;280:19401–9.
139. Wang W, Wang S, Song X, Sima N, Xu X, Luo A, Chen G, Deng D, Xu Q, Meng L, Lu Y, Ma D. The relationship between c-FLIP expression and human papillomavirus E2 gene disruption in cervical carcinogenesis. Gynecol Oncol. 2007;105:571–7.
140. Hashimoto T, Kikkawa U, Kamada S. Contribution of caspase(s) to the cell cycle regulation at mitotic phase. PLoS One. 2011;6, e18449.
141. Hashimoto T, Yamauchi L, Hunter T, Kikkawa U, Kamada S. Possible involvement of caspase-7 in cell cycle progression at mitosis. Genes Cells. 2008;13:609–21.
142. Eymin B, Sordet O, Droin N, Munsch B, Haugg M, van de Craen M, Vandenabeele P, Solary E. Caspase-induced proteolysis of the cyclin-dependent kinase inhibitor p27Kip1 mediates its anti-apoptotic activity. Oncogene. 1999;18:4839–47.
143. Khalil H, Peltzer N, Walicki J, Yang JY, Dubuis G, Gardiol N, Held W, Bigliardi P, Marsland B, Liaudet L, Widmann C. Caspase-3 protects stressed organs against cell death. Mol Cell Biol. 2012;32:4523–33.
144. Yang JY, Michod D, Walicki J, Murphy BM, Kasibhatla S, Martin SJ, Widmann C. Partial cleavage of RasGAP by caspases is required for cell survival in mild stress conditions. Mol Cell Biol. 2004;24:10425–36.
145. Yang JY, Widmann C. The RasGAP N-terminal fragment generated by caspase cleavage protects cells in a Ras/PI3K/Akt-dependent manner that does not rely on NFkappa B activation. J Biol Chem. 2002;277:14641–6.
146. Geiger TR, Peeper DS. Metastasis mechanisms. Biochim Biophys Acta. 2009;1796:293–308.
147. Geisbrecht ER, Montell DJ. A role for Drosophila IAP1-mediated caspase inhibition in Rac-dependent cell migration. Cell. 2004;118:111–25.
148. Kang TB, Ben-Moshe T, Varfolomeev EE, Pewzner-Jung Y, Yogev N, Jurewicz A, Waisman A, Brenner O, Haffner R, Gustafsson E, Ramakrishnan P, Lapidot T, Wallach D. Caspase-8 serves both apoptotic and nonapoptotic roles. J Immunol. 2004;173:2976–84.
149. Portela M, Richardson HE. Death takes a holiday—non-apoptotic role for caspases in cell migration and invasion. EMBO Rep. 2013;14:107–8.
150. Zhao X, Wang D, Zhao Z, Xiao Y, Sengupta S, Xiao Y, Zhang R, Lauber K, Wesselborg S, Feng L, Rose TM, Shen Y, Zhang J, Prestwich G, Xu Y. Caspase-3-dependent activation of calcium-independent phospholipase A2 enhances cell migration in non-apoptotic ovarian cancer cells. J Biol Chem. 2006;281:29357–68.
151. Brown M, Roulson JA, Hart CA, Tawadros T, Clarke NW. Arachidonic acid induction of Rho-mediated transendothelial migration in prostate cancer. Br J Cancer. 2014;110:2099–108.
152. Cheng YJ, Lee CH, Lin YP, Huang JY, Su CC, Chang WT, Yang BC. Caspase-3 enhances lung metastasis and cell migration in a protease-independent mechanism through the ERK pathway. Int J Cancer. 2008;123:1278–85.

153. Rudrapatna VA, Bangi E, Cagan RL. Caspase signalling in the absence of apoptosis drives Jnk-dependent invasion. EMBO Rep. 2013;14:172–7.
154. Na HJ, Hwang JY, Lee KS, Choi YK, Choe J, Kim JY, Moon HE, Kim KW, Koh GY, Lee H, Jeoung D, Won MH, Ha KS, Kwon YG, Kim YM. TRAIL negatively regulates VEGF-induced angiogenesis via caspase-8-mediated enzymatic and non-enzymatic functions. Angiogenesis. 2014;17:179–94.
155. Li F, He Z, Shen J, Huang Q, Li W, Liu X, He Y, Wolf F, Li CY. Apoptotic caspases regulate induction of iPSCs from human fibroblasts. Cell Stem Cell. 2010;7:508–20.
156. Chang DW, Xing Z, Pan Y, Algeciras-Schimnich A, Barnhart BC, Yaish-Ohad S, Peter ME, Yang X. c-FLIP(L) is a dual function regulator for caspase-8 activation and CD95-mediated apoptosis. EMBO J. 2002;21:3704–14.
157. Kim RJ, Nam JS. OCT4 expression enhances features of cancer stem cells in a mouse model of breast cancer. Lab Anim Res. 2011;27:147–52.
158. Fulda S, Debatin KM. Extrinsic versus intrinsic apoptosis pathways in anticancer chemotherapy. Oncogene. 2006;25:4798–811.
159. Dick JE. Stem cell concepts renew cancer research. Blood. 2008;112:4793–807.
160. Huang Q, Li F, Liu X, Li W, Shi W, Liu FF, O'Sullivan B, He Z, Peng Y, Tan AC, Zhou L, Shen J, Han G, Wang XJ, Thorburn J, Thorburn A, Jimeno A, Raben D, Bedford JS, Li CY. Caspase 3-mediated stimulation of tumor cell repopulation during cancer radiotherapy. Nat Med. 2011;17:860–6.
161. Donato AL, Huang Q, Liu X, Li F, Zimmerman MA, Li CY. Caspase 3 promotes surviving melanoma tumor cell growth after cytotoxic therapy. J Invest Dermatol. 2014;134:1686–92.
162. Nakopoulou L, Alexandrou P, Stefanaki K, Panayotopoulou E, Lazaris AC, Davaris PS. Immunohistochemical expression of caspase-3 as an adverse indicator of the clinical outcome in human breast cancer. Pathobiology. 2001;69:266–73.
163. Hu Q, Peng J, Liu W, He X, Cui L, Chen X, Yang M, Liu H, Liu S, Wang H. Elevated cleaved caspase-3 is associated with shortened overall survival in several cancer types. Int J Clin Exp Pathol. 2014;7:5057–70.
164. Labi V, Erlacher M, Krumschnabel G, Manzl C, Tzankov A, Pinon J, Egle A, Villunger A. Apoptosis of leukocytes triggered by acute DNA damage promotes lymphoma formation. Genes Dev. 2010;24:1602–7.
165. Michalak EM, Vandenberg CJ, Delbridge AR, Wu L, Scott CL, Adams JM, Strasser A. Apoptosis-promoted tumorigenesis: gamma-irradiation-induced thymic lymphomagenesis requires Puma-driven leukocyte death. Genes Dev. 2010;24:1608–13.
166. Qian BZ, Li J, Zhang H, Kitamura T, Zhang J, Campion LR, Kaiser EA, Snyder LA, Pollard JW. CCL2 recruits inflammatory monocytes to facilitate breast-tumour metastasis. Nature. 2011;475:222–5.
167. Steidl C, Lee T, Shah SP, Farinha P, Han G, Nayar T, Delaney A, Jones SJ, Iqbal J, Weisenburger DD, Bast MA, Rosenwald A, Muller-Hermelink HK, Rimsza LM, Campo E, Delabie J, Braziel RM, Cook JR, Tubbs RR, Jaffe ES, Lenz G, Connors JM, Staudt LM, Chan WC, Gascoyne RD. Tumor-associated macrophages and survival in classic Hodgkin's lymphoma. N Engl J Med. 2010;362:875–85.
168. Willems JJ, Arnold BP, Gregory CD. Sinister self-sacrifice: the contribution of apoptosis to malignancy. Front Immunol. 2014;5:299.
169. Ford CA, Petrova S, Pound JD, Voss JJ, Melville L, Paterson M, Farnworth SL, Gallimore AM, Cuff S, Wheadon H, Dobbin E, Ogden CA, Dumitriu IE, Dunbar DR, Murray PG, Ruckerl D, Allen JE, Hume DA, van Rooijen N, Goodlad JR, Freeman TC, Gregory CD. Oncogenic properties of apoptotic tumor cells in aggressive B cell lymphoma. Curr Biol. 2015;25:577–88.
170. Lauber K, Bohn E, Krober SM, Xiao YJ, Blumenthal SG, Lindemann RK, Marini P, Wiedig C, Zobywalski A, Baksh S, Xu Y, Autenrieth IB, Schulze-Osthoff K, Belka C, Stuhler G, Wesselborg S. Apoptotic cells induce migration of phagocytes via caspase-3-mediated release of a lipid attraction signal. Cell. 2003;113:717–30.
171. Ravichandran KS. Beginnings of a good apoptotic meal: the find-me and eat-me signaling pathways. Immunity. 2011;35:445–55.

Chapter 5
Axl and Mer Receptor Tyrosine Kinases: Distinct and Nonoverlapping Roles in Inflammation and Cancer?

Ian Dransfield and Sarah Farnworth

Abstract The receptor tyrosine kinases Axl and Mer subserve the process of termination of proinflammatory signaling and have key roles in both the resolution of inflammation and restoration of homeostasis. Axl functions prominently under conditions of tissue stress or in response to infection, whereas Mer has a major role in maintenance of homeostasis within tissues. Distinct patterns of expression of Axl and Mer underlie their clearly defined functional roles during the initiation and progression of inflammation. Axl and Mer are expressed by tumor cells and by infiltrating inflammatory cells and the regulation of cellular function via Axl and Mer signaling is also important for tumorigenesis, tumor progression, and metastasis. In this review, we consider the divergent functions of Axl and Mer in the context of inflammatory processes within tumors and the implications for development of therapeutic agents targeting these receptors.

Keywords Receptor tyrosine kinase • Axl • Mer • Gas6 • Protein S • Inflammation • Macrophage • Tumor • Homeostasis • Tumor microenvironment

5.1 Introduction

Following injury or infection, the inflammatory response is geared to provide defense against invading microorganisms, repair of injury, and restoration of tissue architecture that is required for normal organ function [1]. It is now apparent that inflammation and establishment of an inflammatory tissue microenvironment is closely linked to both tumorigenesis and tumor progression and ultimately, the potential for metastasis [2]. The development and progression of inflammatory responses is coordinated by the precise spatiotemporal release of inflammatory

I. Dransfield (✉) • S. Farnworth
MRC and University of Edinburgh Centre for Inflammation Research, Queen's Medical Research Institute, 47, Little France Crescent, Edinburgh EH16 4TJ, UK
e-mail: i.dransfield@ed.ac.uk

C.D. Gregory (ed.), *Apoptosis in Cancer Pathogenesis and Anti-cancer Therapy*, Advances in Experimental Medicine and Biology 930,
DOI 10.1007/978-3-319-39406-0_5

mediators that act to guide the recruitment and activation of inflammatory cells and regulation of stromal cell function necessary for tissue repair [3]. In particular, the association between tumor progression and altered inflammatory homeostasis suggests that failed resolution of inflammation may represent an important underlying mechanism [4]. Thus, therapeutics developed for the treatment of persistent or chronic inflammation associated with debilitating inflammatory or autoimmune diseases that target either initiation, progression, or ultimate termination of inflammatory responses may also find utility in the treatment of cancer.

Restoration of balance of inflammatory or immune responses is one of the key determinants of the process by which inflammation normally resolves. Inhibition of proinflammatory signaling via the Tyro3, Axl, and Mer subfamily of receptor tyrosine kinases (RTKs) is pivotal for resolution of inflammatory responses and restoration of tissue homeostasis [5]. When these regulatory feedback inhibition loops are removed, hyperresponsiveness to inflammatory stimuli or commensal microorganisms may ultimately lead to the development of a persistent inflammatory microenvironment within tissues that favors tumor development and progression. Overexpression and/or mutation of Tyro3, Axl, or Mer has been reported in several different cancers and aberrant intracellular signaling via these receptors likely contributes to tumor progression and metastasis [6]. The ligands for Tyro3, Axl, and Mer bind to phosphatidylserine (PtdSer) which is exposed on the membranes of cells undergoing apoptosis or necrosis, cell-derived microparticles, and many envelope viruses [7]. Importantly, ligand binding to Tyro3/Axl/Mer in the context of PtdSer initiates RTK signaling that provides feedback inhibition of proinflammatory responses [8]. Although these receptors share many molecular properties, Axl and Mer have well-defined, nonoverlapping functional roles during the development and resolution of inflammatory responses [9]. In this article, we discuss this emerging distinction between Axl and Mer in an inflammatory setting that may affect tumor progression and metastasis.

5.2 Axl and Mer Receptor Tyrosine Kinases

Tyro3, Axl, and Mer were identified as a subfamily of related receptor tyrosine kinases (RTKs) in a PCR-based screen of transcripts enriched in preparations of sciatic nerves isolated from rat [10]. The same year, O'Bryan and colleagues identified a transforming gene isolated from two human chronic myelogenous leukemia patients that they termed *Axl* [11]. Expression of *Axl* was found in a wide range of tissues suggesting an important role for this receptor in normal cellular function. Mer (gene name: *Mertk*) was originally cloned from a B lymphoblastoid cell library [12] and is closely related to Axl, with 44 % sequence identity and a similar overall domain structure. RT-PCR analysis revealed widespread expression with message present in testis, ovary, prostate, lung, kidney, and monocytes in bone marrow, with lower expression in the spleen, placenta, thymus, small intestine, colon, and liver.

A careful analysis of the *Mertk−/−* mice revealed increased susceptibility to septic shock in response to LPS with evidence of a heightened inflammatory response in the small intestine [13]. Excessive production of TNF-α by Mer-deficient macrophages suggested that Mer functions as a critical inhibitory pathway to guard against excessive tissue damage in responses to bacterial endotoxin by regulating activation of NFkB [13]. Although *Tyro3/Axl/Mertk* triple knockout animals are viable [14], Lemke and colleagues demonstrated that these animals displayed multiple major organ defects and neurological abnormalities with increased apoptosis and cellular degeneration in a variety of adult tissues, including the brain, prostate epithelium, liver parenchyma, and blood vessel walls [14]. In addition, postnatal degeneration of rods and cones in the retina was evident and apoptotic cells were prominent in the massively enlarged spleens of adult *Tyro3/Axl/Mertk* triple knockout mice consistent with aberrant homeostasis within lymphoid tissue. Hyperproliferation and constitutive activation of T and B lymphocyte populations present in lymphoid tissues from *Tyro3/Axl/Mertk* triple mutant mice was associated with development of elevated antibody titers to nuclear proteins, double-stranded DNA, cardiolipin, and lipids, including phosphatidylserine similar to that observed in human autoimmune diseases such as arthritis and systemic lupus erythematosus [15]. Importantly, hyperactivation of macrophages and dendritic cells was identified as being responsible for the initiation of lymphoproliferation and autoimmunity observed in the triple mutant mice [15]. Together these findings indicated that concerted activation of the Tyro3/Axl/Mer family of receptors is pivotal for maintenance and homeostatic balance of a wide variety of mature mammalian tissues.

5.3 Ligands

Two principal ligands for the Tyro3/Axl/Mer family of receptors have been identified, Gas6 and Protein S, initially based upon their ability to induce tyrosine phosphorylation of Tyro3 [16]. Protein S was originally identified as a vitamin K-dependent protein present in plasma [17] that was able to bind to a Tyro3-Fc construct, but not to an Axl-Fc construct. Axl activation could be induced by Gas6 [16], a gene product originally identified as a growth arrest-specific protein in arrested fibroblasts [18]. Gas6 is a ligand for Tyro3, Axl, and Mer, whereas Protein S binds only Tyro3 and Mer [19]. Both Protein S and Gas6 have an N-terminal Gla-domain containing ~10 gamma-carboxylated glutamic acid residues similar to those found in vitamin K-dependent coagulation proteins which show strong Ca^{2+}-dependent binding to negatively charged phospholipids. However, Protein S differs from Gas6 in having a thrombin-sensitive cleavage site proximal to the Gla-domain [18]. Gas6 and Protein S bind extremely rapidly to PtdSer exposed on the apoptotic cell surface following loss of membrane lipid asymmetry, thus providing a molecular linkage between the apoptotic cell membrane and Axl or Mer expressed on the phagocyte. Axl and Mer have been shown to rapidly localize to sites of contact with apoptotic cells in a ligand-dependent manner [9] and RTK activity is absolutely required for subsequent internalization of tethered apoptotic cells via control of cytoskeletal regulation [20].

5.4 Soluble Receptors

Binding of Protein S or Gas6 has been reported to induce down-regulation of Axl and Mer expression, most likely via a mechanism involving either ubiquitin-mediated degradation or ADAM17-dependent shedding. Proteolytic cleavage of the extracellular domain of Axl and Mer from the cell surface represents a mechanism for inhibition of Axl and Mer function and signaling as inflammation progresses. The soluble forms of Axl [21] and Mer [22] may act to block the function of cell-associated receptors as they bind ligand with high affinity. Increased plasma levels of soluble Mer and Axl were found in chronic kidney disease and this was associated with higher levels of ADAM17 expression in chronic kidney disease patients [23] and in established multiple sclerosis lesions [24]. Elevated plasma levels of sAxl have been reported in a variety of inflammatory disease settings, including community-acquired pneumonia [25], limb ischemia [26], lupus flares [27], obesity and insulin resistance [28] and preeclampsia [29], acute coronary syndromes [30], aortic aneurism [31], and intracranial aneurism rupture [32]. In contrast, increased levels of soluble forms of Mer are found in autoimmune conditions. For example, elevated levels of soluble Mer were present in plasma from patients with primary Sjorgen's Syndrome [33] and levels of soluble Mer present in plasma of patients with juvenile onset systemic lupus erythematosus may be a correlate of disease activity [34]. High levels of sAxl were reported to be present in exudates from transplanted Fibroblast growth factor-transformed tumors [35] and sAxl is associated with disease severity and poor prognosis in renal cancer [36] and in hepatocellular carcinoma [37] and correlates with tumor burden in patients with neurofibromatosis [38]. The presence of soluble forms of Axl may represent a useful biomarker for the presence of systemic inflammation [31] and thus for tumor progression, since sAxl may be derived from tumor cells in some cases.

5.5 Axl- and Mer-Mediated Clearance of Apoptotic Cells

In adults, the process of self-renewal that takes place continuously within diverse tissue settings generates large numbers of apoptotic cells daily. Rapid and efficient phagocytosis of apoptotic cells is important for preventing the release of potentially proinflammatory intracellular contents. In addition, macrophage recognition of apoptotic cells suppresses proinflammatory cytokine release stimulated by inflammatory triggers such as LPS and induces the release of IL-10 and TGF-β [39, 40]. The repertoire of surface receptors that are expressed by phagocytes determines the capacity for apoptotic cell recognition (reviewed elsewhere [41, 42]). Myeloid cell populations exhibit distinct patterns of expression and function of Axl and Mer, both in in vitro settings and in immune tissues in vivo, which will constrain engagement of these receptors in the process of apoptotic cell clearance. Although *Mertk* mRNA transcripts are expressed by virtually all macrophage populations that have been examined [43], Mer is especially prominent in homeostatic tissue settings (e.g., retinal pigment epithelial cells of the eye [44], Sertoli cells of the testis [14], and tingible body macrophages of the spleen [45]). Mer-mediated phagocytic clearance of apoptotic cells may be particularly important for maintaining an anti-inflammatory tissue microenvironment. In some tissues, e.g., the retina and the bone marrow, Mer-mediated apoptotic cell clearance exhibits circadian regulation as part of a homeostatic tissue response to environmental challenge [46, 47].

In contrast, Axl is expressed by tissue cells, e.g., Langerhans cells of the skin [48], alveolar macrophages [49], and splenic dendritic cells [9] that are poised to respond to injury or infectious stimuli and exposure of macrophages to diverse pro-inflammatory stimuli specifically induces Axl expression [9]. In vivo, there is evidence for selective expression of Axl in macrophages exposed to continual environmental triggers, as reported for macrophages present in the airways [49] and elevated expression of Axl may define an inflammatory activation state for tissue phagocytes. During an inflammatory response, Axl-dependent clearance of apoptotic cells may specifically provide a mechanism for initiation of the process of resolution of inflammation.

5.6 The Axl/Gas6 Axis

Lew and colleagues have elegantly demonstrated that Axl signaling is dependent on the context of a PtdSer scaffold, since although Axl binds with high affinity to Gas6 lacking the Gla domain, tyrosine phosphorylation of Axl is not induced [19]. Thus, ligand-induced down-regulation of Axl expression would be predicted to require a source of PtdSer such as apoptotic cells. Furthermore, data from the Lemke laboratory demonstrated that Axl was required to maintain levels of Gas6 present in plasma and Gas6 expression within many tissues [9] which is likely to have important consequences in terms of regulation of Tyro3/Axl/Mer function. Interaction of Gas6 with Axl may be required to prevent rapid tissue clearance of Gas6 and therapeutic use of inhibitors of Axl or genetic knockdown of Axl may also impact upon Gas6 expression within tissues. Whether occupancy of Axl by Gas6 could effectively "arm" tissue phagocytes for rapid Axl-dependent recognition of PtdSer that becomes exposed on apoptotic cells, exosomes or enveloped viruses remains to be determined.

5.7 Tumor Microenvironment and Expression of Mer and Axl

Divergent profiles of expression of Axl and Mer are also seen in human cancers, in which Axl and Mer RTK activity may contribute to tumor progression and metastasis [50]. In an analysis of tumor-associated changes in *Tyro3*, *Axl*, and *Mertk* mRNA expression across multiple tumor types, Zagorska and colleagues identified evidence for exclusivity of expression which would be consistent with distinct roles for these receptors in tumor progression [9]. Consistent with this suggestion, Axl and Mer exhibit restricted expression profiles in specific tumor types. Axl is expressed in acute myelogenous leukemia [51] and in B cell Chronic Lymphocytic Leukemia [52], but not in acute lymphoblastic leukemia [53]. Mer is ectopically expressed in human lymphoblastic [54] and T-ALL cell lines [55] and myeloid leukemias [56] when compared with normal B- and T-lymphocytes [12]. Overexpression of Mer has also been shown in Acute Myeloid Leukemia, Mantle Cell Lymphomas [57], pituitary adenomas [58], prostate cancer [59], and melanoma [60]. Expression of Mer and Axl is differentially regulated in normal melanocytes when compared with melanoma cells with increased expression of Axl on melanoma cell lines compared to normal melanocytes [61].

5.8 Macrophages and the Tumor Microenvironment

The presence of monocyte/macrophages within tumors is a defining feature of the tumor microenvironment and has been associated with poor survival in ovarian, thyroid, and hepatocellular carcinomas [62–64]. Experimental depletion of macrophages with liposomal clodronate reduces growth in a variety of tumor types, including melanoma, lung, and prostate tumors [65–67]. However, there are notable exceptions to this protumor role for tumor-associated macrophages, notably in the bone marrow, the liver, and in the pancreas [68]. For example, depletion of Kupffer cells within liver enhances metastasis and in a xenograft model of tumor metastasis to the liver, worsens prognosis [69]. Whether differential expression of Mer and Axl in these different macrophage populations contributes to their pro- or antitumor effects has not been determined.

Macrophages within tumors are thought to derive mostly from infiltrating monocytes [70, 71] under the control of a combination of chemokines (CCL2), growth factors (CSF-1), and other inflammatory mediators, including complement [72], and hypoxia [73], although local proliferative responses may also contribute [74]. The tumor microenvironment induces a distinct transcriptional profile in the recruited monocytes that is associated with enhanced survival and proliferation of tumor cells, angiogenenic/lymphangiogenic responses, and dampened antitumor responses of both macrophages and cytotoxic T lymphocytes [75]. These effects are likely to require interaction with other immune cells, cancer stem cells, and stromal cells that are present within the tumor microenvironment, as well as tumor cells themselves. In general, tumor-associated macrophage phenotype has been reported to be more regulatory/wound healing or "M2-like" [76]. However, tumor-specific factors, such as tissue origin, stage of tumor progression, and tumor size may also critically influence macrophage phenotype [77] and differential regulation of Axl and Mer is associated with this phenotype [78].

5.9 Tumor Macrophage Heterogeneity

Heterogeneity within the microenvironment of individual tumors may further act to condition macrophages to exhibit site-specific phenotypic and functional differences that may result in differential expression of Axl and Mer. Macrophages with an M2-like phenotype accumulate within the hypoxic/necrotic areas of many different tumors including prostate [79], breast [80], and ovarian carcinomas [81] and release proangiogenic factors such as Vascular Endothelial Growth Factor (VEGF). Macrophages were found to exhibit distinct phenotypes in hypoxic or less hypoxic regions of tumors [82]. Tumor infiltrating monocytes may initially differentiate independently of hypoxia, with subsequent recruitment of macrophages expressing low levels of MHC class II to hypoxic regions and acquisition of a pro-angiogenic phenotype [83].

A proinflammatory or hypoxic [84] tumor microenvironment might favor induction of expression of Axl on both tumor cells or infiltrating monocytes. High levels of Gas6 and Axl were reported in hypoxia-inducible factor-1α expressing prostate cancer cells and in bone metastases compared with normal tissues. A hypoxic tumor microenvironment may inhibit Gas6-mediated downregulation of Axl and lead to sustained Gas6/Axl signaling [85]. Furthermore, oxidative stress could activate Axl phosphorylation to synergistically enhance cell migration in an Akt-dependent manner. Axl was found to be required for neuronal migration in Gonadotropin-releasing hormone responses [86] and promotes cell invasion through induction of Matrix metalloproteinase-9 expression [87]. Activation of Rac1 via Axl elicits reactive oxygen species accumulation, which is associated with malignant cancer phenotypes, resistance to chemotherapy, and metastasis [88]. Elevated expression and activation of Axl may therefore be important in tumors that are characterized by extensive inflammatory cell infiltrates, or those lacking extensive vascularization, which may have important implications for the use of antiangiogenic therapeutics [89]. In contrast, expression of Mer may be induced by exposure of phagocytes to apoptotic cells in a nuclear receptor LXRalpha/beta-dependent manner [90], thereby amplifying Mer-mediated apoptotic cell clearance mechanisms. Macrophage expression of Mer is associated with an anti-inflammatory and tissue remodeling phenotype induced by M-CSF and IL-10, glucocorticoids, and PPARγ antagonists [91, 92]. Mer-dependent clearance of dying cells has been shown to be critical in shaping the cytokine micro-environment of developing tumors [93]. Macrophages associated with human Burkitt lymphoma xenografts, a tumor which exhibits high levels of constitutive apoptosis, show upregulated expression of Mer [94].

5.10 Microvesicles

In addition to soluble mediators and the unique extracellular matrix composition associated with tumors that may alter macrophage phenotype and function, tumor cells may release microvesicles which could influence tumor development in a number of ways. First, microvesicles derived from some tumors, e.g., chronic lymphocytic leukemia (CLL) B-cells, express constitutively phosphorylated Axl, which correlates with activation of multiple signaling intermediates including Lyn, phosphoinositide-3 kinase, Syk/ζ-associated protein of 70 kDa, phospholipase Cγ2. Small molecule inhibition of Axl with R428 or TP-0903 induced rapid CLL B cell apoptosis [52], possibly as a consequence of reduced expression of the antiapoptotic proteins Mcl-1, Bcl-2, and XIAP and upregulation of BIM [95]. Second, microvesicles expose PtdSer on their outer membrane and can be opsonized with PtdSer binding proteins, including Gas6 and Protein S. Macrophages and epithelial cells present within tumors that express Axl or Mer may then specifically interact with opsonized microvesicles using the same molecular mechanisms that are involved in the phagocytosis of apoptotic cells. Finally, it is becoming clear that internalization of microvesicle "cargo," which includes miRNA and a variety of proteins, by macrophages can directly affect macrophage functional responses [96].

5.11 Axl- and Mer-Mediated Signaling

In a homeostatic or inflammatory setting, signaling via Mer and Axl forms part of a critical feedback loop that inhibits cellular responses to diverse proinflammatory stimuli including Toll-like receptor (TLR) ligands, type I interferons, and hypoxia. For example, ligation of Axl in the context of PtdSer exposure [19] acts to attenuate TLR- or cytokine receptor-mediated activation via inhibition of both MyD88-dependent and MyD88-independent pathways. As a consequence, Axl-mediated signaling inhibits activation of ERK1/2, p38MAPK, and NF-kB pathways [8]. Ligation of Axl leads to specific induction of SOCS1 and SOCS3, E3 ubiquitin ligases that control degradation of pivotal adaptors for NF-kB and TLR signaling such as TRAF6 and TIRAP. In STAT1−/− dendritic cells, upregulation of SOCS1 and SOCS3 transcripts was markedly reduced in response to addition of Gas6, consistent with a requirement for type I IFN receptor/STAT1 in upregulation of Axl and subsequent inhibition of proinflammatory signaling [8]. However, in the context of a developing tumor, Mer or Axl signaling may activate cellular survival/antiapoptotic, proliferative (PI3K/Akt, p38, Erk) and migration (FAK, RhoA) responses. Mer-dependent activation of the NF-kB and induction of STAT5 and STAT6 phosphorylation and transcriptional activity was reported in acute myeloid leukemia [56], melanoma cell lines [60], and in T-ALL cell lines [55]. High levels of expression of Axl also offer a survival advantage to tumor cells, conferring drug resistance in AML [97]; enhanced survival, adhesion, and proliferation in Schwannoma cells [98]; tumor progression in breast cancer [99]; and reduced survival in patients with head and neck cancer [100] or Ewing Sarcoma [101].

5.12 Interplay Between Axl/Mer and Other Signaling Pathways

Increased expression of Axl and associated Axl-dependent signaling pathways may induce chemoresistance that is induced following repeated use of chemotherapeutic agents [97]. Axl is upregulated in imatinib-resistant CML [102] and Axl expression is associated with resistance to cetuximab therapy [103] and EFGR therapy [104]. Hyperactivation of Axl has been reported in lapatinib-resistant breast cancer and siRNA knockdown of Axl expression restores sensitivity to lapatinib in these cells [105]. Downregulation of Axl by shRNA in hepatocellular carcinoma cell lines resulted in the inhibition of invasive capacity in vitro and in vivo [106]. Overexpression of Axl in squamous cell carcinomas of head and neck is associated with persistent mTOR activation and a lack of response to PI3Kα inhibition. Phospholipase Cγ-dependent activation of mTOR may also occur via Axl-induced phosphorylation of epidermal growth factor receptor (EGFR) [107]. However, Gas6-induced activation of Axl may result in restoring migratory defects and inhibition of apoptosis in glioblastoma cells, rendering them more sensitive to sunitinib treatment [108]. Axl-mediated signaling in breast cancer cells leads to the phosphorylation of the scaffolding proteins ELMO1/2 resulting in the formation of a complex with Axl. ELMO knockdown prevented Gas6-induced Rac1 activation in breast cancer cells, reducing proliferation and abolishing breast cancer cell invasion [109].

Signaling downstream of Tyro3/Axl and Mer may impact upon a range of oncogenic pathways, as described earlier. The specific pathways activated are likely to be dependent on the receptor, ligand, and cell type. Mer phosphorylation was found to be reduced in *Axl/Tyro3–/–* mice, suggesting that signaling may involve both homodimerization and heterodimerization of Tyro3/Axl/Mer to induce subsequent receptor auto-phosphorylation in some cell types and processes [110]. Furthermore, there may be crosstalk between Tyro3/Axl/Mer and other RTKs. For example, Gas6/Axl crosstalk with HGF/MET is important for migration and survival in Hypothalamic Gonadotropin-releasing hormone neurons [111].

5.13 Strategies for Altering Expression and/or Function of Axl and Mer

The impact of different experimental strategies designed to dissect the function of Axl and Mer in tumorigenesis and progression may require careful interpretation. Use of small molecule inhibitors, receptor mAb, shRNA knockdown, or the use of knockout mice may also affect Axl- or Mer-mediated responses of tumor-associated cells, particularly macrophages or NK cells. An interesting recent study showed that inhibition of Tyro3/Axl/Mer kinase activation in wild-type NK cells was associated with enhanced antimetastatic NK cell activity in vivo and reduced murine mammary cancer and melanoma metastases in an NK cell-dependent manner [112]. NK cell-mediated rejection of metastatic tumors was induced following deletion/inactivation of the E3 ubiquitin ligase Cbl-b (casitas B-lineage lymphoma-b) that acts to ubiquitylate Tyro3, Axl, and Mer. In the absence of Cbl-b, Gas6-induced downregulation of Axl in NK cells was not observed. Binding of Axl to Gas6 induces interaction with the ubiquitin ligase c-Cbl and ubiquitylation of Axl leading to endocytic/lysosomal degradation that was independent of proteosomal activity [113]. Furthermore, differential Axl/Mer signaling in neoplastic cells present within tumors [114] may provide protumor effects that in combination with Axl/Mer-mediated inhibition of tumor-associated macrophage inflammatory cytokines and chemokine production [13, 115, 116] act to favor tumor progression.

5.14 Axl/Mer Inhibitors

Due to the important role of Mer in homeostatic apoptotic cell clearance within tissues, long-term inhibition of Mer may have an impact on vision impairment, male fertility, and autoimmune disease. Whether lower expression of Axl within tissues will result in fewer side effects following therapeutic blockade remains to be determined. Excessive cytokine production following Axl inhibition during infection could predispose to autoimmunity or susceptibility to septic shock. Small molecule ATP mimetics that specifically inhibit Tyro3/Axl/Mer RTK activity are in development as cancer therapies or treatment of enveloped virus infections [7]. In general, these inhibitors have been reported to reverse protumor effects, most notably when used in combination with other agents. Treatment with the Mer-specific UNC569 increased the sensitivity of acute lymphoblastic leukemia (ALL) cells and a pediatric tumor cell line (BT12) to chemotherapy and decreased colony formation, possibly via induction of apoptosis. UNC569 inhibition of Mer in ALL xenografts in vivo resulted in reduced tumor infiltration into the central nervous system [117] and regressed disease in a transgenic zebrafish model of T-ALL [118]. Similarly, inhibition of Mer with an UNC569 derivative, UNC1062 promoted apoptosis and inhibited colony formation in melanoma cells [60], BT12 cells, and nonsmall cell lung cancer cell lines A549 and Colo699 [119]. The Axl-specific inhibitor R428 was found to reduce metastatic burden and extend survival in mouse models of breast cancer metastasis [120]. Inhibition with the small molecule inhibitor BMS777607 (which inhibits Met, Ron, Axl, and Mer RTKs) was found to attenuate breast cancer cell migration, an effect that was more marked when Lyn and p130Cas were also targeted [121]. Inhibition of Axl was also found to reduce growth and proliferation of head and neck cancer tumor cells and led to resensitization of EGFR inhibitor (erlotinib) resistant tumor cells to therapy [100]. Synergistic effects of inhibition of Axl with antimitotic agents in killing tyrosine kinase inhibitor-resistant cancer cells that had undergone EMT have been reported [122] and also synergistic effects of Axl inhibition with cisplatin treatment in the suppression of liver micro-metastasis [120].

5.15 Decoy Receptors

Administration of an exogenous source of soluble Axl was found to inhibit progression of established metastatic ovarian cancer in vivo [123]. Development of "decoy" receptors that act as potent inhibitors of Axl or Mer-mediated signaling could reduce metastasis and disease progression in vivo [124]. NK cells with high expression of Axl/Tyro3 exhibit potent cytotoxic activity and NK cell activity in Axl−/− mice is markedly reduced [8]. A key role for Axl in CD56+ NK cell development from CD34+ hematopoietic stem cells has been demonstrated using Axl-Fc constructs, revealing a positive regulatory effect of the Axl/Gas6 pathway on FLT3 signaling [125].

5.16 Knockdown or Antibody Treatment

An alternative approach to inhibition of Tyro3/Axl/Mer receptor tyrosine kinase activity is to suppress protumor effects via reduction of cellular expression. Function blocking antibodies or shRNA knockdown of Mer was found to reduce glioblastoma migration in vitro [126] and delay progression of a mouse model of human leukemia (B-ALL) [127]. Mer knockdown also inhibited melanoma proliferation, migration, Akt signaling, colony formation and cell survival in vitro [61], and melanoma tumorigenesis in vivo [60]. Knockdown of Mer with shRNA was found to reduce colony formation in vitro in acute myeloma cell lines, significantly reduced the rate of myoblast apoptosis in response to serum starvation and delayed leukemia development in vivo [56], and enhanced apoptosis and chemosensitivity of NSCLC and astrocytoma [128, 129]. A monoclonal antibody to Mer (Mer590), which prevents Mer phosphorylation and signaling via receptor internalization, has been shown to impede glioblastoma migration in vitro [126] and increase chemosensitivity in nonsmall cell lung cancer [130]. A combination of Mer shRNA with Mer590 mAb may have additive effects on reduction of Mer expression levels, significantly increasing cell death of nonsmall cell lung cancer cell line Colo699 [130]. Similarly, antagonistic anti-Axl antibodies down-regulate Axl expression, reduce growth in a xenograft model by inhibiting tumor cell proliferation, and promoting apoptosis [116]. Axl knockdown with siRNA inhibits angiogenesis, with impaired endothelial tubule formation [131] and also VEGF signaling, thereby potentiating the effect of anti-VEGF in several different tumor models [132]. These effects would be consistent with vascular defects and increased vessel permeability seen in *Axl−/−* mice that suggest a critical role for Axl in maintenance of normal vascular architecture [133].

5.17 miRNAs and Axl/Mer

Recent data suggests that Axl and Mer expression is downregulated by specific microRNAs. In response to DNA damage or oncogenic stress, p53 activation results in transcriptional regulation of miR-34s. In a p53-regulated miRNA-deficient mouse tumor model, miR-34 target genes, including Axl, were overexpressed. Exogenous miR-34 reduced proliferation and invasion of epithelial tumor cells and prevented tumor formation and progression in mice [134]. miR-34a and miR-199a have been reported to decrease Axl expression [135, 136] and targeting of Axl via miR-34a was reported to suppress ovarian cancer, reducing both proliferation and motility [137]. Axl was also targeted by miR-199a, acting to negatively regulate progression of osteosarcoma [138]. Further studies will likely reveal a key role for miR-34 regulation of Axl in both inflammatory and tumor contexts. The levels of soluble Mer may be increased following knockdown of miR-126, which acts to suppress metastatic initiation and colonization by negatively regulating endothelial recruitment [139]. MiR-126 interacts with *MERTK* 3′ untranslated regions, reducing levels of soluble Mer, which enhances endothelial recruitment through binding and inhibition of Gas6.

5.18 Axl and Epithelial–Mesenchymal Transition

Recent studies have also revealed an intriguing role for Axl in the control of epithelial mesenchymal transition (EMT), which may enhance cell migration and survival associated with malignant progression. EMT regulates the generation of cancer stem cells that are capable of tumor initiation and self-renewal and contribute to resistance to treatment and development of metastases. Axl expression shows a strong association with mesenchymal phenotype, for example, in nonsmall cell lung cancer and triple-negative breast cancer [122]. Axl expression was found to be correlated with cell invasiveness and mesenchymal-like tumor formation in mammary epithelial cells. In nonsmall cell lung cancer cell lines, genes related to the epithelial-to-mesenchymal transition, including Axl were differentially methylated between epithelial and mesenchymal cells. In a pancreatic cancer cell line, stable knockdown of *Axl* resulted in reduced viability and anchorage-independent growth and attenuated migration/invasion which was associated with downregulation of EMT-associated transcription factors, slug, snail, and twist [140]. *Axl* knockdown resulted in reduced metastatic spread of breast cancer cells from the mammary gland to lymph nodes and increased overall survival [141]. Similarly, *Axl* knockdown was found to reduce migration and invasion of breast cancer cells to the lung following injection into the tail vein in vivo [142]. EMT could be reversed by Axl downregulation in both human mammary epithelial cells and murine breast CSCs, resulting in attenuation of self-renewal capacity, restored chemosensitivity, and reduced tumor formation in vivo [143].

5.19 Concluding Remarks

Mer and Axl have distinct functional roles within tissues that are dependent on the inflammatory state, with Mer generally being associated with homeostatic regulation. Axl and Mer exhibit differential patterns of tissue expression, ligand-binding

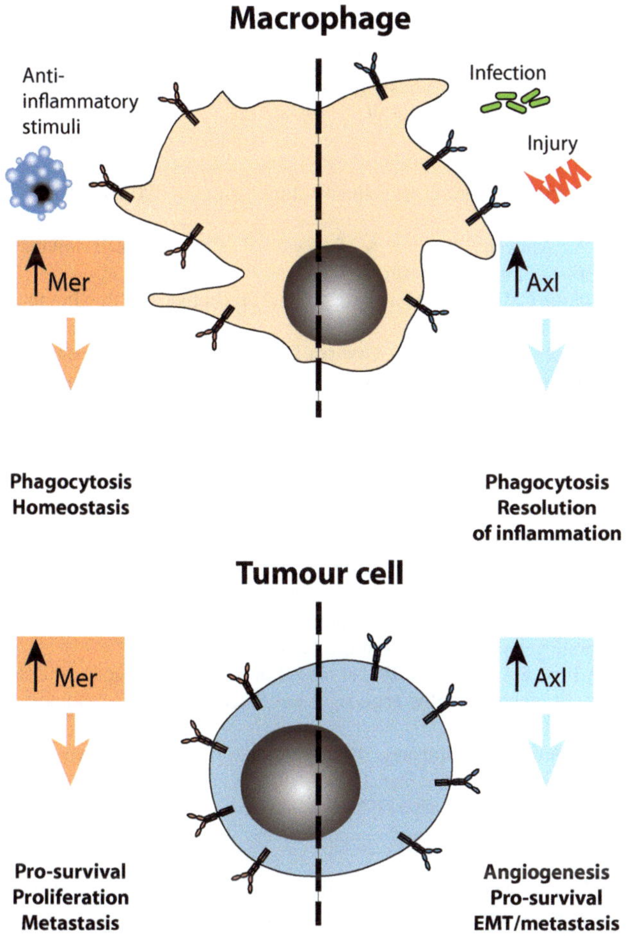

Fig. 5.1 Schematic of the divergent functional roles of Mer and Axl within the tumor microenvironment. Mer expressed on macrophages is generally associated with homeostatic regulation, whereas Axl is upregulated upon exposure of macrophages to hypoxia or proinflammatory stimuli and initiates the process of resolution of inflammation. Mer-expressing tumor cells demonstrate enhanced survival, proliferation, and metastasis, whereas Axl-expressing tumor cells demonstrate increased angiogenesis and EMT as well as enhanced survival and metastasis. These distinct functions of Mer and Axl in homeostasis and inflammation may be important for development of new therapeutics to target the receptor tyrosine kinases for the treatment of tumors

specificities, and signal transduction potential. As a consequence, these two receptors are likely to make distinct contributions to tumor development and progression (summarized in Fig. 5.1). Understanding the mechanisms underlying regulation of expression and function of Axl and Mer will be critical for maximizing the therapeutic efficacy of reagents that have been developed to target these molecules for the treatment of cancer.

References

1. Chovatiya R, Medzhitov R. Stress, inflammation, and defense of homeostasis. Mol Cell. 2014;54:281–8. doi:10.1016/j.molcel.2014.03.030.
2. Balkwill FR, Mantovani A. Cancer-related inflammation: common themes and therapeutic opportunities. Semin Cancer Biol. 2012;22:33–40. doi:10.1016/j.semcancer.2011.12.005.
3. Nathan C. Points of control in inflammation. Nature. 2002;420:846–52. doi:10.1038/nature01320.
4. Mantovani A, Allavena P, Sica A, Balkwill F. Cancer-related inflammation. Nature. 2008;454:436–44. doi:10.1038/nature07205.
5. Lemke G, Rothlin CV. Immunobiology of the TAM receptors. Nat Rev Immunol. 2008;8:327–36. doi:10.1038/nri2303.
6. Linger RM, Keating A, Earp HS, Graham DK. TAM receptor tyrosine kinase: biological functions, signaling, and potential therapeutics targeting in human cancer. Adv Cancer Res. 2008;100:35–83. doi:10.1016/S0065-230X(08)00002-X.TAM.
7. Bhattacharyya S, Zagórska A, Lew ED, et al. Enveloped viruses disable innate immune responses in dendritic cells by direct activation of TAM receptors. Cell Host Microbe. 2013;14:136–47. doi:10.1016/j.chom.2013.07.005.
8. Rothlin CV, Ghosh S, Zuniga EI, et al. TAM receptors are pleiotropic inhibitors of the innate immune response. Cell. 2007;131:1124–36. doi:10.1016/j.cell.2007.10.034.
9. Zagórska A, Través PG, Lew ED, et al. Diversification of TAM receptor tyrosine kinase function. Nat Immunol. 2014. doi:10.1038/ni.2986.
10. Lai C, Lemke G. An extended family of protein-tyrosine kinase genes differentially expressed in the vertebrate nervous system. Neuron. 1991;6:691–704.
11. O'Bryan JP, Frye RA, Cogswell PC, et al. axl, a transforming gene isolated from primary human myeloid leukemia cells, encodes a novel receptor tyrosine kinase. Mol Cell Biol. 1991;11:5016–31.
12. Graham DK, Dawson T, Mullaney D, et al. Cloning and mRNA expression analysis of a novel human protooncogene, c-mer. 1994;5:647–57.
13. Camenisch TD, Koller BH, Earp HS, Matsushima GK. A novel receptor tyrosine kinase, Mer, inhibits TNF-alpha production and lipopolysaccharide-induced endotoxic shock. J Immunol. 1999;162:3498–503.
14. Lu Q, Gore M, Zhang Q, et al. Tyro-3 family receptors are essential regulators of mammalian spermatogenesis. Nature. 1999;398:723–8. doi:10.1038/19554.
15. Lu Q, Lemke G. Homeostatic regulation of the immune system by receptor tyrosine kinases of the Tyro 3 family. Science. 2001;293:306–11. doi:10.1126/science.1061663.
16. Stitt TN, Conn G, Gore M, et al. The anticoagulation factor protein S and its relative, Gas6, are ligands for the Tyro 3/Axl family of receptor tyrosine kinases. Cell. 1995;80:661–70.
17. Di Scipio RG, Hermodson MA, Yates SG, Davie EW. A comparison of human prothrombin, factor IX (Christmas factor), factor X (Stuart factor), and protein S. Biochemistry (Mosc). 1977;16:698–706.
18. Manfioletti G, Brancolini C, Avanzi G, Schneider C. The protein encoded by a growth arrest-specific gene (gas6) is a new member of the vitamin K-dependent proteins related to protein S, a negative coregulator in the blood coagulation cascade. Mol Cell Biol. 1993;13:4976–85.
19. Lew ED, Oh J, Burrola PG, et al. Differential TAM receptor-ligand-phospholipid interactions delimit differential TAM bioactivities. eLife. 2014. doi:10.7554/eLife.03385.
20. Dransfield I, Zagórska A, Lew ED, et al. Mer receptor tyrosine kinase mediates both tethering and phagocytosis of apoptotic cells. Cell Death Dis. 2015;6, e1646. doi:10.1038/cddis.2015.18.
21. O'Bryan JP, Fridell YW, Koski R, et al. The transforming receptor tyrosine kinase, Axl, is post-translationally regulated by proteolytic cleavage. J Biol Chem. 1995;270:551–7.
22. Sather S, Kenyon KD, Lefkowitz JB, et al. A soluble form of the Mer receptor tyrosine kinase inhibits macrophage clearance of apoptotic cells and platelet aggregation. Blood. 2007;109:1026–33. doi:10.1182/blood-2006-05-021634.

23. Lee IJ, Hilliard BA, Ulas M, et al. Monocyte and plasma expression of TAM ligand and receptor in renal failure: Links to unregulated immunity and chronic inflammation. Clin Immunol. 2015;158:231–41. doi:10.1016/j.clim.2015.01.012.
24. Weinger JG, Omari KM, Marsden K, et al. Up-regulation of soluble Axl and Mer receptor tyrosine kinases negatively correlates with Gas6 in established multiple sclerosis lesions. Am J Pathol. 2009;175:283–93. doi:10.2353/ajpath.2009.080807.
25. Ko C-P, Yu Y-L, Hsiao P-C, et al. Plasma levels of soluble Axl correlate with severity of community-acquired pneumonia. Mol Med Rep. 2014;9:1400–4. doi:10.3892/mmr.2014.1933.
26. Ekman C, Gottsäter A, Lindblad B, Dahlbäck B. Plasma concentrations of Gas6 and soluble Axl correlate with disease and predict mortality in patients with critical limb ischemia. Clin Biochem. 2010;43:873–6. doi:10.1016/j.clinbiochem.2010.04.006.
27. Zhu H, Sun X, Zhu L, et al. Different expression patterns and clinical significance of mAxl and sAxl in systemic lupus erythematosus. Lupus. 2014. doi:10.1177/0961203314520839.
28. Hsiao F-C, Lin Y-F, Hsieh P-S, et al. Circulating growth arrest-specific 6 protein is associated with adiposity, systemic inflammation, and insulin resistance among overweight and obese adolescents. J Clin Endocrinol Metab. 2013;98:E267–74. doi:10.1210/jc.2012-3179.
29. Liu X, Gong Y, Jia J, et al. Plasma concentrations of sAxl are associated with severe pre-eclampsia. Clin Biochem. 2014;47:173–6. doi:10.1016/j.clinbiochem.2013.11.001.
30. Liu Y-W, Yang Q-F, Zuo P-Y, et al. Elevated serum levels of soluble Axl in acute coronary syndrome. Am J Med Sci. 2015;349:124–9. doi:10.1097/MAJ.0000000000000362.
31. Ekman C, Site DF, Gottsäter A, et al. Plasma concentrations of growth arrest specific protein 6 and the soluble form of its tyrosine kinase receptor Axl as markers of large abdominal aortic aneurysms. Clin Biochem. 2010;43:110–4. doi:10.1016/j.clinbiochem.2009.07.025.
32. Xu J, Ma F, Yan W, et al. Identification of the soluble form of tyrosine kinase receptor Axl as a potential biomarker for intracranial aneurysm rupture. BMC Neurol. 2015;15:23. doi:10.1186/s12883-015-0282-8.
33. Qin B, Wang J, Ma N, et al. The association of Tyro3/Axl/Mer signaling with inflammatory response, disease activity in patients with primary Sjögren's syndrome. Joint Bone Spine. 2015. doi:10.1016/j.jbspin.2015.01.008.
34. Ballantine L, Midgley A, Harris D, et al. Increased soluble phagocytic receptors sMer, sTyro3 and sAxl and reduced phagocytosis in juvenile-onset systemic lupus erythematosus. Pediatr Rheumatol Online J. 2015;13:10. doi:10.1186/s12969-015-0007-y.
35. Costa M, Bellosta P, Basilico C. Cleavage and release of a soluble form of the receptor tyrosine kinase ARK in vitro and in vivo. J Cell Physiol. 1996;168:737–44. doi:10.1002/(SICI)1097-4652(199609)168:3<737::AID-JCP27>3.0.CO;2-U.
36. Gustafsson A, Martuszewska D, Johansson M, et al. Differential expression of Axl and Gas6 in renal cell carcinoma reflecting tumor advancement and survival. Clin Cancer Res. 2009;15:4742–9. doi:10.1158/1078-0432.CCR-08-2514.
37. Reichl P, Fang M, Starlinger P, et al. Multicenter analysis of soluble Axl reveals diagnostic value for very early stage hepatocellular carcinoma. Int J Cancer. 2015;137:385–94. doi:10.1002/ijc.29394.
38. Johansson G, Peng P-C, Huang P-Y, et al. Soluble AXL: a possible circulating biomarker for neurofibromatosis type 1 related tumor burden. PLoS One. 2014;9, e115916. doi:10.1371/journal.pone.0115916.
39. Fadok VA, Bratton DL, Konowal A, et al. Macrophages that have ingested apoptotic cells in vitro inhibit proinflammatory cytokine production through autocrine/paracrine mechanisms involving TGF-beta, PGE2, and PAF. J Clin Invest. 1998;101:890–8. doi:10.1172/JCI1112.
40. Voll RE, Herrmann M, Roth EA, et al. Immunosuppressive effects of apoptotic cells. Nature. 1997;390:350–1. doi:10.1038/37022.
41. Hochreiter-Hufford A, Ravichandran KS. Clearing the dead: apoptotic cell sensing, recognition, engulfment, and digestion. Cold Spring Harb Perspect Biol. 2013;5:a008748. doi:10.1101/cshperspect.a008748.
42. Savill J, Dransfield I, Gregory C, Haslett C. A blast from the past: clearance of apoptotic cells regulates immune responses. Nat Rev Immunol. 2002;2:965–75. doi:10.1038/nri957.

43. Gautier EL, Shay T, Miller J, et al. Gene-expression profiles and transcriptional regulatory pathways that underlie the identity and diversity of mouse tissue macrophages. Nat Immunol. 2012;13:1118–28. doi:10.1038/ni.2419.
44. Gal A, Li Y, Thompson DA, et al. Mutations in MERTK, the human orthologue of the RCS rat retinal dystrophy gene, cause retinitis pigmentosa. Nat Genet. 2000;26:270–1. doi:10.1038/81555.
45. Rahman ZSM, Shao W-H, Khan TN, et al. Impaired apoptotic cell clearance in the germinal center by Mer-deficient tingible body macrophages leads to enhanced antibody-forming cell and germinal center responses. J Immunol. 2010;185:5859–68. doi:10.4049/jimmunol.1001187.
46. Casanova-Acebes M, Pitaval C, Weiss LA, et al. Rhythmic modulation of the hematopoietic niche through neutrophil clearance. Cell. 2013;153:1025–35. doi:10.1016/j.cell.2013.04.040.
47. Irschick EU, Haas G, Geiger M, et al. Phagocytosis of human retinal pigment epithelial cells: evidence of a diurnal rhythm, involvement of the cytoskeleton and interference of antiviral drugs. Ophthalmic Res. 2006;38:164–74. doi:10.1159/000091476.
48. Bauer T, Zagórska A, Jurkin J, et al. Identification of Axl as a downstream effector of TGF-β1 during Langerhans cell differentiation and epidermal homeostasis. J Exp Med. 2012;209:2033–47. doi:10.1084/jem.20120493.
49. Fujimori T, Grabiec AM, Kaur M, et al. The Axl receptor tyrosine kinase is a discriminator of macrophage function in the inflamed lung. Mucosal Immunol. 2015;8:1021–30. doi:10.1038/mi.2014.129.
50. Graham DK, Deryckere D, Davies KD, Earp HS. The TAM family: phosphatidylserine sensing receptor tyrosine kinases gone awry in cancer. Nat Publ Group. 2014;14:769–85. doi: 10.1038/nrc3847.
51. Rochlitz C, Lohri A, Bacchi M, et al. Axl expression is associated with adverse prognosis and with expression of Bcl-2 and CD34 in de novo acute myeloid leukemia (AML): results from a multicenter trial of the Swiss Group for Clinical Cancer Research (SAKK). Leukemia. 1999;13:1352–8.
52. Ghosh AK, Secreto C, Boysen J, et al. The novel receptor tyrosine kinase Axl is constitutively active in B-cell chronic lymphocytic leukemia and acts as a docking site of nonreceptor kinases: implications for therapy. Blood. 2011;117:1928–37. doi:10.1182/blood-2010-09-305649.
53. Neubauer A, Fiebeler A, Graham DK, et al. Expression of axl, a transforming receptor tyrosine kinase, in normal and malignant hematopoiesis. Blood. 1994;84:1931–41.
54. Graham DK, Salzberg DB, Kurtzberg J, et al. Ectopic expression of the proto-oncogene Mer in pediatric T-cell acute lymphoblastic leukemia. Clin Cancer Res. 2006;12:2662–9. doi:10.1158/1078-0432.CCR-05-2208.
55. Brandao LN, Winges A, Christoph S, et al. Inhibition of MerTK increases chemosensitivity and decreases oncogenic potential in T-cell acute lymphoblastic leukemia. Blood Cancer J. 2013;3, e101. doi:10.1038/bcj.2012.46.
56. Lee-Sherick AB, Eisenman KM, Sather S, et al. Aberrant Mer receptor tyrosine kinase expression contributes to leukemogenesis in acute myeloid leukemia. Oncogene. 2013;32:5359–68. doi:10.1038/onc.2013.40.
57. Ek S, Högerkorp C-M, Dictor M, et al. Mantle cell lymphomas express a distinct genetic signature affecting lymphocyte trafficking and growth regulation as compared with subpopulations of normal human B cells. Cancer Res. 2002;62:4398–405.
58. Evans CO, Young AN, Brown MR, et al. Novel patterns of gene expression in pituitary adenomas identified by complementary deoxyribonucleic acid microarrays and quantitative reverse transcription-polymerase chain reaction. J Clin Endocrinol Metab. 2001;86:3097–107. doi:10.1210/jcem.86.7.7616.
59. Wu Y-M, Robinson DR, Kung H-J. Signal pathways in up-regulation of chemokines by tyrosine kinase MER/NYK in prostate cancer cells. Cancer Res. 2004;64:7311–20. doi:10.1158/0008-5472.CAN-04-0972.
60. Schlegel J, Sambade MJ, Sather S, et al. MERTK receptor tyrosine kinase is a therapeutic target in melanoma. J Clin Invest. 2013;123:2257–67. doi:10.1172/JCI67816.
61. Tworkoski KA, Platt JT, Bacchiocchi A, et al. MERTK controls melanoma cell migration and survival and differentially regulates cell behavior relative to AXL. Pigment Cell Melanoma Res. 2013;26:527–41. doi:10.1111/pcmr.12110.

62. Reinartz S, Schumann T, Finkernagel F, et al. Mixed-polarization phenotype of ascites-associated macrophages in human ovarian carcinoma: correlation of CD163 expression, cytokine levels and early relapse. Int J Cancer. 2014;134:32–42. doi:10.1002/ijc.28335.
63. Ryder M, Ghossein RA, Ricarte-Filho JCM, et al. Increased density of tumor-associated macrophages is associated with decreased survival in advanced thyroid cancer. Endocr Relat Cancer. 2008;15:1069–74. doi:10.1677/ERC-08-0036.
64. Zhu X-D, Zhang J-B, Zhuang P-Y, et al. High expression of macrophage colony-stimulating factor in peritumoral liver tissue is associated with poor survival after curative resection of hepatocellular carcinoma. J Clin Oncol. 2008;26:2707–16. doi:10.1200/JCO.2007.15.6521.
65. Gazzaniga S, Bravo AI, Guglielmotti A, et al. Targeting tumor-associated macrophages and inhibition of MCP-1 reduce angiogenesis and tumor growth in a human melanoma xenograft. J Invest Dermatol. 2007;127:2031–41. doi:10.1038/sj.jid.5700827.
66. Kimura YN, Watari K, Fotovati A, et al. Inflammatory stimuli from macrophages and cancer cells synergistically promote tumor growth and angiogenesis. Cancer Sci. 2007;98:2009–18. doi:10.1111/j.1349-7006.2007.00633.x.
67. Zeisberger SM, Odermatt B, Marty C, et al. Clodronate-liposome-mediated depletion of tumor-associated macrophages: a new and highly effective antiangiogenic therapy approach. Br J Cancer. 2006;95:272–81. doi:10.1038/sj.bjc.6603240.
68. Qian B-Z, Pollard JW. Macrophage diversity enhances tumor progression and metastasis. Cell. 2010;141:39–51. doi:10.1016/j.cell.2010.03.014.
69. Heuff G, Oldenburg HS, Boutkan H, et al. Enhanced tumor growth in the rat liver after selective elimination of Kupffer cells. Cancer Immunol Immunother. 1993;37:125–30.
70. Franklin RA, Liao W, Sarkar A, et al. The cellular and molecular origin of tumor-associated macrophages. Science. 2014;344:921–5. doi:10.1126/science.1252510.
71. Shand FHW, Ueha S, Otsuji M, et al. Tracking of intertissue migration reveals the origins of tumor-infiltrating monocytes. Proc Natl Acad Sci U S A. 2014;111:7771–6. doi:10.1073/pnas.1402914111.
72. Bonavita E, Gentile S, Rubino M, et al. PTX3 is an extrinsic oncosuppressor regulating complement-dependent inflammation in cancer. Cell. 2015;160:700–14. doi:10.1016/j.cell.2015.01.004.
73. Tripathi C, Tewari BN, Kanchan RK, et al. Macrophages are recruited to hypoxic tumor areas and acquire a pro-angiogenic M2-polarized phenotype via hypoxic cancer cell derived cytokines Oncostatin M and Eotaxin. Oncotarget. 2014;5:5350–68.
74. Bottazzi B, Erba E, Nobili N, et al. A paracrine circuit in the regulation of the proliferation of macrophages infiltrating murine sarcomas. J Immunol. 1990;144:2409–12.
75. Mantovani A, Biswas SK, Galdiero MR, et al. Macrophage plasticity and polarization in tissue repair and remodelling. J Pathol. 2013;229:176–85. doi:10.1002/path.4133.
76. Mantovani A, Germano G, Marchesi F, et al. Cancer-promoting tumor-associated macrophages: new vistas and open questions. Eur J Immunol. 2011;41:2522–5. doi:10.1002/eji.201141894.
77. Movahedi K, Laoui D, Gysemans C, et al. Different tumor microenvironments contain functionally distinct subsets of macrophages derived from Ly6C(high) monocytes. Cancer Res. 2010;70:5728–39. doi:10.1158/0008-5472.CAN-09-4672.
78. Zizzo G, Hilliard BA, Monestier M, Cohen PL. Efficient clearance of early apoptotic cells by human macrophages requires M2c polarization and MerTK induction. J Immunol. 2012;189:3508–20. doi:10.4049/jimmunol.1200662.
79. Chiang C-S, Fu SY, Wang S-C, et al. Irradiation promotes an m2 macrophage phenotype in tumor hypoxia. Front Oncol. 2012;2:89. doi:10.3389/fonc.2012.00089.
80. Leek RD, Talks KL, Pezzella F, et al. Relation of hypoxia-inducible factor-2 alpha (HIF-2 alpha) expression in tumor-infiltrative macrophages to tumor angiogenesis and the oxidative thymidine phosphorylase pathway in Human breast cancer. Cancer Res. 2002;62:1326–9.
81. Negus RP, Turner L, Burke F, Balkwill FR. Hypoxia down-regulates MCP-1 expression: implications for macrophage distribution in tumors. J Leukoc Biol. 1998;63:758–65.
82. Chittezhath M, Dhillon MK, Lim JY, et al. Molecular profiling reveals a tumor-promoting phenotype of monocytes and macrophages in human cancer progression. Immunity. 2014;41:815–29. doi:10.1016/j.immuni.2014.09.014.

83. Laoui D, Van Overmeire E, Di Conza G, et al. Tumor hypoxia does not drive differentiation of tumor-associated macrophages but rather fine-tunes the M2-like macrophage population. Cancer Res. 2014;74:24–30. doi:10.1158/0008-5472.CAN-13-1196.
84. Rankin EB, Fuh KC, Castellini L, et al. Direct regulation of GAS6/AXL signaling by HIF promotes renal metastasis through SRC and MET. Proc Natl Acad Sci U S A. 2014;111:13373–8. doi:10.1073/pnas.1404848111.
85. Mishra A, Wang J, Shiozawa Y, et al. Hypoxia stabilizes GAS6/Axl signaling in metastatic prostate cancer. Mol Cancer Res. 2012;10:703–12. doi:10.1158/1541-7786.MCR-11-0569.
86. Pierce A, Bliesner B, Xu M, et al. Axl and Tyro3 modulate female reproduction by influencing gonadotropin-releasing hormone neuron survival and migration. Mol Endocrinol. 2008;22:2481–95. doi:10.1210/me.2008-0169.
87. Tai K-Y, Shieh Y-S, Lee C-S, et al. Axl promotes cell invasion by inducing MMP-9 activity through activation of NF-kappaB and Brg-1. Oncogene. 2008;27:4044–55. doi:10.1038/onc.2008.57.
88. Huang J-S, Cho C-Y, Hong C-C, et al. Oxidative stress enhances Axl-mediated cell migration through an Akt1/Rac1-dependent mechanism. Free Radic Biol Med. 2013;65:1246–56. doi:10.1016/j.freeradbiomed.2013.09.011.
89. McIntyre A, Harris AL. Metabolic and hypoxic adaptation to anti-angiogenic therapy: a target for induced essentiality. EMBO Mol Med. 2015;7:368–79. doi:10.15252/emmm.201404271.
90. A-Gonzalez N, Bensinger SJ, Hong C, et al. Apoptotic cells promote their own clearance and immune tolerance through activation of the nuclear receptor LXR. Immunity. 2009;31:245–58. doi:10.1016/j.immuni.2009.06.018.
91. Hodrea J, Majai G, Doró Z, et al. The glucocorticoid dexamethasone programs human dendritic cells for enhanced phagocytosis of apoptotic neutrophils and inflammatory response. J Leukoc Biol. 2012;91:127–36. doi:10.1189/jlb.0511243.
92. McColl A, Michlewska S, Dransfield I, Rossi AG. Effects of glucocorticoids on apoptosis and clearance of apoptotic cells. ScientificWorldJournal. 2007;7:1165–81. doi:10.1100/tsw.2007.224.
93. Stanford JC, Young C, Hicks D, et al. Efferocytosis produces a prometastatic landscape during postpartum mammary gland involution. J Clin Invest. 2014;124:4737–52. doi:10.1172/JCI76375.
94. Ford CA, Petrova S, Pound JD, et al. Oncogenic properties of apoptotic tumor cells in aggressive B cell lymphoma. Curr Biol. 2015;25:577–88. doi:10.1016/j.cub.2014.12.059.
95. Sinha S, Boysen J, Nelson M, et al. Targeted Axl inhibition primes chronic lymphocytic leukemia B cells to apoptosis and shows synergistic/additive effects in combination with BTK inhibitors. Clin Cancer Res. 2015;21:2115–26. doi:10.1158/1078-0432.CCR-14-1892.
96. Hulsmans M, Holvoet P. MicroRNA-containing microvesicles regulating inflammation in association with atherosclerotic disease. Cardiovasc Res. 2013;100:7–18. doi:10.1093/cvr/cvt161.
97. Hong C-C, Lay J-D, Huang J-S, et al. Receptor tyrosine kinase AXL is induced by chemotherapy drugs and overexpression of AXL confers drug resistance in acute myeloid leukemia. Cancer Lett. 2008;268:314–24. doi:10.1016/j.canlet.2008.04.017.
98. Ammoun S, Provenzano L, Zhou L, et al. Axl/Gas6/NFkB signaling in Schwannoma pathological proliferation, adhesion and survival. Oncogene. 2014;33:336–46. doi:10.1038/onc.2012.587.
99. Zhang Y-X, Knyazev PG, Cheburkin YV, et al. AXL is a potential target for therapeutic intervention in breast cancer progression. Cancer Res. 2008;68:1905–15. doi:10.1158/0008-5472.CAN-07-2661.
100. Giles KM, Kalinowski FC, Candy PA, et al. Axl mediates acquired resistance of head and neck cancer cells to the epidermal growth factor receptor inhibitor erlotinib. Mol Cancer Ther. 2013;12:2541–58. doi:10.1158/1535-7163.MCT-13-0170.
101. Fleuren EDG, Hillebrandt-Roeffen MHS, Flucke UE, et al. The role of AXL and the in vitro activity of the receptor tyrosine kinase inhibitor BGB324 in Ewing sarcoma. Oncotarget. 2014;5:12753–68.
102. Mahadevan D, Cooke L, Riley C, et al. A novel tyrosine kinase switch is a mechanism of imatinib resistance in gastrointestinal stromal tumors. Oncogene. 2007;26:3909–19. doi:10.1038/sj.onc.1210173.

103. Brand TM, Iida M, Stein AP, et al. AXL mediates resistance to cetuximab therapy. Cancer Res. 2014;74:5152–64. doi:10.1158/0008-5472.CAN-14-0294.
104. Zhang Z, Lee JC, Lin L, et al. Activation of the AXL kinase causes resistance to EGFR-targeted therapy in lung cancer. Nat Genet. 2012;44:852–60. doi:10.1038/ng.2330.
105. Liu L, Greger J, Shi H, et al. Novel mechanism of lapatinib resistance in HER2-positive breast tumor cells: activation of AXL. Cancer Res. 2009;69:6871–8. doi:10.1158/0008-5472.CAN-08-4490.
106. Xu J, Jia L, Ma H, et al. Axl gene knockdown inhibits the metastasis properties of hepatocellular carcinoma via PI3K/Akt-PAK1 signal pathway. Tumor Biol. 2014;35:3809–17. doi:10.1007/s13277-013-1521-5.
107. Elkabets M, Pazarentzos E, Juric D, et al. AXL mediates resistance to PI3Kα inhibition by activating the EGFR/PKC/mTOR axis in head and neck and esophageal squamous cell carcinomas. Cancer Cell. 2015;27:533–46. doi:10.1016/j.ccell.2015.03.010.
108. Martinho O, Zucca LE, Reis RM. AXL as a modulator of sunitinib response in glioblastoma cell lines. Exp Cell Res. 2015;332:1–10. doi:10.1016/j.yexcr.2015.01.009.
109. Abu-Thuraia A, Gauthier R, Chidiac R, et al. Axl phosphorylates Elmo scaffold proteins to promote Rac activation and cell invasion. Mol Cell Biol. 2015;35:76–87. doi:10.1128/MCB.00764-14.
110. Seitz HM, Camenisch TD, Lemke G, et al. Macrophages and dendritic cells use different Axl/Mertk/Tyro3 receptors in clearance of apoptotic cells. J Immunol. 2007;178:5635–42.
111. Salian-Mehta S, Xu M, Wierman ME. AXL and MET crosstalk to promote gonadotropin releasing hormone (GnRH) neuronal cell migration and survival. Mol Cell Endocrinol. 2013;374:92–100. doi:10.1016/j.mce.2013.04.018.
112. Paolino M, Choidas A, Wallner S, et al. The E3 ligase Cbl-b and TAM receptors regulate cancer metastasis via natural killer cells. Nature. 2014;507:508–12. doi:10.1038/nature12998.
113. Valverde P. Effects of Gas6 and hydrogen peroxide in Axl ubiquitination and downregulation. Biochem Biophys Res Commun. 2005;333:180–5. doi:10.1016/j.bbrc.2005.05.086.
114. Georgescu MM, Kirsch KH, Shishido T, et al. Biological effects of c-Mer receptor tyrosine kinase in hematopoietic cells depend on the Grb2 binding site in the receptor and activation of NF-kappaB. Mol Cell Biol. 1999;19:1171–81.
115. Lee Y-J, Han J-Y, Byun J, et al. Inhibiting Mer receptor tyrosine kinase suppresses STAT1, SOCS1/3, and NF-kB activation and enhances inflammatory responses in lipopolysaccharide-induced acute lung injury. J Leukoc Biol. 2012;91:921–32. doi:10.1189/jlb.0611289.
116. Ye X, Li Y, Stawicki S, et al. An anti-Axl monoclonal antibody attenuates xenograft tumor growth and enhances the effect of multiple anticancer therapies. Oncogene. 2010;29:5254–64. doi:10.1038/onc.2010.268.
117. Krause S, Pfeiffer C, Strube S, et al. Mer tyrosine kinase promotes the survival of t(1;19)-positive acute lymphoblastic leukemia (ALL) in the central nervous system (CNS). Blood. 2015;125:820–30. doi:10.1182/blood-2014-06-583062.
118. Christoph S, Deryckere D, Schlegel J, et al. UNC569, a novel small-molecule mer inhibitor with efficacy against acute lymphoblastic leukemia in vitro and in vivo. Mol Cancer Ther. 2013;12:2367–77. doi:10.1158/1535-7163.MCT-13-0040.
119. Liu J, Zhang W, Stashko MA, et al. UNC1062, a new and potent Mer inhibitor. Eur J Med Chem. 2013;65:83–93. doi:10.1016/j.ejmech.2013.03.035.
120. Holland SJ, Pan A, Franci C, et al. R428, a selective small molecule inhibitor of Axl kinase, blocks tumor spread and prolongs survival in models of metastatic breast cancer. Cancer Res. 2010;70:1544–54. doi:10.1158/0008-5472.CAN-09-2997.
121. Pénzes K, Baumann C, Szabadkai I, et al. Combined inhibition of AXL, Lyn and p130Cas kinases block migration of triple negative breast cancer cells. Cancer Biol Ther. 2014;15:1571–82. doi:10.4161/15384047.2014.956634.
122. Wilson C, Ye X, Pham T, et al. AXL inhibition sensitizes mesenchymal cancer cells to antimitotic drugs. Cancer Res. 2014;74:5878–90. doi:10.1158/0008-5472.CAN-14-1009.
123. Rankin EB, Fuh KC, Taylor TE, et al. AXL is an essential factor and therapeutic target for metastatic ovarian cancer. Cancer Res. 2010;70:7570–9. doi:10.1158/0008-5472.CAN-10-1267.

124. Kariolis MS, Miao YR, Jones DS, et al. An engineered Axl "decoy receptor" effectively silences the Gas6-Axl signaling axis. Nat Chem Biol. 2014;10:977–83. doi:10.1038/nchembio.1636.
125. Park I-K, Trotta R, Yu J, Caligiuri MA. Axl/Gas6 pathway positively regulates FLT3 activation in human natural killer cell development. Eur J Immunol. 2013;43:2750–5. doi:10.1002/eji.201243116.
126. Rogers AEJ, Le JP, Sather S, et al. Mer receptor tyrosine kinase inhibition impedes glioblastoma multiforme migration and alters cellular morphology. Oncogene. 2012;31:4171–81. doi:10.1038/onc.2011.588.
127. Linger RMA, Lee-Sherick AB, DeRyckere D, et al. Mer receptor tyrosine kinase is a therapeutic target in pre-B-cell acute lymphoblastic leukemia. Blood. 2013;122:1599–609. doi:10.1182/blood-2013-01-478156.
128. Keating AK, Kim GK, Jones AE, et al. Inhibition of Mer and Axl receptor tyrosine kinases in astrocytoma cells leads to increased apoptosis and improved chemosensitivity. Mol Cancer Ther. 2010;9:1298–307. doi:10.1158/1535-7163.MCT-09-0707.
129. Linger RMA, Cohen RA, Cummings CT, et al. Mer or Axl receptor tyrosine kinase inhibition promotes apoptosis, blocks growth and enhances chemosensitivity of human non-small cell lung cancer. Oncogene. 2012. doi:10.1038/onc.2012.355.
130. Cummings CT, Linger RMA, Cohen RA, et al. Mer590, a novel monoclonal antibody targeting MER receptor tyrosine kinase, decreases colony formation and increases chemosensitivity in non-small cell lung cancer. Oncotarget. 2014;5:10434–45.
131. Li Y, Ye X, Tan C, et al. Axl as a potential therapeutic target in cancer: role of Axl in tumor growth, metastasis and angiogenesis. Oncogene. 2009;28:3442–55. doi:10.1038/onc.2009.212.
132. Ruan G-X, Kazlauskas A. Axl is essential for VEGF-A-dependent activation of PI3K/Akt. EMBO J. 2012;31:1692–703. doi:10.1038/emboj.2012.21.
133. Burstyn-Cohen T, Heeb MJ, Lemke G. Lack of protein S in mice causes embryonic lethal coagulopathy and vascular dysgenesis. J Clin Invest. 2009;119:2942–53. doi:10.1172/JCI39325.
134. Kasinski AL, Slack FJ. miRNA-34 prevents cancer initiation and progression in a therapeutically resistant K-ras and p53-induced mouse model of lung adenocarcinoma. Cancer Res. 2012;72:5576–87. doi:10.1158/0008-5472.CAN-12-2001.
135. Mackiewicz M, Huppi K, Pitt JJ, et al. Identification of the receptor tyrosine kinase AXL in breast cancer as a target for the human miR-34a microRNA. Breast Cancer Res Treat. 2011;130:663–79. doi:10.1007/s10549-011-1690-0.
136. Mudduluru G, Ceppi P, Kumarswamy R, et al. Regulation of Axl receptor tyrosine kinase expression by miR-34a and miR-199a/b in solid cancer. Oncogene. 2011;30:2888–99. doi:10.1038/onc.2011.13.
137. Li R, Shi X, Ling F, et al. MiR-34a suppresses ovarian cancer proliferation and motility by targeting AXL. Tumor Biol. 2015;36:7277–83. doi:10.1007/s13277-015-3445-8.
138. Tian R, Xie X, Han J, et al. miR-199a-3p negatively regulates the progression of osteosarcoma through targeting AXL. Am J Cancer Res. 2014;4:738–50.
139. Png KJ, Halberg N, Yoshida M, Tavazoie SF. A microRNA regulon that mediates endothelial recruitment and metastasis by cancer cells. Nature. 2012;481:190–4. doi:10.1038/nature10661.
140. Koorstra J-BM, Karikari CA, Feldmann G, et al. The Axl receptor tyrosine kinase confers an adverse prognostic influence in pancreatic cancer and represents a new therapeutic target. Cancer Biol Ther. 2009;8:618–26.
141. Gjerdrum C, Tiron C, Høiby T, et al. Axl is an essential epithelial-to-mesenchymal transition-induced regulator of breast cancer metastasis and patient survival. Proc Natl Acad Sci U S A. 2010;107:1124–9. doi:10.1073/pnas.0909333107.
142. Vuoriluoto K, Haugen H, Kiviluoto S, et al. Vimentin regulates EMT induction by Slug and oncogenic H-Ras and migration by governing Axl expression in breast cancer. Oncogene. 2011;30:1436–48. doi:10.1038/onc.2010.509.
143. Asiedu MK, Beauchamp-Perez FD, Ingle JN, et al. AXL induces epithelial-to-mesenchymal transition and regulates the function of breast cancer stem cells. Oncogene. 2014;33:1316–24. doi:10.1038/onc.2013.57.

Chapter 6
Immunogenic Apoptotic Cell Death and Anticancer Immunity

Peter Vandenabeele, Katrien Vandecasteele, Claus Bachert, Olga Krysko, and Dmitri V. Krysko

Abstract For many years it has been thought that apoptotic cells rapidly cleared by phagocytic cells do not trigger an immune response but rather have anti-inflammatory properties. However, accumulating experimental data indicate that certain anticancer therapies can induce an immunogenic form of apoptosis associated with the emission of damage-associated molecular patterns (DAMPs), which function as adjuvants to activate host antitumor immune responses. In this review, we will first discuss recent advances and the significance of danger signaling pathways involved in the emission of DAMPs, including calreticulin, ATP, and HMGB1. We will also emphasize that switching on a particular signaling pathway depends on the immunogenic cell death stimulus. Further, we address the role of ER stress in danger signaling and the classification of immunogenic cell death inducers in relation to how ER stress is triggered. In the final part, we discuss the role of radiotherapy-induced immunogenic apoptosis and the relationship of its immunogenicity to the fraction dose and concomitant chemotherapy.

P. Vandenabeele
Molecular Signalling and Cell Death Unit, Inflammation Research Center, VIB, Ghent, Belgium

Department of Biomedical Molecular Biology, Ghent University, Ghent, Belgium

Methusalem program, Ghent University, Ghent, Belgium

K. Vandecasteele
Department of Radiation Oncology and Experimental Cancer Research, Ghent University, Ghent, Belgium

C. Bachert • O. Krysko
The Upper Airway Research Laboratory, Department of Oto-Rhino-Laryngology, Ghent University Hospital, Ghent, Belgium

D.V. Krysko (✉)
Molecular Signalling and Cell Death Unit, Inflammation Research Center, VIB, Ghent, Belgium

Department of Biomedical Molecular Biology, Ghent University, Ghent, Belgium
e-mail: dmitri.krysko@ugent.be

C.D. Gregory (ed.), *Apoptosis in Cancer Pathogenesis and Anti-cancer Therapy*, Advances in Experimental Medicine and Biology 930,
DOI 10.1007/978-3-319-39406-0_6

Keywords ATP • HMGB1 • Calreticulin • DAMPs • Cancer • Radiotherapy • ER stress • Tumors • Chemotherapeutics

6.1 Introduction

In the human body close to 500 billion cells die each day by apoptosis, and they are continuously recognized and removed by the phagocytic system without causing inflammation or scars. The process of clearing—dying cells play a critical role in development, maintenance of tissue homeostasis, control of immune responses, and resolution of inflammation. Immunological responses elicited by apoptotic cells have been studied extensively in the last two decades. Back in the nineties it was shown that uptake of apoptotic neutrophils or eosinophils by human monocyte-derived macrophages does not induce secretion of granulocyte macrophage colony stimulating factor (GM-CSF) or thromboxane B2 [1, 2]. In later studies it was shown that apoptotic cells actually inhibit the production of many proinflammatory cytokines by antigen-presenting cells (Fig. 6.1) [1, 3–10]. Cells undergoing apoptosis are known

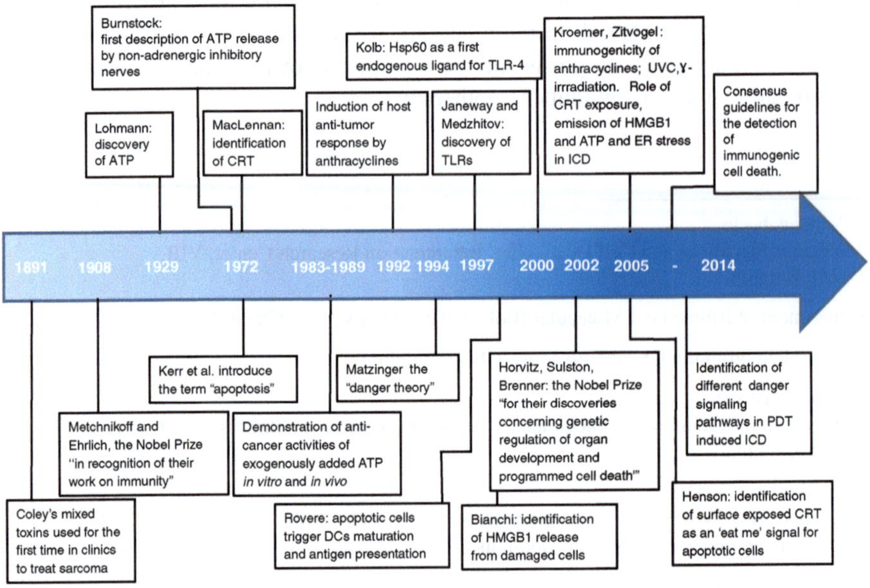

Fig. 6.1 Timeline of the key milestones in the development of the immunogenic cell death concept. Of note that immunotherapy in the treatment of cancer was first successfully used in 1891 by William B. Coley, who injected streptococcal products into patients with inoperable cancer. These products became known as Coley's Toxins. The following references are used to make this figure: [14–16, 20, 25, 32, 33, 42, 43, 46, 92–104]

to modulate their tissue microenvironments either by acting on phagocytes and thereby inhibiting immunological and inflammatory responses and promoting "healing" signaling pathways and/or by releasing immunomodulatory signals. Indeed, in the context of anticancer therapy it is generally accepted that most chemotherapeutic drugs elicit apoptotic cell death. Phagocytosis of apoptotic cells maintains an anti-inflammatory state in the extracellular environment and thereby contributes to an immunosuppressive network in a primary tumor site to promote further tumorigenesis [11]. Several studies have confirmed this notion. It has been shown that apoptotic tumor cells promote coordinated tumor growth, angiogenesis, and accumulation of tumor-associated macrophages (TAMs) in aggressive B cell lymphomas [12]. It has also been demonstrated that radiotherapy induces caspase-3-dependent release from apoptotic cells of arachidonic acid and prostaglandin E_2, which then promote the growth of the tumor cells that survive radiation activation [13]. This correlates with observations in cancer patients that tumors with elevated levels of activated caspase-3 are associated with a poor disease outcome [13]. All these studies indeed demonstrate that cancer cells undergoing apoptosis can promote tumor progression. However, in the late nineties it was reported that dendritic cells (DCs) internalize apoptotic cells and process them for presentation to both MHC class I- and class II-restricted T cells with an efficiency that is dependent on the number of apoptotic cells [14]. Later, it was discovered that certain types of anticancer treatments, such as chemotherapeutics (e.g., anthracyclines) [15], γ-irradiation [16, 17], and photodynamic therapy [18–21] (Table 6.1) can induce a specific form of apoptosis, which was named immunogenic apoptosis (IA) due to its immunostimulatory or adjuvant-like properties (Fig. 6.1). When cancer cell lines exposed to lethal doses of inducers of immunogenic apoptosis in vitro are used to vaccinate syngenic mice, they protect them against a subsequent challenge with live cancer cells of the same type. The immunogenicity of apoptotic cancerous cells relies on the spatiotemporal emission of specific signals called danger-associated molecular patterns (DAMPs), such as calreticulin (CRT), ATP, and HMGB-1. Most of these molecules have predominantly nonimmunological functions inside the cell but they become immunogenic after they are emitted extracellularly. DAMPs are derived from different subcellular compartments, including the plasma membrane, nucleus, ER, and cytosol, and they can often be modified by the proteolysis and/or oxidation associated with cell death mechanisms [22, 23]. DAMPs exert their immunostimulatory effects upon their recognition by membrane-bound or cytoplasmic pattern-recognition receptors (PRRs, e.g., Toll-like Receptor-4, TLR4), phagocytic receptors or scavenger receptors (e.g., LDL-receptor-related protein, LRP1/CD91), and purinergic receptors (e.g., P_2RX_7/P_2RY_2). These danger signals, in combination with cancer antigens, induce maturation of dendritic cells (DCs) and can lead to an adaptive immune response against tumor cells, thereby mediating anticancer immunity. This review covers recent advances in our understanding of the molecular mechanisms involved in danger signaling, DAMPs emission, the role of ER stress, and classification of immunogenic cell death inducers in relation to the way ER stress is triggered. In the final part, we discuss the role of radiotherapy-induced immunogenic apoptosis and the relationship of its immunogenicity to the fraction dose and concomitant chemotherapy.

Table 6.1 An overview of immunogenic cell death inducers and emission of DAMPs related to the stage of cell death

ICD inducers	Cellular target for ICD inducers	Surface exposed DAMPs and the stage of apoptosis	Secreted or released DAMPs and the stage of apoptosis	Refs
Type I				
Mitoxantrone, doxorubicin, idarubicin, oxaliplatin, UVC, γ-irradiation	Nucleus (DNA or DNA proteins related to cell mitosis	Preapoptotic: CRT/ERp57	Early apoptotic secreted: ATP	[33, 42, 46]
		Mid to late apoptotic: HSP-70	Late apoptotic passive release: HMGB1	
Cyclophosphamide	Nucleus (DNA)	Preapoptotic: CRT	Late apoptotic passive release: HMGB1	[105]
Bortezomib	Cytosol (26S proteasome, CIP2A and ERAD machinery)	Early to mid apoptotic: HSP90	Late apoptotic passive release: HMGB1	[106–110]
Cardiac glycosides	Cell surface (Na^+/K^+ ATPase)	Preapoptotic: CRT	Early to mid apoptotic ATP	[111]
			Late apoptotic passive release HMGB1	
Shikonin	Cytosol (tumor-specific pyruvate kinase-M2 protein)	Early to mid apoptotic: CRT, HSP90, GRP78	ND	[112, 113]
7A7 (EFR-specific antibody)	Cell surface receptor (EGFR)	Preapoptotic: CRT and ERp57	ND	[114]
		Early to mid apoptotic: HSP70 and HSP90		
Wogonin	Mitochondria	Early apoptotic: CRT	Late passive release ATP and HMGB1	[115]
High hydrostatic pressure	Cellular proteins	Preapoptotic (?): CRT, HSP70, HSP90	Late passive release ATP and HMGB1	[116]
Vorinostat (histone deacetylase inhibitor)	Nucleus (chromatin structure))	Early to mid apoptotic: CRT	Late passive release ATP and HMGB1	[117–119]
Bleomycin	Nucleus (DNA)	Early to mid apoptotic: CRT and ERp57	Early apoptotic secreted: ATP	[120]
			Late apoptotic passive release: HMGB1	

(continued)

Table 6.1 (continued)

ICD inducers	Cellular target for ICD inducers	Surface exposed DAMPs and the stage of apoptosis	Secreted or released DAMPs and the stage of apoptosis	Refs
Electrical pulses[a]	Cellular proteins	Early to mid apoptotic: CRT	Early to mid apoptotic ATP	[121]
			Late apoptotic passive release: HMGB1	
Septacidin	Cellular proteins	Early to mid apoptotic: CRT	Early to mid apoptotic ATP	[122]
			Late apoptotic passive release: HMGB1	
Honokiol	Cellular proteins (possibly)	CRT (stage is ND)	ND	[123]
Type II				
Hypericin-based PDT	Endoplasmic reticulum	Preapoptotic: CRT, HSP70	Preapoptotic secreted ATP	[19, 20, 43, 124]
			Late apoptotic passive release HSP70, HSP90, and CRT	
Oncolytic viruses (e.g., CVB3)	Endoplasmic reticulum	Early apoptotic: CRT	Early apoptotic secreted ATP late apoptotic passive release HMGB1	[125, 126]

CRT calreticulin, *DAMP* damage-associated molecular pattern, *ND* not determined, *EGFR* epidermal growth factor receptor, *ERAD* endoplasmic reticulum-associated degradation, *GRP* glucose-regulated protein, *HMGB1* high mobility group protein B1, *HSP* heat shock protein, *ICD* immunogenic cell death, *PDT* photodynamic therapy, *UVC* ultraviolet C, *CVB3* coxsackievirus B

[a]Combining electric pulses with the chemotherapeutic agent bleomycin was required for HMGB1 release

6.2 ER Stress and ROS: Crucial Players in Danger Signaling

Immunogenic anticancer drugs and treatments can trigger IA in dying cancer cells via the combined action of ER stress and ROS production, which activate danger signaling pathways and mediate the trafficking of DAMPs to the extracellular space [20, 24, 25]. ER stress was proposed to be a crucial component because the emission of DAMPs (e.g., calreticulin and ATP) and subsequent immunogenicity of cell death in vivo was found to be diminished when molecular effectors of the ER stress pathway were silenced [20, 25]. Anticancer drugs that do not induce ER stress (e.g., cisplatin) are poor inducers of IA [26]. Notably, the immunogenicity of drugs such as cisplatin could be restored by combining it with thapsigargin or tunicamycin [26] or by expression of the ER resident protein reticulon-1 [27]. ROS was also proposed to be required for immunogenicity of cell death because antioxidants (*N*-Acetyl cysteine, glutathione ethyl ester, and L-histidine) decrease its immunogenicity [20, 25]. As many immunogenic cell death inducers are diverse both biologically and chemically (reviewed in detail in [24, 28]), there seems to be no simple structure–function relationship that could explain the ability of these agents to induce IA. Therefore, we proposed that immunogenic cell death inducers can be classified into two categories (Type I and Type II) based on their distinct mode of action in the induction of ER stress and apoptosis [24]. Most of immunogenic cell death inducers (Table 6.1) are categorized as type I immunogenic cell death inducers that primarily trigger cell death via targeting cytosolic proteins, plasma membranes, or nucleic proteins rather than primary targeting ER mechanisms [24, 29, 30]. The type II immunogenic cell death inducers preferentially target the ER and include hypericin-based PDT and oncolytic coxsackievirus B3 (CVB3, Table 6.1). Although ER stress and ROS are essential in the immunogenicity of cell death, it is still not clear how these two signaling modules cooperate to efficiently induce immunogenic cell death. Therefore, further studies to elucidate the precise interplay between the ER stress and ROS is required to modulate antitumor immune responses.

6.3 Main Effectors of Immunogenic Cell Death: CRT, ATP, and HMGB1

Calreticulin (CRT) is an ER chaperone and its function is usually linked with Ca^{2+} homeostasis [31]. The role of CRT in the clearance of apoptotic cells was first described by Gardai et al. [32], who showed that CRT acts as a recognition ligand ("eat me" signal) on the surface of apoptotic cells by binding and activating LRP1/CD91 on the engulfing cell (Fig. 6.1). However, a new life was given to CRT by studies showing that CRT exposure is a key determinant of immunogenicity of dying cells and anticancer immune responses [33]. In that study, the authors found that anthracyclines induce rapid preapoptotic translocation of CRT to the cell surface and that blockade or knockdown of CRT suppresses the immunogenicity of apoptotic cancerous cells in mice. Several signaling pathways triggered by immunogenic cell death inducers have been described (Fig. 6.2). One pathway is induced by anthracyclines and relies on the phosphorylation of eukaryotic initiation factor 2a (eIF2a) by the ER stress-sensing kinase, PKR-related ER kinase (PERK), the activation of caspase-8, BAX and BAK, the transport of ER-derived vesicles through the Golgi apparatus, and the SNAP receptor (SNARE)-dependent exocytosis of these vesicles [25]. It has also been shown that paracrine signals that involve the chemokine CXCL8 contribute to CRT exposure on the cell surface [34]. The second pathway for CRT exposure is more rapid and relies on PERK-mediated trafficking of ecto-CRT by regulation of the proximal secretory pathway [20]. In this signaling pathway, eIF2a phosphorylation and caspase-8 signaling were not required for CRT exposure. Vaccination of mice with cells deficient in any of the proteins

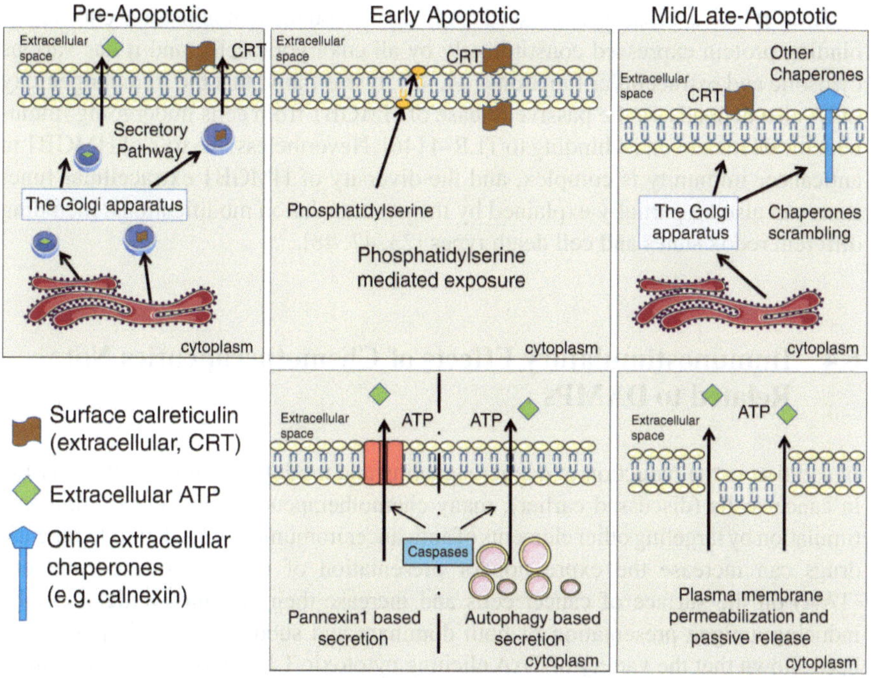

Fig. 6.2 An overview of the danger signaling pathways involved in surface CRT exposure and ATP secretion and their relation to different apoptotic stages. Signaling pathways responsible for surface exposure of CRT and secretion of ATP depend on immunogenic cell death stimuli [24]

required for CRT exposure or with cells in which CRT was knocked down reduced the immunogenicity of the cancer cells [20, 33]. All these results underline the key role of CRT exposure on the cell surface to the efficacy of anticancer therapy.

ATP is involved in various cellular metabolic processes and intracellular responses. However, it has become clear that APT is also actively secreted or passively released from dying cancerous cells, and that it is modulating the immunogenicity of dying cancerous cells (Fig. 6.1) [22, 23, 35, 36] via activation of purinergic P_2X_7 and P_2X_2 receptors [37]. The mechanisms of ATP secretion are strongly dependent on the type of immunogenic cell death inducer. Anthracyclines induce ATP secretion by a mechanism involving the caspase-dependent activation of pannexin 1 channels, lysosomal exocytosis, and plasma membrane blebbing [36, 38, 39]. Moreover, cancer cells undergoing IA in response to anthracycline secrete ATP in an autophagy-dependent manner [40–42]. Autophagy-deficient tumors exposed to chemotherapy cannot attract tumor-infiltrating leukocytes and therefore do not induce therapeutic anticancer immune responses [42]. However, in contrast to anthracyclines, hypericin-based PDT-induced ATP secretion is independent of autophagy [43] and involves the classical and PERK-regulated proximal secretory pathway, as well as PI3K-dependent exocytosis [20]. All these studies suggest that the mechanisms of ATP secretion might vary from one immunogenic cancer cell death inducer to another (Fig. 6.2).

HMGB1 is a broadly expressed and highly abundant nonhistone chromatin-binding protein expressed constitutively by all eukaryotic cells, and it has various cytosolic and extracellular functions [44, 45]. It was found that the immunogenicity of IA also depends on the passive release of HMGB1 from cells undergoing immunogenic death and on its binding to TLR-4 [46]. Nevertheless, the role of HMGB1 in anticancer immunity is complex, and the diversity of HMGB1 extracellular functions can also be partially explained by the posttranslation modifications, including different redox states and cell death types [23, 47, 48].

6.4 Immunostimulatory Effects of Chemotherapeutics Not Related to DAMPs

In addition to the induction of danger signaling and modulation of DAMPs emission in cancer cells (discussed earlier), many chemotherapeutics can induce immunostimulation by targeting other elements of anticancer immunity [36]. Chemotherapeutic drugs can increase the expression or presentation of tumor-associated antigens (TAA) on the surface of cancer cells and increase their so-called antigenicity by inducing antigen presentation of both dominant and subdominant epitopes. It has been shown that the variety of TAA eliciting cytotoxic T lymphocytes (CTL) can be increased by cisplatin and gemcitabine [49]. The authors showed that chemotherapy reveals weaker tumor antigens to the immune system, resulting in the induction of specific CTLs. The antigenicity of cancer cells can be enhanced by increasing the expression of MHC class I molecules (e.g., cyclophosphamide, gemcitabine, oxaliplatin, paclitaxel, and γ-irradiation) [36, 50, 51]. In addition, some anticancer drugs can increase the expression of TAA, including carcinoembryonic antigen (induced by 5-fluorouracil), multiple cancer testis antigens (increased by 5-aza-20deoxycytidine and γ-irradiation), and melanoma-associated antigens (increased by vemurafenib) [36, 50, 52, 53]. It is of interest that subtoxic doses of paclitaxel and doxorubicin increased the expression of components of the MHC class I antigen processing machinery (calmodulin, LMP2, LMP7, TAP1, and tapasin) in cancer cells [54]. Chemotherapeutic agents also cause immunopotentiation by directly stimulating immune cells. It has been shown that low doses of paclitaxel, doxorubicin, mitomycin C, and methotrexate that do not cause cell death up-regulate the ability of DCs to present antigens to antigen-specific T cells [55]. Recently, we demonstrated that intraperitoneal injection of doxorubicin in mice triggers the signs of acute inflammatory response (accumulation of neutrophils and increased levels of IL6, TNF, and MCP-1) [56–58]. Of interest is that the inflammatory response was significantly reduced in mice deficient in myeloid differentiation primary response gene 88 (MyD88), TLR-2 or TLR-9 [58], or tumor necrosis factor receptor-1 (TNFR1) [57]. These studies provide important new insights into how the innate immune system is modulated by immunogenic drugs such as doxorubicin (Table 6.1). It was also shown that the percentage of regulatory T cells among the $CD4^+$ lymphocytes was decreased by cyclophosphamide, which allowed a whole tumor cell vaccine or costimulatory receptor OX40 (OX86) immunotherapy to eradicate established tumors in colon carcinoma or melanoma models [59, 60]. The number of myeloid-derived suppressor cells (MDSCs) was reduced by gemcitabine in the spleen of mice bearing large tumors but did not affect CD4 and CD8 T cells, NK cells, macrophages, and B cells [61–63]. The bisphosphonate zoledronate, a drug that has been approved by the FDA for the treatment of bone metastases, was shown to induce caspase-1 activation in DC-like cells, which then provide mature IL-18 and IL-1β for the activation of IL-2-primed NK cells [64]. All these data suggest that some chemotherapeutics can directly stimulate immune cell functions and that their therapeutic efficacy could be at least partly explained by their ability to modulate the host immune system.

6.5 Radiotherapy-Induced Immunogenic Cell Death: Fraction Dose and Concomitant Chemotherapy

Together with surgery and chemotherapy, gamma-irradiation (RT) is important in the treatment of cancer. For decades, its main antitumor activity was believed to result from a direct and local cytotoxic effect on malignant cells within the irradiated area [65]. Nowadays, there is growing evidence for the occurrence of immune-mediated systemic effects resulting from local RT. Clinical proof of principle for such abscopal effects is provided by regression of distant metastases after local RT. Abscopal effects have been observed with various dose and fractionation regimens in melanoma (3×8 Gy to 3×18 Gy) [66–68] and lung adenocarcinoma (5×6 Gy) [69]. The necessity of combining RT with immunotherapy (in these cases CTLA4 blockade) to achieve these abscopal effects indicates that proimmunogenic effects are often dampened by the immune-suppressive microenvironment that characterizes cancer [70–73].

As for other immunogenic agents [74], radiation-induced immunogenic cell death is characterized, in cell cultures, by preapoptotic exposure on the extracellular surface of the "eat-me" signal CRT [25, 75, 76] and emission of ATP [75, 77, 78], and by late-apoptotic release of the "find-me" signal HMGB-1 [46, 75, 77, 79]. Animal and clinical experimental evidence supporting the ability of RT to induce immunogenic cell death remains scarce [77], and the clinical relevance of these pathways to the therapeutic efficacy of RT has yet to be validated.

Induction of immunogenic cell death is most likely highly dependent on total dose and fractionation. Golden et al. showed, in cell cultures, that the clinically used single doses between 2 and 20 Gy (1×2–20 Gy) effectively induce the signals for each individual component of immunogenic cell death in a dose-dependent manner [75]. Gameiro et al. showed the same, albeit with a clinically irrelevant single dose of 100 Gy [77]. Demaria et al. overviewed the literature and found immunogenic cell death to be often detected in tumor cell cultures exposed to mid-to-high doses of RT (1×>5–10 Gy) [80]. They initiated animal experiments using three RT regimens (1×20 Gy, 3×8 Gy and 5×6 Gy) combined with CTLA-4 antibody treatment in syngeneic mice with breast and colorectal carcinoma. While anti-CTLA-4 treatment on its own and its combination with a single-dose RT were not able to induce an abscopal effect, the fractioned regimens did [81]. This could explain why a single 8-Gy fraction treatment of bone metastases in prostate cancer patients failed to induce an abscopal effect when combined with anti-CTLA-4 treatment [82], whereas the above described clinical trials succeeded [66–69].

In addition to the induction of immunogenic cell death, other components up- or downregulated in response to RT are involved in antitumor immunity [71]. Tumor cell surface expression of MHC Class I molecules increases and CD47 (a "don't eat-me" signal for DCs) decreases in a dose-dependent manner in cell cultures [83–85]. Additionally, it was shown in a murine model that RT (2 × 12 Gy) increases the expression on tumor cell surface of RAE-1, a ligand for natural killer cell group 2D [86]. Distinct radiation fraction doses also have a direct effect on the irradiated tumor microenvironment. Clinical observations showed that immune-suppressing Treg cells are more radioresistant than $CD8^+$ T cells [87, 88]. In a xenotransplant mouse model, a lower RT dose (1 × 2 Gy) reprograms macrophages toward an iNOS+/M1 phenotype, allowing them to recruit tumor-specific T cells [89].

The above-mentioned data support the growing consensus that hypofractionated regimens (a limited number but >1 fraction high doses per fraction) are more effective at inducing the proimmunogenic effects of RT than single high doses or normofractionation (2 Gy per fraction or "x" times × 2 Gy) [90]. The hypofractionated regimens are mostly used to treat small (often oligo-) metastatic lesions, whereas for treatment of the primary tumor, normofractionation combined with chemotherapy is often the standard treatment. Concomitant use of both treatments has been shown to be superior to sequential chemo-RT in numerous clinical trials. It should be considered that concomitant chemo-RT causes a tumor cell death that is both qualitatively and quantitatively different from that achieved by each therapy alone [83]. Frey et al. showed that combining 5-FU, oxaliplatin, and irinotecan with RT could induce immunogenic cell death in colorectal cancer cells [91]. Golden et al. designed a cell culture assay to examine the effect on immunogenic cell death when combining RT (1 × 2 Gy) with paclitaxel and found that all three components of immunogenic cell death (i.e., CRT, ATP, and HMGB1; discussed earlier) to be increased significantly when chemotherapy and RT were used together as compared to separate treatments [75, 83]. Animal and clinical experiments are awaited to validate these interesting findings.

6.6 Conclusions

Only one decade ago, apoptotic cell death was presented as anti-inflammatory and tolerogenic, or even as a silent mode of cell death. However, insights over the last decade increasingly support the view that under specific conditions certain types and regimens of anticancer therapy can induce an immunogenic form of apoptosis that can be beneficial for the induction of anticancer immunity and long-lasting remission in cancer patients. Many questions remain regarding what determines the difference between immunogenic aspects of apoptosis and the danger signaling subroutines in the various types of cancers. Deeper insight into the molecular mechanisms of immunogenicity of apoptotic cells will lead to novel experimental immunotherapies for cancer, and is therefore a challenging research area. This work highlights the need for careful preclinical testing of the immunological effects of chemotherapies, alone and in combination with partner cytotoxic agents and immunotherapies, before proceeding to clinical investigations.

Acknowledgments We thank A. Bredan for editing the manuscript. This work was supported by project grants from the Fund for Scientific Research Flanders (FWO-Vlaanderen: G060713N, 3G067512, G0A5413N to DVK) and UGent Special Research Fund (BOF14/GOA/019 to D.V.K. and C.B.). Vandenabeele's group is supported by VIB, Ghent University (GROUP-ID Consortium of the UGent MRP initiative), FWO-Vlaanderen (G.0875.11, G.0973.11, G.0A45.12N, FWO G.0787.13N, FWO G.0C31.14N, FWO G0E04.16N), Federal Research Program (IAP 7/32). PV holds a Methusalem grant BOF09/01M00709 and BOF16/MET_V/007 from the Flemish Government, by Belgian grants (Interuniversity Attraction Poles, IAP 7/32) and grant from the 'Foundation against Cancer, 2012-188'. DVK is paid by the Methusalem grant.

References

1. Meagher LC, Savill JS, Baker A, Fuller RW, Haslett C. Phagocytosis of apoptotic neutrophils does not induce macrophage release of thromboxane B2. J Leukoc Biol. 1992;52:269–73.
2. Stern M, Savill J, Haslett C. Human monocyte-derived macrophage phagocytosis of senescent eosinophils undergoing apoptosis. Mediation by alpha v beta 3/CD36/thrombospondin recognition mechanism and lack of phlogistic response. Am J Pathol. 1996;149:911–21.
3. Brouckaert G, Kalai M, Krysko DV, Saelens X, Vercammen D, Ndlovu M, Haegeman G, D'Herde K, Vandenabeele P. Phagocytosis of necrotic cells by macrophages is phosphatidylserine dependent and does not induce inflammatory cytokine production. Mol Biol Cell. 2004;15:1089–100.
4. Cocco RE, Ucker DS. Distinct modes of macrophage recognition for apoptotic and necrotic cells are not specified exclusively by phosphatidylserine exposure. Mol Biol Cell. 2001;12:919–30.
5. Cvetanovic M, Ucker DS. Innate immune discrimination of apoptotic cells: repression of proinflammatory macrophage transcription is coupled directly to specific recognition. J Immunol. 2004;172:880–9.
6. Fadok VA, Bratton DL, Konowal A, Freed PW, Westcott JY, Henson PM. Macrophages that have ingested apoptotic cells in vitro inhibit proinflammatory cytokine production through autocrine/paracrine mechanisms involving TGF-beta, PGE2, and PAF. J Clin Invest. 1998;101:890–8.
7. Krysko DV, D'Herde K, Vandenabeele P. Clearance of apoptotic and necrotic cells and its immunological consequences. Apoptosis. 2006;11:1709–26.
8. Krysko DV, Vandenabeele P. From regulation of dying cell engulfment to development of anti-cancer therapy. Cell Death Differ. 2008;15:29–38.
9. Krysko DV, Vandenabeele P. Clearance of dead cells: mechanisms, immune responses and implication in the development of diseases. Apoptosis. 2010;15:995–7.
10. Voll RE, Herrmann M, Roth EA, Stach C, Kalden JR, Girkontaite I. Immunosuppressive effects of apoptotic cells. Nature. 1997;390:350–1.
11. Gregory CD, Pound JD. Cell death in the neighbourhood: direct microenvironmental effects of apoptosis in normal and neoplastic tissues. J Pathol. 2011;223:177–94.

12. Ford CA, Petrova S, Pound JD, Voss JJ, Melville L, Paterson M, Farnworth SL, Gallimore AM, Cuff S, Wheadon H, et al. Oncogenic properties of apoptotic tumor cells in aggressive B cell lymphoma. Curr Biol. 2015;25:577–88.
13. Huang Q, Li F, Liu X, Li W, Shi W, Liu FF, O'Sullivan B, He Z, Peng Y, Tan AC, et al. Caspase 3-mediated stimulation of tumor cell repopulation during cancer radiotherapy. Nat Med. 2011;17:860–6.
14. Rovere P, Vallinoto C, Bondanza A, Crosti MC, Rescigno M, Ricciardi-Castagnoli P, Rugarli C, Manfredi AA. Bystander apoptosis triggers dendritic cell maturation and antigen-presenting function. J Immunol. 1998;161:4467–71.
15. Casares N, Pequignot MO, Tesniere A, Ghiringhelli F, Roux S, Chaput N, Schmitt E, Hamai A, Hervas-Stubbs S, Obeid M, et al. Caspase-dependent immunogenicity of doxorubicin-induced tumor cell death. J Exp Med. 2005;202:1691–701.
16. Obeid M, Panaretakis T, Joza N, Tufi R, Tesniere A, van Endert P, Zitvogel L, Kroemer G. Calreticulin exposure is required for the immunogenicity of gamma-irradiation and UVC light-induced apoptosis. Cell Death Differ. 2007;14:1848–50.
17. Scheithauer H, Belka C, Lauber K, Gaipl US. Immunological aspects of radiotherapy. Radiat Oncol. 2014;9:185.
18. Garg AD, Krysko DV, Vandenabeele P, Agostinis P. The emergence of phox-ER stress induced immunogenic apoptosis. Oncoimmunology. 2012;1:786–8.
19. Garg AD, Krysko DV, Vandenabeele P, Agostinis P. Hypericin-based photodynamic therapy induces surface exposure of damage-associated molecular patterns like HSP70 and calreticulin. Cancer Immunol Immunother. 2012;61:215–21.
20. Garg AD, Krysko DV, Verfaillie T, Kaczmarek A, Ferreira GB, Marysael T, Rubio N, Firczuk M, Mathieu C, Roebroek AJ, et al. A novel pathway combining calreticulin exposure and ATP secretion in immunogenic cancer cell death. EMBO J. 2012;31:1062–79.
21. Panzarini E, Inguscio V, Fimia GM, Dini L. Rose Bengal acetate photodynamic therapy (RBAc-PDT) induces exposure and release of Damage-Associated Molecular Patterns (DAMPs) in human HeLa cells. PLoS One. 2014;9, e105778.
22. Krysko DV, Agostinis P, Krysko O, Garg AD, Bachert C, Lambrecht BN, Vandenabeele P. Emerging role of damage-associated molecular patterns derived from mitochondria in inflammation. Trends Immunol. 2011;32:157–64.
23. Krysko O, Love Aaes T, Bachert C, Vandenabeele P, Krysko DV. Many faces of DAMPs in cancer therapy. Cell Death Dis. 2013;4, e631.
24. Krysko DV, Garg AD, Kaczmarek A, Krysko O, Agostinis P, Vandenabeele P. Immunogenic cell death and DAMPs in cancer therapy. Nat Rev Cancer. 2012;12:860–75.
25. Panaretakis T, Kepp O, Brockmeier U, Tesniere A, Bjorklund AC, Chapman DC, Durchschlag M, Joza N, Pierron G, van Endert P, et al. Mechanisms of pre-apoptotic calreticulin exposure in immunogenic cell death. EMBO J. 2009;28:578–90.
26. Martins I, Kepp O, Schlemmer F, Adjemian S, Tailler M, Shen S, Michaud M, Menger L, Gdoura A, Tajeddine N, et al. Restoration of the immunogenicity of cisplatin-induced cancer cell death by endoplasmic reticulum stress. Oncogene. 2011;30:1147–58.
27. Michaud M, Sukkurwala AQ, Di Sano F, Zitvogel L, Kepp O, Kroemer G. Synthetic induction of immunogenic cell death by genetic stimulation of endoplasmic reticulum stress. Oncoimmunology. 2014;3, e28276.
28. Dudek AM, Garg AD, Krysko DV, De Ruysscher D, Agostinis P. Inducers of immunogenic cancer cell death. Cytokine Growth Factor Rev. 2013;24:319–33.
29. Garg AD, Martin S, Golab J, Agostinis P. Danger signalling during cancer cell death: origins, plasticity and regulation. Cell Death Differ. 2014;21:26–38.
30. Inoue H, Tani K. Multimodal immunogenic cancer cell death as a consequence of anticancer cytotoxic treatments. Cell Death Differ. 2014;21:39–49.
31. Gelebart P, Opas M, Michalak M. Calreticulin, a Ca2+-binding chaperone of the endoplasmic reticulum. Int J Biochem Cell Biol. 2005;37:260–6.
32. Gardai SJ, McPhillips KA, Frasch SC, Janssen WJ, Starefeldt A, Murphy-Ullrich JE, Bratton DL, Oldenborg PA, Michalak M, Henson PM. Cell-surface calreticulin initiates clearance of

viable or apoptotic cells through trans-activation of LRP on the phagocyte. Cell. 2005;123:321–34.
33. Obeid M, Tesniere A, Ghiringhelli F, Fimia GM, Apetoh L, Perfettini JL, Castedo M, Mignot G, Panaretakis T, Casares N, et al. Calreticulin exposure dictates the immunogenicity of cancer cell death. Nat Med. 2007;13:54–61.
34. Sukkurwala AQ, Martins I, Wang Y, Schlemmer F, Ruckenstuhl C, Durchschlag M, Michaud M, Senovilla L, Sistigu A, Ma Y, et al. Immunogenic calreticulin exposure occurs through a phylogenetically conserved stress pathway involving the chemokine CXCL8. Cell Death Differ. 2014;21:59–68.
35. Wang Y, Martins I, Ma Y, Kepp O, Galluzzi L, Kroemer G. Autophagy-dependent ATP release from dying cells via lysosomal exocytosis. Autophagy. 2013;9:1624–5.
36. Zitvogel L, Galluzzi L, Smyth MJ, Kroemer G. Mechanism of action of conventional and targeted anticancer therapies: reinstating immunosurveillance. Immunity. 2013;39:74–88.
37. Ghiringhelli F, Apetoh L, Tesniere A, Aymeric L, Ma Y, Ortiz C, Vermaelen K, Panaretakis T, Mignot G, Ullrich E, et al. Activation of the NLRP3 inflammasome in dendritic cells induces IL-1beta-dependent adaptive immunity against tumors. Nat Med. 2009;15:1170–8.
38. Elliott MR, Chekeni FB, Trampont PC, Lazarowski ER, Kadl A, Walk SF, Park D, Woodson RI, Ostankovich M, Sharma P, et al. Nucleotides released by apoptotic cells act as a find-me signal to promote phagocytic clearance. Nature. 2009;461:282–6.
39. Martins I, Wang Y, Michaud M, Ma Y, Sukkurwala AQ, Shen S, Kepp O, Metivier D, Galluzzi L, Perfettini JL, et al. Molecular mechanisms of ATP secretion during immunogenic cell death. Cell Death Differ. 2014;21:79–91.
40. Martins I, Michaud M, Sukkurwala AQ, Adjemian S, Ma Y, Shen S, Kepp O, Menger L, Vacchelli E, Galluzzi L, et al. Premortem autophagy determines the immunogenicity of chemotherapy-induced cancer cell death. Autophagy. 2012;8:413–5.
41. Martins I, Tesniere A, Kepp O, Michaud M, Schlemmer F, Senovilla L, Seror C, Metivier D, Perfettini JL, Zitvogel L, Kroemer G. Chemotherapy induces ATP release from tumor cells. Cell Cycle. 2009;8:3723–8.
42. Michaud M, Martins I, Sukkurwala AQ, Adjemian S, Ma Y, Pellegatti P, Shen S, Kepp O, Scoazec M, Mignot G, et al. Autophagy-dependent anticancer immune responses induced by chemotherapeutic agents in mice. Science. 2011;334:1573–7.
43. Garg AD, Dudek AM, Ferreira GB, Verfaillie T, Vandenabeele P, Krysko DV, Mathieu C, Agostinis P. ROS-induced autophagy in cancer cells assists in evasion from determinants of immunogenic cell death. Autophagy. 2013;9:1292–307.
44. Kang R, Chen R, Zhang Q, Hou W, Wu S, Cao L, Huang J, Yu Y, Fan XG, Yan Z, et al. HMGB1 in health and disease. Mol Aspects Med. 2014;40:1–116.
45. Yu Y, Tang D, Kang R. Oxidative stress-mediated HMGB1 biology. Front Physiol. 2015;6:93.
46. Apetoh L, Ghiringhelli F, Tesniere A, Obeid M, Ortiz C, Criollo A, Mignot G, Maiuri MC, Ullrich E, Saulnier P, et al. Toll-like receptor 4-dependent contribution of the immune system to anticancer chemotherapy and radiotherapy. Nat Med. 2007;13:1050–9.
47. Kazama H, Ricci JE, Herndon JM, Hoppe G, Green DR, Ferguson TA. Induction of immunological tolerance by apoptotic cells requires caspase-dependent oxidation of high-mobility group box-1 protein. Immunity. 2008;29:21–32.
48. Venereau E, Casalgrandi M, Schiraldi M, Antoine DJ, Cattaneo A, De Marchis F, Liu J, Antonelli A, Preti A, Raeli L, et al. Mutually exclusive redox forms of HMGB1 promote cell recruitment or proinflammatory cytokine release. J Exp Med. 2012;209:1519–28.
49. Jackaman C, Majewski D, Fox SA, Nowak AK, Nelson DJ. Chemotherapy broadens the range of tumor antigens seen by cytotoxic CD8(+) T cells in vivo. Cancer Immunol Immunother. 2012;61:2343–56.
50. Chen G, Emens LA. Chemoimmunotherapy: reengineering tumor immunity. Cancer Immunol Immunother. 2013;62:203–16.
51. Liu WM, Fowler DW, Smith P, Dalgleish AG. Pre-treatment with chemotherapy can enhance the antigenicity and immunogenicity of tumours by promoting adaptive immune responses. Br J Cancer. 2010;102:115–23.

52. Frederick DT, Piris A, Cogdill AP, Cooper ZA, Lezcano C, Ferrone CR, Mitra D, Boni A, Newton LP, Liu C, et al. BRAF inhibition is associated with enhanced melanoma antigen expression and a more favorable tumor microenvironment in patients with metastatic melanoma. Clin Cancer Res. 2013;19:1225–31.
53. Sharma A, Bode B, Wenger RH, Lehmann K, Sartori AA, Moch H, Knuth A, Boehmer L, Broek M. gamma-Radiation promotes immunological recognition of cancer cells through increased expression of cancer-testis antigens in vitro and in vivo. PLoS One. 2011;6, e28217.
54. Kaneno R, Shurin GV, Kaneno FM, Naiditch H, Luo J, Shurin MR. Chemotherapeutic agents in low noncytotoxic concentrations increase immunogenicity of human colon cancer cells. Cell Oncol. 2011;34:97–106.
55. Shurin GV, Tourkova IL, Kaneno R, Shurin MR. Chemotherapeutic agents in noncytotoxic concentrations increase antigen presentation by dendritic cells via an IL-12-dependent mechanism. J Immunol. 2009;183:137–44.
56. Kaczmarek A, Brinkman BM, Heyndrickx L, Vandenabeele P, Krysko DV. Severity of doxorubicin-induced small intestinal mucositis is regulated by the TLR-2 and TLR-9 pathways. J Pathol. 2012;226:598–608.
57. Kaczmarek A, Krysko O, Heyndrickx L, Love Aaes T, Delvaeye T, Bachert C, Leybaert L, Vandenabeele P, Krysko DV. TNF/TNF-R1 pathway is involved in doxorubicin-induced acute sterile inflammation. Cell Death Dis. 2013;4, e961.
58. Krysko DV, Kaczmarek A, Krysko O, Heyndrickx L, Woznicki J, Bogaert P, Cauwels A, Takahashi N, Magez S, Bachert C, Vandenabeele P. TLR-2 and TLR-9 are sensors of apoptosis in a mouse model of doxorubicin-induced acute inflammation. Cell Death Differ. 2011;18:1316–25.
59. Ghiringhelli F, Larmonier N, Schmitt E, Parcellier A, Cathelin D, Garrido C, Chauffert B, Solary E, Bonnotte B, Martin F. CD4+CD25+ regulatory T cells suppress tumor immunity but are sensitive to cyclophosphamide which allows immunotherapy of established tumors to be curative. Eur J Immunol. 2004;34:336–44.
60. Hirschhorn-Cymerman D, Rizzuto GA, Merghoub T, Cohen AD, Avogadri F, Lesokhin AM, Weinberg AD, Wolchok JD, Houghton AN. OX40 engagement and chemotherapy combination provides potent antitumor immunity with concomitant regulatory T cell apoptosis. J Exp Med. 2009;206:1103–16.
61. Bunt SK, Yang L, Sinha P, Clements VK, Leips J, Ostrand-Rosenberg S. Reduced inflammation in the tumor microenvironment delays the accumulation of myeloid-derived suppressor cells and limits tumor progression. Cancer Res. 2007;67:10019–26.
62. Ko HJ, Kim YJ, Kim YS, Chang WS, Ko SY, Chang SY, Sakaguchi S, Kang CY. A combination of chemoimmunotherapies can efficiently break self-tolerance and induce antitumor immunity in a tolerogenic murine tumor model. Cancer Res. 2007;67:7477–86.
63. Suzuki E, Kapoor V, Jassar AS, Kaiser LR, Albelda SM. Gemcitabine selectively eliminates splenic Gr-1+/CD11b+ myeloid suppressor cells in tumor-bearing animals and enhances antitumor immune activity. Clin Cancer Res. 2005;11:6713–21.
64. Nussbaumer O, Gruenbacher G, Gander H, Thurnher M. DC-like cell-dependent activation of human natural killer cells by the bisphosphonate zoledronic acid is regulated by gammadelta T lymphocytes. Blood. 2011;118:2743–51.
65. Bernier J, Hall EJ, Giaccia A. Radiation oncology: a century of achievements. Nat Rev Cancer. 2004;4:737–47.
66. Hiniker SM, Chen DS, Reddy S, Chang DT, Jones JC, Mollick JA, Swetter SM, Knox SJ. A systemic complete response of metastatic melanoma to local radiation and immunotherapy. Transl Oncol. 2012;5:404–7.
67. Postow MA, Callahan MK, Barker CA, Yamada Y, Yuan J, Kitano S, Mu Z, Rasalan T, Adamow M, Ritter E, et al. Immunologic correlates of the abscopal effect in a patient with melanoma. N Engl J Med. 2012;366:925–31.
68. Stamell EF, Wolchok JD, Gnjatic S, Lee NY, Brownell I. The abscopal effect associated with a systemic anti-melanoma immune response. Int J Radiat Oncol Biol Phys. 2013;85:293–5.

69. Golden EB, Demaria S, Schiff PB, Chachoua A, Formenti SC. An abscopal response to radiation and ipilimumab in a patient with metastatic non-small cell lung cancer. Cancer Immunol Res. 2013;1:365–72.
70. De Meerleer G, Khoo V, Escudier B, Joniau S, Bossi A, Ost P, Briganti A, Fonteyne V, Van Vulpen M, Lumen N, et al. Radiotherapy for renal-cell carcinoma. Lancet Oncol. 2014;15:e170–7.
71. De Wolf K, Vermaelen K, De Meerleer G, Lambrecht BN, Ost P. The potential of radiotherapy to enhance the efficacy of renal cell carcinoma therapy. Oncoimmunology. 2015;4, e1042198.
72. Formenti SC, Demaria S. Combining radiotherapy and cancer immunotherapy: a paradigm shift. J Natl Cancer Inst. 2013;105:256–65.
73. Golden EB, Apetoh L. Radiotherapy and immunogenic cell death. Semin Radiat Oncol. 2015;25:11–7.
74. Kepp O, Senovilla L, Vitale I, Vacchelli E, Adjemian S, Agostinis P, Apetoh L, Aranda F, Barnaba V, Bloy N, et al. Consensus guidelines for the detection of immunogenic cell death. Oncoimmunology. 2014;3, e955691.
75. Golden EB, Frances D, Pellicciotta I, Demaria S, Helen Barcellos-Hoff M, Formenti SC. Radiation fosters dose-dependent and chemotherapy-induced immunogenic cell death. Oncoimmunology. 2014;3, e28518.
76. Obeid M, Tesniere A, Panaretakis T, Tufi R, Joza N, van Endert P, Ghiringhelli F, Apetoh L, Chaput N, Flament C, et al. Ecto-calreticulin in immunogenic chemotherapy. Immunol Rev. 2007;220:22–34.
77. Gameiro SR, Jammeh ML, Wattenberg MM, Tsang KY, Ferrone S, Hodge JW. Radiation-induced immunogenic modulation of tumor enhances antigen processing and calreticulin exposure, resulting in enhanced T-cell killing. Oncotarget. 2014;5:403–16.
78. Ohshima Y, Tsukimoto M, Takenouchi T, Harada H, Suzuki A, Sato M, Kitani H, Kojima S. gamma-Irradiation induces P2X(7) receptor-dependent ATP release from B16 melanoma cells. Biochim Biophys Acta. 2010;1800:40–6.
79. Suzuki Y, Mimura K, Yoshimoto Y, Watanabe M, Ohkubo Y, Izawa S, Murata K, Fujii H, Nakano T, Kono KA. Immunogenic tumor cell death induced by chemoradiotherapy in patients with esophageal squamous cell carcinoma. Cancer Res. 2012;72:3967–76.
80. Demaria S, Formenti SC. Radiation as an immunological adjuvant: current evidence on dose and fractionation. Front Oncol. 2012;2:153.
81. Dewan MZ, Galloway AE, Kawashima N, Dewyngaert JK, Babb JS, Formenti SC, Demaria S. Fractionated but not single-dose radiotherapy induces an immune-mediated abscopal effect when combined with anti-CTLA-4 antibody. Clin Cancer Res. 2009;15:5379–88.
82. Kwon ED, Drake CG, Scher HI, Fizazi K, Bossi A, van den Eertwegh AJ, Krainer M, Houede N, Santos R, Mahammedi H, et al. Ipilimumab versus placebo after radiotherapy in patients with metastatic castration-resistant prostate cancer that had progressed after docetaxel chemotherapy (CA184-043): a multicentre, randomised, double-blind, phase 3 trial. Lancet Oncol. 2014;15:700–12.
83. Formenti SC, Demaria S. Effects of chemoradiation on tumor-host interactions: the immunologic side. J Clin Oncol. 2008;26:1562–3; author reply 1563.
84. Hauser SH, Calorini L, Wazer DE, Gattoni-Celli S. Radiation-enhanced expression of major histocompatibility complex class I antigen H-2Db in B16 melanoma cells. Cancer Res. 1993;53:1952–5.
85. Reits EA, Hodge JW, Herberts CA, Groothuis TA, Chakraborty M, Wansley EK, Camphausen K, Luiten RM, de Ru AH, Neijssen J, et al. Radiation modulates the peptide repertoire, enhances MHC class I expression, and induces successful antitumor immunotherapy. J Exp Med. 2006;203:1259–71.
86. Ruocco MG, Pilones KA, Kawashima N, Cammer M, Huang J, Babb JS, Liu M, Formenti SC, Dustin ML, Demaria S. Suppressing T cell motility induced by anti–CTLA-4 monotherapy improves antitumor effects. J Clin Invest. 2012;122:3718–30.

87. Kachikwu EL, Iwamoto KS, Liao YP, DeMarco JJ, Agazaryan N, Economou JS, McBride WH, Schaue D. Radiation enhances regulatory T cell representation. Int J Radiat Oncol Biol Phys. 2011;81:1128–35.
88. Qinfeng S, Depu W, Xiaofeng Y, Shah W, Hongwei C, Yili W. In situ observation of the effects of local irradiation on cytotoxic and regulatory T lymphocytes in cervical cancer tissue. Radiat Res. 2013;179:584–9.
89. Klug F, Prakash H, Huber PE, Seibel T, Bender N, Halama N, Pfirschke C, Voss RH, Timke C, Umansky L, et al. Low-dose irradiation programs macrophage differentiation to an iNOS+/M1 phenotype that orchestrates effective T cell immunotherapy. Cancer Cell. 2013;24:589–602.
90. Formenti SC. Is classical stereotactic radiotherapy the optimal partner for immunotherapy? Oncology 2015;29:340, 347, 387.
91. Frey B, Stache C, Rubner Y, Werthmoller N, Schulz K, Sieber R, Semrau S, Rodel F, Fietkau R, Gaipl US. Combined treatment of human colorectal tumor cell lines with chemotherapeutic agents and ionizing irradiation can in vitro induce tumor cell death forms with immunogenic potential. J Immunotoxicol. 2012;9:301–13.
92. Burnstock G. Purinergic nerves. Pharmacol Rev. 1972;24:509–81.
93. Kaufmann SH. Immunology's foundation: the 100-year anniversary of the Nobel Prize to Paul Ehrlich and Elie Metchnikoff. Nat Immunol. 2008;9:705–12.
94. Kerr JF, Wyllie AH, Currie AR. Apoptosis: a basic biological phenomenon with wide-ranging implications in tissue kinetics. Br J Cancer. 1972;26:239–57.
95. Lohmann K. Über die Pyrophosphatfraktion im Muskel. Naturwissenschaften. 1929;17:624–5.
96. Maccubbin DL, Wing KR, Mace KF, Ho RL, Ehrke MJ, Mihich E. Adriamycin-induced modulation of host defenses in tumor-bearing mice. Cancer Res. 1992;52:3572–6.
97. MacLennan DH, Yip CC, Iles GH, Seeman P. Isolation of sarcoplasmic reticulum proteins. Cold Spring Harb Symp Quant Biol. 1972;37:469–77.
98. Matzinger P. Tolerance, danger, and the extended family. Annu Rev Immunol. 1994;12:991–1045.
99. McCarthy EF. The toxins of William B. Coley and the treatment of bone and soft-tissue sarcomas. Iowa Orthop J. 2006;26:154–8.
100. Medzhitov R, Preston-Hurlburt P, Janeway Jr CA. A human homologue of the Drosophila Toll protein signals activation of adaptive immunity. Nature. 1997;388:394–7.
101. Ohashi K, Burkart V, Flohe S, Kolb H. Cutting edge: heat shock protein 60 is a putative endogenous ligand of the toll-like receptor-4 complex. J Immunol. 2000;164:558–61.
102. Rapaport E, Fishman RF, Gercel C. Growth inhibition of human tumor cells in soft-agar cultures by treatment with low levels of adenosine 5′-triphosphate. Cancer Res. 1983;43:4402–6.
103. Rapaport E, Fontaine J. Anticancer activities of adenine nucleotides in mice are mediated through expansion of erythrocyte ATP pools. Proc Natl Acad Sci U S A. 1989;86:1662–6.
104. Scaffidi P, Misteli T, Bianchi ME. Release of chromatin protein HMGB1 by necrotic cells triggers inflammation. Nature. 2002;418:191–5.
105. Schiavoni G, Sistigu A, Valentini M, Mattei F, Sestili P, Spadaro F, Sanchez M, Lorenzi S, D'Urso MT, Belardelli F, et al. Cyclophosphamide synergizes with type I interferons through systemic dendritic cell reactivation and induction of immunogenic tumor apoptosis. Cancer Res. 2011;71:768–78.
106. Davies AM, Lara Jr PN, Mack PC, Gandara DR. Incorporating bortezomib into the treatment of lung cancer. Clin Cancer Res. 2007;13:s4647–51.
107. Tseng LM, Liu CY, Chang KC, Chu PY, Shiau CW, Chen KF. CIP2A is a target of bortezomib in human triple negative breast cancer cells. Breast Cancer Res. 2012;14:R68.
108. Ling YH, Liebes L, Zou Y, Perez-Soler R. Reactive oxygen species generation and mitochondrial dysfunction in the apoptotic response to Bortezomib, a novel proteasome inhibitor, in human H460 non-small cell lung cancer cells. J Biol Chem. 2003;278:33714–23.
109. Spisek R, Charalambous A, Mazumder A, Vesole DH, Jagannath S, Dhodapkar MV. Bortezomib enhances dendritic cell (DC)-mediated induction of immunity to human myeloma via exposure of cell surface heat shock protein 90 on dying tumor cells: therapeutic implications. Blood. 2007;109:4839–45.

110. Cirone M, Di Renzo L, Lotti LV, Conte V, Trivedi P, Santarelli R, Gonnella R, Frati L, Faggioni A. Primary effusion lymphoma cell death induced by bortezomib and AG 490 activates dendritic cells through CD91. PLoS One. 2012;7, e31732.
111. Menger L, Vacchelli E, Adjemian S, Martins I, Ma Y, Shen S, Yamazaki T, Sukkurwala AQ, Michaud M, Mignot G, et al. Cardiac glycosides exert anticancer effects by inducing immunogenic cell death. Sci Transl Med. 2012;4:143ra199.
112. Chen J, Xie J, Jiang Z, Wang B, Wang Y, Hu X. Shikonin and its analogs inhibit cancer cell glycolysis by targeting tumor pyruvate kinase-M2. Oncogene. 2011;30:4297–306.
113. Chen HM, Wang PH, Chen SS, Wen CC, Chen YH, Yang WC, Yang NS. Shikonin induces immunogenic cell death in tumor cells and enhances dendritic cell-based cancer vaccine. Cancer Immunol Immunother. 2012;61:1989–2002.
114. Garrido G, Rabasa A, Sanchez B, Lopez MV, Blanco R, Lopez A, Hernandez DR, Perez R, Fernandez LE. Induction of immunogenic apoptosis by blockade of epidermal growth factor receptor activation with a specific antibody. J Immunol. 2011;187:4954–66.
115. Yang Y, Li XJ, Chen Z, Zhu XX, Wang J, Zhang LB, Qiang L, Ma YJ, Li ZY, Guo QL, You QD. Wogonin induced calreticulin/annexin A1 exposure dictates the immunogenicity of cancer cells in a PERK/AKT dependent manner. PLoS One. 2012;7, e50811.
116. Fucikova J, Moserova I, Truxova I, Hermanova I, Vancurova I, Partlova S, Fialova A, Sojka L, Cartron PF, Houska M, et al. High hydrostatic pressure induces immunogenic cell death in human tumor cells. Int J Cancer. 2014;135:1165–77.
117. Sonnemann J, Gressmann S, Becker S, Wittig S, Schmudde M, Beck JF. The histone deacetylase inhibitor vorinostat induces calreticulin exposure in childhood brain tumour cells in vitro. Cancer Chemother Pharmacol. 2010;66:611–6.
118. West AC, Mattarollo SR, Shortt J, Cluse LA, Christiansen AJ, Smyth MJ, Johnstone RW. An intact immune system is required for the anticancer activities of histone deacetylase inhibitors. Cancer Res. 2013;73:7265–76.
119. Guillot F, Boutin B, Blanquart C, Fonteneau JF, Robard M, Gregoire M, Pouliquen D. Vaccination with epigenetically treated mesothelioma cells induces immunisation and blocks tumour growth. Vaccine. 2011;29:5534–43.
120. Bugaut H, Bruchard M, Berger H, Derangere V, Odoul L, Euvrard R, Ladoire S, Chalmin F, Vegran F, Rebe C, et al. Bleomycin exerts ambivalent antitumor immune effect by triggering both immunogenic cell death and proliferation of regulatory T cells. PLoS One. 2013;8, e65181.
121. Calvet CY, Famin D, Andre FM, Mir LM. Electrochemotherapy with bleomycin induces hallmarks of immunogenic cell death in murine colon cancer cells. Oncoimmunology. 2014;3, e28131.
122. Sukkurwala AQ, Adjemian S, Senovilla L, Michaud M, Spaggiari S, Vacchelli E, Baracco EE, Galluzzi L, Zitvogel L, Kepp O, Kroemer G. Screening of novel immunogenic cell death inducers within the NCI Mechanistic Diversity Set. Oncoimmunology. 2014;3, e28473.
123. Liu SH, Lee WJ, Lai DW, Wu SM, Liu CY, Tien HR, Chiu CS, Peng YC, Jan YJ, Chao TH, et al. Honokiol confers immunogenicity by dictating calreticulin exposure, activating ER stress and inhibiting epithelial-to-mesenchymal transition. Mol Oncol. 2015;9:834–49.
124. Sanovic R, Verwanger T, Hartl A, Krammer B. Low dose hypericin-PDT induces complete tumor regression in BALB/c mice bearing CT26 colon carcinoma. Photodiagnosis Photodyn Ther. 2011;8:291–6.
125. Liu Z, Zhang HM, Yuan J, Ye X, Taylor GA, Yang D. The immunity-related GTPase Irgm3 relieves endoplasmic reticulum stress response during coxsackievirus B3 infection via a PI3K/Akt dependent pathway. Cell Microbiol. 2012;14:133–46.
126. Miyamoto S, Inoue H, Nakamura T, Yamada M, Sakamoto C, Urata Y, Okazaki T, Marumoto T, Takahashi A, Takayama K, et al. Coxsackievirus B3 is an oncolytic virus with immunostimulatory properties that is active against lung adenocarcinoma. Cancer Res. 2012;72:2609–21.

Chapter 7
Cancer Cell Death-Inducing Radiotherapy: Impact on Local Tumour Control, Tumour Cell Proliferation and Induction of Systemic Anti-tumour Immunity

Benjamin Frey, Anja Derer, Heike Scheithauer, Roland Wunderlich, Rainer Fietkau, and Udo S. Gaipl

Abstract Radiotherapy (RT) predominantly is aimed to induce DNA damage in tumour cells that results in reduction of their clonogenicity and finally in tumour cell death. Adaptation of RT with higher single doses has become necessary and led to a more detailed view on what kind of tumour cell death is induced and which immunological consequences result from it. RT is capable of rendering tumour cells immunogenic by modifying the tumour cell phenotype and the microenvironment. Danger signals are released as well as the senescence-associated secretory phenotype. This results in maturation of dendritic cells and priming of cytotoxic T cells as well as in activation of natural killer cells. However, RT on the other hand can also result in immune suppressive events including apoptosis induction and foster tumour cell proliferation. That's why RT is nowadays increasingly combined with selected immunotherapies.

Keywords Radiotherapy • DNA damage • Apoptosis • Necrosis • Autophagy • Danger signals • Senescence-associated secretory phenotype • Immunogenic cell death • Immunotherapy

---

B. Frey • A. Derer • R. Fietkau • U.S. Gaipl (✉)
Department of Radiation Oncology, Universitätsklinikum Erlangen, Friedrich-Alexander-Universität Erlangen-Nürnberg (FAU), Universitätsstrasse 27, 91054 Erlangen, Germany
e-mail: udo.gaipl@uk-erlangen.de

H. Scheithauer
Department of Radiotherapy and Radiation Oncology, Ludwig Maximilian University Munich, 81377 Munich, Germany

R. Wunderlich
Research Unit Radiation Cytogenetics, Helmholtz Centre Munich, Ingolstaedter Landstrasse 1, 85764 Neuherberg, Germany

C.D. Gregory (ed.), *Apoptosis in Cancer Pathogenesis and Anti-cancer Therapy*, Advances in Experimental Medicine and Biology 930,
DOI 10.1007/978-3-319-39406-0_7

7.1 Introduction

Two months after the announcement of the discovery of X-rays by Conrad Röntgen on November 30 1895, E. H. Grubbé, a medical student living in Chicago at that time, applied the X-rays therapeutically for the treatment of breast cancer and inflammatory lesions. He was provident and protected the surrounding healthy tissues by a sheet of lead taken from a tea chest. This was the hour of birth of radiotherapy (RT) [1]. The second classical cytotoxic treatment option for cancer disease is chemotherapy. The latter was ultimately discovered by physicians to treat cancer in the First World War. They observed that leukocytes disappeared in humans who survived mustard gas (dichloroethyl sulphide) exposure. They concluded that every poison could be also a potential efficacious remedy [2]. Until today, the three classical columns of cancer therapy are still chemotherapy (CT), RT and, the oldest form of tumour treatment, surgery.

During the last decades, immunotherapy (IT) accrued and multimodal therapies make nowadays more and more their way into clinical practice [3]. These cancer treatment modalities were formerly classified as those acting locally (surgery and RT) and those systemically (CT, IT). However, local modification of tumour cells might also result in secondary systemic responses. The focus of this article is therefore set on the ability of RT to induce distinct forms of tumour cell death and on the subsequent systemic consequences.

7.2 DNA Damage Induction and Repair Capacity as Basis for Local Efficacy of Radiotherapy

The most sensitive cellular structure for radiation is the deoxyribonucleic acid (DNA). X-rays as exogenous DNA damaging source can induce DNA single-strand breaks (SSB), double-strand breaks (DSB), oxidation of DNA bases and non-DSB clustered DNA lesions [4]. The damage is induced either by direct action of radiation on the DNA or mostly secondary by reactive oxygen species (ROS) or reactive nitrogen species (RNS) [5]. Irrespective of the DNA damage sources, the DNA damage response (DDR) is activated consecutively. Several DNA repair pathways have evolved like homologous recombination (HR), non-homologous end-joining (NHEJ), back-up NHEJ (B-NHEJ) nucleotide (NER) and base excision repair (BER) as well as mismatch repair (MMR) dependent on size and modality of the DNA damage [6].

The success or failure of standard clinical radiation treatment has mainly been determined by the four R's of radiobiology: repair of DNA damage, reoxygenation of hypoxic tumour areas, redistribution of cells in the cell cycle and repopulation [7, 8]. Tumour cells usually less effectively repair sublethal DNA damage compared to healthy tissue cells. This is one reason why repeated irradiation, namely fractionated irradiation, is beneficial since the healthy tissue can regenerate during the radiation break. Furthermore, time is created to allow reoxygenation of hypoxic tumour areas. This highly enhances the radiosensitivity of the tumour cells [9]. The latter also exit the radioresistant S-phase of the cell cycle during radiation breaks and become more sensitive for re-irradiation [10]. However, the breaks should not be too long to avoid repopulation of tumour cells. These are the reasons for delivering radiation in lower doses but repeated fractions.

One has to keep always in mind that the local irradiation of the tumour has to fulfil two main requirements: On the one hand the tumour control probability (TCP) must be as high as possible, but on the other hand the normal tissue complication probability (NTCP) has to be as small as possible [11]. Therefore, the applied dose is finely balanced between minimal, justifiable NTCP matched with a maximal TCP. The linear quadratic model is still the basis for clinicians to estimate the total dose and fractions of irradiation for the respective tumour entities. The dose of irradiation that is necessary to destroy tumour cells and the tolerance dose for healthy tissue is known by clinicians based on long-lasting experience with classical fractionated RT with a single dose of 1.8–2.0 Gy. α/β values were defined long ago for tissues. This was based on observations in mice, namely when and to what extent irradiation causes damage in certain organs [12, 13]. High values characterise early reacting tissue with rare repair and fast repopulation, as e.g. the skin (α/β: 9–19 Gy) and many tumours. Late reacting tissues such as kidney have α/β values <5 Gy and high repair capacity. During fractionated irradiation, the late reacting tissue can regenerate during the radiation breaks and is thereby spared.

Adaption of radiation schemes is necessary for distinct tumour entities since, e.g. prostate cancer has exceptionally low values of α/β. Here, the use of a higher dose per fraction is indicated on this radiobiological basis as it is also currently intensively discussed for breast cancer [14].

It has become feasible to deliver higher single doses due to technical advancements in planning procedures (e.g. intensity-modulated RT), accuracy of dose application (e.g. image-guided RT) and application of protons and heavy ions for RT. How novel techniques in RT change the standards for cancer treatment has recently been comprehensively summarised by Durante et al. and Orth et al. [15, 16].

7.3 Radiotherapy Induces Different Cell Death Modalities

7.3.1 Mitotic Catastrophe

If the DNA damage cannot be properly repaired by the radiation-exposed cells, they execute cell death. Mitotic catastrophe, a type of cell death that occurs during mitosis, was considered for a long time by radiobiologists to be the only way cells die after irradiation. In mammalian cells it is the failure to undergo complete mitosis after DNA damage. This results in multi-ploidy and counting of multinucleated cells is the basis for detection of mitotic catastrophe [17]. The combination of cell cycle checkpoint deficiencies and specific types of DNA damage most likely lead to mitotic catastrophe and cancer cells are especially prone to that [18]. Nevertheless, there is no consensus on the distinctive morphological appearance of mitotic catastrophe as far as the extent of chromatin condensation. The latter is, however, also the morphological hallmark of apoptosis [19].

7.3.2 Senescence

Cells also evolved a bypass to deal with persistent DNA damage, namely senescence. It was first described by Hayflick and colleagues, who demonstrated that as a consequence of telomere shortening with each cycle of DNA replication human fibroblasts do not proliferate until infinity in culture [20]. Senescent cells are characterised by low expression of proteins driving proliferation, morphological changes as increase in volume and, if adherent, flattered morphology. They further highly express senescence-associated acidic lysosomal β-galactosidase. The latter is a manifestation of residual lysosomal activity at a suboptimal pH and it becomes detectable due to the increased lysosomal content in senescent cells [21]. Telomere erosion, DNA damage and oncogenic signalling induce senescence, the so-called replicative, stress and oncogene-induced senescence, respectively. It has always been in the attention of oncologists since it is the basis for prolonged or ideally permanent growth arrest of tumour cells.

However, senescent cells can regain proliferative capacity in a p53-dependent manner after radiation exposure while cells undergoing apoptosis do not. This was especially demonstrated in vitro, as for p53 wild-type MCF-7 compared to MDA-MB231 breast cancer cells with mutant p53 [22]. One should additionally keep in mind that caspase proficiency might be related to it, since MCF-7 cells are deficient for caspase-3 and MDA-MB231 cells not. We recently showed that the in vitro immunogenic potential of caspase-3 proficient breast cancer cells with basal low immunogenicity is increased by hypofractionated irradiation and that of caspase-3 deficient ones not [23].

Since senescent cells remain in a metabolic active state they cannot be defined as dead [24]. They actively shape the microenvironment and the expression and secretion of immune modulating proteins changes during the induction and establishment of senescence [25]. This has been termed as senescence-associated secretory phenotype (SASP) [26]. Senescent cells activate a self-amplifying secretory network. The SASP includes pro-inflammatory cytokines like Interleukin (IL)-1α, IL-1β, IL-6 and IL-8, chemokines and growth factors and thereby connects local senescent cells with systemic inflammatory events [27, 28].

7.3.3 Autophagy

Not only radiation-induced forms of cell demise and inflammation are interconnected, but also additionally the DNA damage response, as demonstrated for autophagy. The latter is a conserved lysosomal pathway for degrading cytoplasmic proteins, macromolecules and organelles. It is kind of a cellular recycling factory unit that also promotes energy efficiency through adenosine triphosphate (ATP) generation. It further mediates damage control by removing non-functional proteins and organelles. A detailed summary on the molecular and cellular mechanisms of autophagy was provided by Glick and colleagues [29]. Autophagy can be monitored by autophagosome formation, but usage of multiple assays is recommended for its detection [30]. We here focus on the impact of autophagy on radiosensitivity, DNA damage response and inflammation.

Cancer cells exploit autophagy to adapt to nutrient limiting, metabolically stressful and hypoxic tumour microenvironment, since the physiological function of autophagy is related to the maintenance of cellular homeostasis under cellular stress [31]. Additionally, a non-protective form of autophagy does exist. Here, the cell is carrying out autophagy-mediated degrading functions, but autophagy inhibition does not lead to sensitisation for radiation or drugs [32]. Furthermore, autophagy can be cytotoxic [33] or cytostatic. The latter one is characterised by prolonged growth inhibition and reduced clonogenic survival without resulting in cell death induction [34]. Because of cytotoxic and cytostatic autophagy, cancer cells most likely often display a reduced autophagy. Overexpression of Beclin 1, a Bcl-2-interacting coiled-coil protein, inhibits cellular proliferation and has autophagy-promoting activity. Beclin-1 expression is absent or frequently low in cancer, e.g. in prostate, breast and ovarian cancer [35].

The relationship between DNA repair and autophagy in cancer cells is just fragmentarily understood. Autophagy has been shown to regulate some of the DNA repair proteins after DNA damage (summarised in [36]). Furthermore, evidence was provided that a mechanistic link between processing of DNA damage and activation of autophagy does exist [37]. In a mouse model of poly-microbial sepsis it was elegantly demonstrated that DNA damaging chemotherapeutics like anthracyclines improved the survival of the septic mice without affecting bacterial burden. This was not a sole effect of suppression of release of inflammatory cytokines like IL-1β and danger signals like high-mobility group box 1 (HMGB1) that could also be achieved by antibiotics, but also of promoting tissue protection from inflammatory damage. This was achieved by autophagy induction in dependence of the activation of the DNA damage response [38, 39]. Recently, hints were identified that defective autophagy in vivo caused an absence or reduction in regulatory proteins critical to both homologous recombination (HR) and non-homologous end joining (NHEJ) DNA damage repair pathways. Further, a failure to induce these proteins in response to radiation was asserted [40]. Cottone and colleagues have identified the activation of autophagy and the release of HMGB1 as key events how colon carcinoma cells recruit leukocytes. Concomitant induction of autophagy to apoptosis by 5-fluorouracil (5-FU) was necessary to induce the leukocytes attraction. They suggest that HMGB1 is translocated to the cytosol and may there promote the activation of autophagy, which in turn fosters further HMGB1 translocation form the nucleus into the cytosol and its consecutive release in the extracellular milieu [41]. Irradiation of tumours with 2 Gy as other DNA-damaging stressor resulted in recruitment of cytotoxic T cells, here in dependence of macrophage differentiation to an $iNOS^{+}$/M1 phenotype [42]. All these works give on the one hand evidence that after DNA damaging stress not only single cell death forms are induced and that on the other hand interconnections between DNA damage responses, inflammation and systemic immune modulation exist.

7.3.4 Apoptosis

Even though cell death can have many facets the two best known forms are still apoptosis and necrosis. Apoptosis, a form of programmed cell death, is crucial not only during embryonic development, but is present throughout the whole lifetime of multicellular organisms to attain cellular homeostasis. Apoptotic cells are characterised by nuclear and cytoplasmic condensation, nuclear fragmentation and cell shrinkage induced by plasma membrane blebbing [43]. Most importantly and contrary to necrotic cells, apoptotic cells maintain their membrane integrity until late stages of apoptosis execution. Apoptotic cells release and expose a broad range of 'find me' and 'eat me' signals for phagocytes such as macrophages [44]. The uptake of apoptotic cells occurs in a non- or even anti-inflammatory manner [45]. This immune suppressive effect might contribute to the in part unwanted effects of apoptosis induction by radiotherapy [46].

In response to ionising radiation, apoptosis is predominantly observed in cells of the hematopoietic system [47]. In solid tumours, the multicellular architecture may strongly contribute to render individual tumour cells less susceptible to apoptosis [48]. The TP53 gene provides instructions for making a protein called tumour protein p53 (p53) and is together with the PI3KCA gene that encodes for PI 3-kinases (PI3K) the most mutated gene in all types of cancers [49]. The tumour suppressor p53 primarily functions as a transcription factor, but its binding to the nuclear matrix generally increases after genotoxic stress [50]. p53 is involved in damage recognition, cell-cycle arrest, DNA repair, senescence or apoptosis. Of note is that p53 has roles that do not involve its transactivation functions during DNA repair; it modulates DNA repair processes, except for homologous recombination, by both transactivation-dependent and -independent pathways, as well as damage recognition and apoptosis [51]. It links apoptotic signalling pathways to radiation-induced DNA damage and is capable of directly regulating the Bax-dependent mitochondrial pathway to cell death [52]. In addition to intrinsic apoptosis pathways, extrinsic ones exist based on ligation of death receptors. In response to radiation, proteins of the death receptors are upregulated in a p53 dependent and independent manner [53]. Further, p53 controls signalling-mediated phagocytosis of apoptotic cells through its target Death Domain1α (DD1α). The latter functions as an engulfment ligand and thereby ensures a proper clearance of cell corpses. This contributes to the maintenance of immune tolerance [54].

Other members of the p53 tumour suppressor family of genes like p73 might compensate the lack of function of p53 and mediate radiation-induced apoptosis [55]. Therefore, the general statement that is mostly based on p53 functionality, that distinct tumours are sensitive for apoptosis after irradiation or not has to be considered critically.

In addition, distinct stimuli can promote an immunogenic variant of apoptosis [24, 56]. Treatment with tumour necrosis factor (TNF)-related apoptosis inducing ligand (TRAIL), e.g. induces membrane calreticulin (CRT) exposure on cancer cells [57]. The pre-apoptotic exposure of the endoplasmic reticulum (ER)-derived CRT together with the late or post-apoptotic release of danger signals like HMGB1 (see below) renders dying tumour cells immunogenic and can be induced by distinct chemotherapeutic agents like anthracyclines and oxaliplatin and by ionising radiation [58]. The exposure pathway of CRT is activated by pre-apoptotic ER stress and mediated via caspase-8-dependent proteolysis of the ER-sessile protein BAP31 and by activation of the pro-apoptotic proteins Bax and Bak [59]. Another scenario where apoptotic cells become immunogenic is that they proceed to secondary necrosis, meaning that they lose their membrane integrity. This happens when the clearance of apoptotic cells is impaired. This clearance defect is present in certain autoimmune diseases or when massive apoptosis occurs, e.g. after multimodal tumour treatments including RT [60, 61]. Secondary necrotic cells are often termed late apoptotic cells. This naming refers to the fact that the cells already underwent the apoptotic programme for a certain time. However, from the immunological point of view, due to the disturbed plasma membrane they behave like necrotic cells (Fig. 7.1).

7.3.5 *Necrosis*

The overall definition of necrosis is that cells have lost their plasma membrane integrity. In Radiation Oncology, the term necrosis was for a long time just linked with radionecrosis, a late side effect of irradiation with high single doses [62]. Soft tissue and bone changes occur and lead in a small percentage of the patients to tissue necrosis.

Beneficial necrosis of tumour cells induced by RT came into the mind of clinicians when data came up that immunogenic cancer cell death has profound clinical and therapeutic implications. Necrotic cells release danger-associated molecular patterns (DAMPs) like HMGB1, heat shock proteins (HSP), nucleotides or uric acid that trigger the activation of both, the innate and the adaptive immune system [63]. Primary necrosis was considered as a non-physiological form of cell death induced by trauma, ROS, pathogens and massive toxicity in general. However, similar to apoptosis, necrosis can also occur in a regulated fashion, meaning that a genetically encoded molecular machinery runs. The so-called necroptosis, which is dependent on the receptor interacting protein (RIP) kinases RIP1 and RIP3 can be induced by factors such as tumour necrosis factor (TNF), Fas Ligand or TRAIL and utilises the same initial signalling cascade as cell-death receptor-induced apoptosis [64]. Necroptosis further requires the substrate of RIP3K, the mixed lineage kinase like (MLKL). Necroptosis can be manipulated by inhibitors such as necrostatin 1, which blocks RIP1 kinase activity [65, 66]. Mounting evidence exists that many of the currently used anticancer agents are capable of engaging necroptotic signalling pathways. This offers the opportunity to reactivate cell death programmes in human malignancies, especially in those being considered as apoptosis resistant [67].

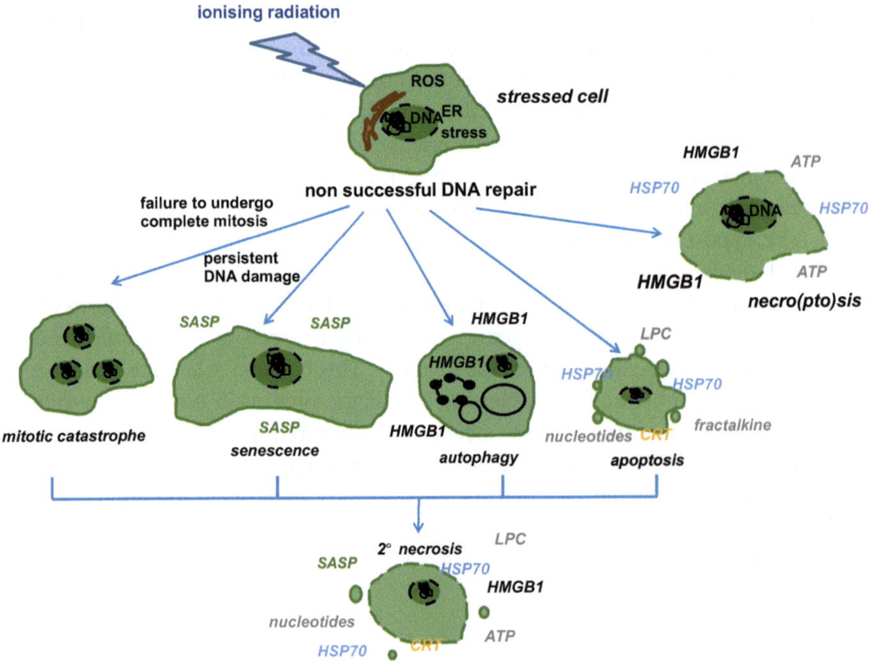

Fig. 7.1 Ionising radiation induces various tumour cell death modalities. The exposure of tumour cells to ionising radiation results in DNA damage, DNA damage response, ER stress response and in the induction of the displayed cell death forms. Radiation hereby not only impacts on the tumour cell phenotype but also on the tumour cell microenvironment. Of note is that all cell death forms can proceed to necrosis when during time their plasma membrane is disturbed. *ATP* adenosine triphosphate, *CRT* calreticulin, *ER* endoplasmic reticulum, *HMGB1* high-mobility group box 1, *HSP* heat shock protein, *LPC* lysophosphatidylcholine, *ROS* reactive oxygen species, *SASP* senescence-associated secretory phenotype/proteins, 2° secondary

In colorectal cancer cell lines, predominantly necrosis was inducible by RT and/or hyperthermia concomitantly with an increased expression of RIP1 [68]. We recently demonstrated that necroptosis is inducible with the pan caspase inhibitor zVAD-fmk in poorly immunogenic B16 melanoma cells [69]. Combination of RT, CT and immune stimulation by hyperthermia and zVAD-fmk resulted in significant tumour growth retardation compared to treatments without zVAD-fmk. This was dependent on the adaptive immune system, HMGB1 and nucleotides. Therapy-induced immunogenic cancer cell death might therefore be the key event in triggering anti-tumour immune responses.

7.4 Immunogenic Cancer Cell Death

The definition of immunogenic cancer cell death is based on molecular and cellular mechanisms as well as certain in vivo characteristics [70]. Non-immunogenic cell death is characterised by PS exposure and swift clearance of the dying and stressed

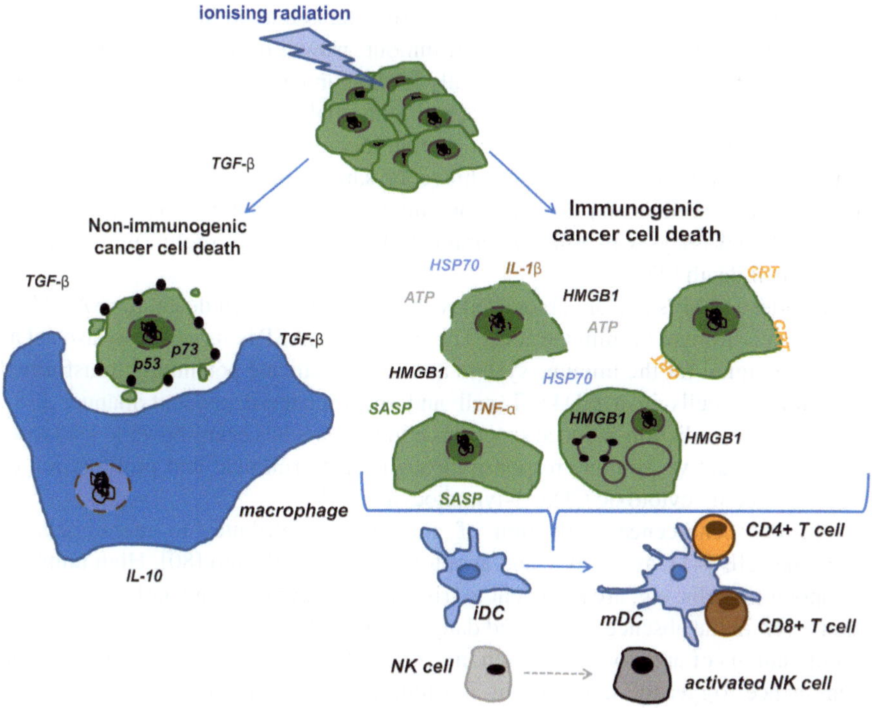

Fig. 7.2 Radiation-induced immunogenic cancer cell death results in activation of the innate and adaptive immune system. Treatment of tumour cells with ionising radiation can induce non-immunogenic cancer cell death, namely apoptotic tumour cells that do expose phosphatidylserine (*black dots*) on the outer membrane leaflet and secrete TGF-β. They are finally cleared by macrophages in an anti-inflammatory manner. On the other hand, very early apoptotic cells that do expose CRT are immunogenic, as well as senescent cells, cells undergoing autophagy and necrotic cells mainly by release of danger signals, inflammatory cytokines and SASP. Especially a mixture of these cell death modalities is highly immunogenic and results in maturation of DC, consecutive priming of T cells and activation of NK cells. *ATP* adenosine triphosphate, *CRT* calreticulin, *HMGB1* high-mobility group box 1, *HSP* heat shock protein, *iDC* immature dendritic cell, *IL* interleukin, *mDC* mature dendritic cell, *NK* natural killer, *SASP* senescence-associated secretory phenotype/proteins, *TGF-β* transforming growth factor beta, *TNF* tumour necrosis factor

cells by macrophages. Concomitantly, apoptotic-cell derived blebs [71] and radiation-induced TGF-beta [72] might result in inhibition of anti-tumour immune responses [73] (Fig. 7.2). In contrast, immunogenic cancer cell death is mostly connected with the release of the DAMPs HMGB1 and ATP and with the exposure of CRT. Additionally, further immune activating danger signals like Hsp70 and immunostimulatory cytokines like TNF-α and IL-1β are released [74].

This results in maturation and activation of DCs and ensuing priming of tumour-specific CD8+ T cells. Furthermore, NK cells can be activated by immunogenic cells including their microenvironment [75] (Fig. 7.2). For the in vivo examination of the immunogenic potential of tumour cells, both an immunisation and a therapeutic assay should be used. Both are based on the comparison of tumour growth in wild type compared to immune deficient mice: treatments that do induce immunogenic tumour cell death do result in retarded tumour growth only in wild-type animals [70]. That the tolerance has been actually broken and a memory immune response has been indeed induced should be tested with challenge experiments in animals that were primarily cured. Of note is that antineoplastic regimens that do engage immune effector mechanisms also achieve the same result without inducing immunogenic cancer cell death [76]. Therefore, multiple additional in vitro testing including functional assays with primary immune cells is mandatory to define immunogenic cancer cell death [77].

Besides DAMPs that are associated with immunogenic cell death, the SASP fosters the recruitment of immune cells. Therefore, the SASP is supposed to also act as a danger signal for the immune system aiming to eradicate potentially transformed or damaged cells in a CD4+ T cell and macrophage-dependent manner [78]. Furthermore, radiation-induced senescence in tumours has been shown to lead to an increased adaptive immune response through the recruitment and proliferation of tumour specific cytotoxic $CD8^{+}$ T-lymphocytes [79].

Besides senescence, activation of autophagy contributes to recruitment of immune cells [41], as necrotic and apoptotic tumour cells, too [80]. High numbers of apoptotic cells, e.g. are sufficient to trigger DC maturation and antigen presentation, even in the absence of released danger signals [81]. This suggests that in vivo, combinations of apoptotic cell death, necrotic cell death, autophagic cell death and senescence trigger the induction of anti-tumour immune responses in a concerted action (Fig. 7.2).

7.5 Systemic Effects of Radiation

The insufficient immunological control of tumours is one hallmark of cancer [82]. Tumours must escape immune surveillance during development and when being established. The cancer immunoediting consists of the elimination, equilibrium and escape phase [83]. In the elimination phase, the immune system is capable of stopping cancer development and destroys tumour cells. In the equilibrium phase a latent state exists, while in the escape phase the immunological defence mechanisms fail and the tumour progresses. The immune system is not only involved in cancer prevention and development but also in cancer therapy [84].

RT might contribute to overcome tumour escape by modifying the phenotype of the tumour cells [85, 86]. In the ideal case, radiation generates an in situ vaccine. However, mostly immune responses to model antigens expressed by tumours have been examined. It remained uncertain whether RT can prime T cells specific for endogenous antigens expressed by poorly immunogenic tumours. Vanpouille-Box and colleagues recently demonstrated that this is also possible, however only when combining RT with blockade of TGF-beta and/or PD-1 [87]. The generated T cells were effective at causing regression of the irradiated tumours but also of non-irradiated metastases.

The so-called out-of-field or abscopal effects of RT are best when RT is combined with further immune activation [88]. To avoid the "mystic" wording abscopal and due to continuously growing numbers of preclinical and clinical studies that immune reactions mediate abscopal responses, they should be better termed RT-induced systemic immune-mediated effects [74]. The key mechanisms involved in ionising radiation-induced systemic effects were recently comprehensively summarised by Mavragani and colleagues [89].

7.6 Immunogenicity of Distinct Doses of RT and of Combination with Immunotherapies

Nowadays, due to technical improvements, RT is delivered in various fractionations. Standard fractionation consists of single doses of 1.8–2.2 Gy (one fraction per day, 5 days a week continuing for 3–7 weeks) and hypofractionation of 3–20 Gy (one fraction a day given for 1–3 days a week) [90]. The available data whether standard fractionation is as immunogenic as fewer applications with higher single doses (hypofractionation) or a very high single dose (radiosurgery) are not conclusive. Irradiation with a high single dose of 10 Gy of glioblastoma mouse tumours induced tumour growth retardation, increased the influx of CD8+ T cells and decreased that of Treg. However, significant improvement of long-term survival was only achieved when combining radiosurgery with blockade of the immune checkpoint molecule programmed cell death protein 1 (PD-1) [91]. While a single high dose of 20 Gy was as effective as 3 × 8 Gy or 5 × 6 Gy in retarding growth of the irradiated tumour, only fractionated irradiation in combination with an antibody against the immune checkpoint protein cytotoxic T-lymphocyte antigen 4 (CTLA-4) induced tumour growth retardation also outside of the irradiation field, as here shown in a mouse breast carcinoma model [92]. In ex vivo assays with human tumour and immune cells, the activation of DCs was similar when getting into contact with norm- or hypofractionated irradiated colorectal cancer cells, but much less after a single irradiation with 15 Gy [93].

Nevertheless, the current hypothesis is that higher doses might impact more strongly on intratumoural induction and production of type I interferon (IFN) with consecutive triggering of innate and adaptive immune mechanisms [94]. Ablative RT dramatically increases T-cell priming in draining lymphoid tissues, leading to both reduction of the primary tumour and of distant metastasis in dependence of CD8+ T cells. These immune responses are greatly amplified by addition of immunotherapy [95]. Lower single doses used in standard fractionation might especially impact on tumour vascularisation and therewith connected infiltration of immune cells [42, 96] (Fig. 7.4). Definite is that combination of RT with further immune activation induces the most striking anti-tumour immune reactions [85]. As already outlined shortly earlier, in response to radiation, tumour cells increase the surface expression of adhesion molecules, death receptors, stress-induced ligands, cryptic antigens and stimulatory molecules, such as MHC I and CD80, thereby becoming more sensitive to T cell-mediated cytotoxicity [86]. In the tumour microenvironment, pro-inflammatory molecules increase and maturation of DCs, antigen presentation and lymph node migration is fostered [97]. On the other hand, the immune cells might also be killed by radiation and pro-tumourigenic factors can be upregulated [98]. Consequently, radiation regimens have to be optimised and adjusted to maximise immunostimulatory functions and for the successful combination with other treatments, including IT [99].

Primarily radiation-induced immune suppression by, e.g. upregulation of PD-L1 on tumour cells has to be exploited for multimodal therapies with checkpoint inhibitors. These are currently the most promising therapies for induction of long-lasting anti-tumour effects as seen by a plateau in the patients' survival curves [100, 101]. Checkpoint-blockade inhibitors improve adaptive immune responses induced by the RT-mediated increase in tumour antigens and tumours with high somatic mutation prevalence do respond best [102]. Nevertheless, not all of these selected patients respond. Therefore, the most beneficial combination with selected RT schemes and the chronological sequence of application of RT and IT has still to be identified [60]. We just recently summarised preclinical and clinical data on how the immune modulating properties of RT can be exploited for the combined treatment of cancer with immune checkpoint inhibitors [74].

7.7 Immune Suppressive and Proliferation Promoting Effects of Radiotherapy

As almost always, two sides of the coin exist. X-rays can also reinforce immunosuppressive pathways (Fig. 7.3).

Treg are intrinsically radioresistant which might lead to their intratumoural enrichment during RT. In the tumour microenvironment, Treg acquire a highly suppressive phenotype which is further increased by RT [103]. This is one rationale for combination of RT with further IT, as already mentioned earlier for checkpoint inhibitors. Short-term ablation of Treg in advanced spontaneous tumours induces both high numbers of dead tumour cells and in combination with RT significantly reduced metastatic tumour progression concomitant with prolonged survival [104].

As Treg, Langerhans cells (LC) are quite resistant immune cells [105]. Recently, it was found that LC resisted damage by irradiation because of their intrinsic expression of the cyclin-dependent kinase inhibitor CDKN1A (p21). Further, the LC-mediated generation of Treg was enhanced by radiation and directly correlated with the growth of the skin tumour [106].

RT might further induce the macrophage colony-stimulating factor CSF1 in tumours and myeloid-derived suppressor cells accumulate in the tumour as well as in spleen, lung, lymph nodes and peripheral blood in a prostate cancer model [107]. This is again a convincing fact why especially combination of RT with immune modulation with CSF1 inhibitors in this case triggers beneficial anti-tumour responses (Fig. 7.4). Therapies have to be optimised in a way that the positive

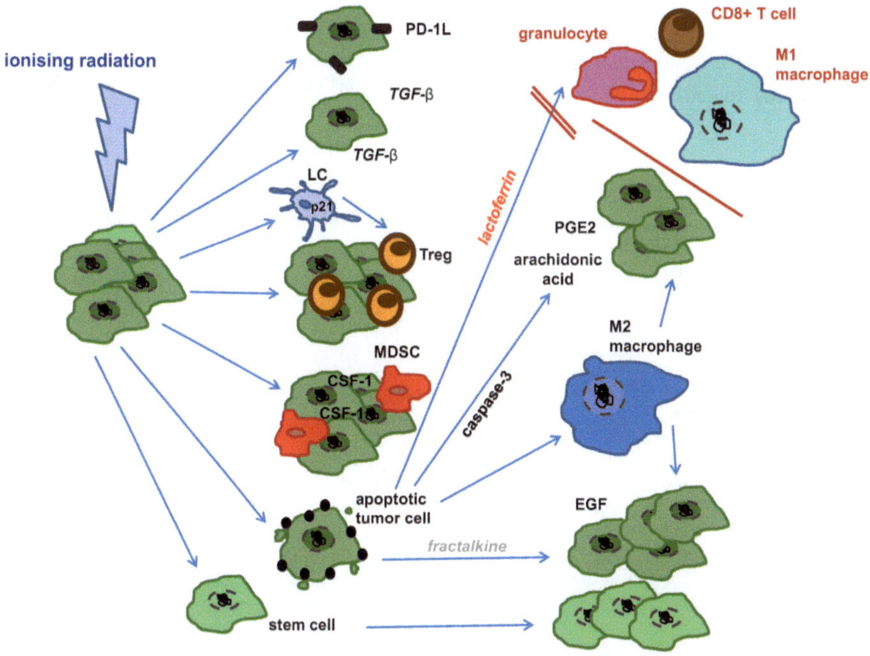

Fig. 7.3 Radiation-induced immune suppression and tumour cell proliferation. Tumour cells exposed to ionising radiation can acquire an immune suppressive phenotype characterised by the expression of checkpoint inhibitor ligands such as PD-L1, the secretion of TGF-β, the infiltration of regulatory T cells and myeloid-derived suppressor cells into the tumour, and by inducing immune suppressive apoptosis. The latter is connected to reduced infiltration of eosinophils into the tumour, by M2 macrophage polarisation and by caspase-3, fractalkine and EGF-dependent increased tumour cell proliferation. Further, during and after RT, radioresistant cancer stem cells could be selected as well as Langerhans cells that generate in turn immune suppressive Treg. *CSF-1* macrophage colony-stimulating factor 1, *EGF* epidermal growth factor, *LC* Langerhans cell, *MDSC* myeloid-derived suppressor cell, *PD-L1* programmed cell death protein 1 ligand, *PGE2* prostaglandin E2, *TGF-β* transforming growth factor beta, *Treg* regulatory T cell

immunological impact of RT on anti-cancer responses outweighs the negative ones [108, 109]. Recently, it was demonstrated that granulocyte-macrophage colony-stimulating factor as a potent stimulator of DC maturation in combination with local RT generates abscopal responses in patients with metastatic solid tumours such as non-small cell lung and breast cancer [110].

Again, we should also have in mind the local as well as systemic consequences of RT. Apoptosis induction by RT is beneficial with regard to local tumour cell killing, but not inevitably from the immunological point of view [111]. Ford et al. recently demonstrated for B cell lymphomas that apoptotic tumour cells promote tumour growth, angiogenesis and accumulation of tumour-associated macrophages (TAM) resulting from in situ macrophage proliferation [112]. TAM are one of the major inflammatory cells that infiltrate tumours and epidemiological studies depict a correlation

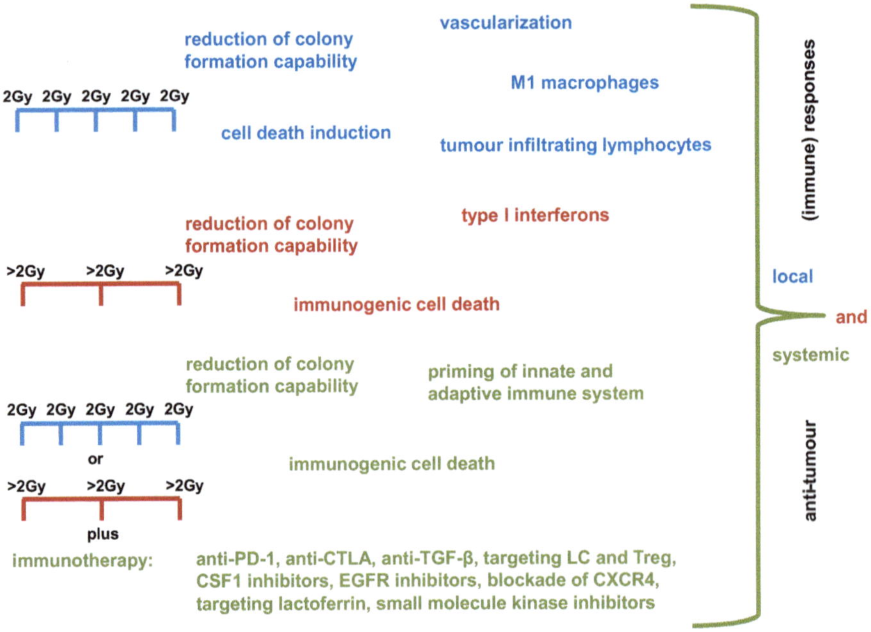

Fig. 7.4 Therapeutic exploitation of norm- and hypofractionated radiotherapy in combination with immune therapies for cancer treatment. For the induction of local and systemic tumour control, norm- and/or hypofractionated radiotherapy has to be combined with selected immune therapies to overcome the immune suppressive pathways depicted in Fig. 7.3. The most prominent events related to norm-fractionated radiotherapy, hypofractionated radiotherapy and immunotherapy are depicted in blue, red and green, respectively. *CSF-1* macrophage colony-stimulating factor 1, *CXCR4* chemokine (C-X-C Motif) receptor 4, *EGFR* epidermal growth factor receptor, *LC* Langerhans cell, *PD-1* programmed cell death protein 1, *TGF-β* transforming growth factor beta, *Treg* regulatory T cell

between TAM density and poor cancer prognosis [113]. Tissue destruction, even a small one occurring when taking a biopsy, may result in polarisation of macrophages to an M2 phenotype that could foster accelerated tumour progression [114].

Tumour cell apoptosis does thus not only impact on the immune system but also on proliferation of surrounding cells. Already in 1956 it was described that tumours killed by X-rays stimulate the proliferation of viable tumour cells [115]. It has been suggested that this is dependent on trophic substances derived from the tumour cells but also of the tumour bed, the microenvironment [116]. Recently, Chaurio et al. demonstrated that in an allogenic situation UV-B-irradiated apoptotic cells stimulate the growth of co-implanted viable tumour cells. These experiments were conducted in immune competent mice [117]. Since UV-B induces a mixture of apoptotic and necrotic cells it would be worth to examine in the future how distinct forms of tumour cells death impact on the proliferation of viable tumour cells and what mixture of cell death forms results predominantly in fostering of tumour cell proliferation and/or induction of anti-tumour immunity, respectively.

But what are the radiation-induced trophic substances that stimulate tumour cell proliferation? Apoptotic cells release a variety of "find-me" signalling factors, including nucleotides, the lipid lysophosphatidylcholine (LPC) and proteins such as fractalkine (summarised in [118]). The latter mediates the chemotaxis of macrophages to apoptotic lymphocytes [119]. Therefore, it might indirectly induce viable tumour cell proliferation by attracting macrophages into the tumour that are there polarised to M2 macrophages and directly by transactivation of the epidermal growth factor (EGF) pathway in the tumour cells [120]. This might be a further reason why combined treatments of tumours with RT and EGF receptor tyrosine kinase inhibitors are efficient [121]. Huang et al. demonstrated that caspase-3 is central in regulating the growth-promoting properties of dying cells by inducing the release of arachidonic acid and the production of prostaglandin E2 (PGE2) being a key regulator of tumour growth. Of special note is that caspase-3 was activated during RT [122]. RT-induced apoptosis may indeed lead to caspase 3-dependent tumour cell repopulation [46], but on the other hand caspase-3 is important to trigger immunogenic cancer cell death after hypofractionated irradiation [23]. Since TAM interacting with apoptotic tumour cells are central to activating multiple oncogenic pathways, to promote tumour cell growth and survival, angiogenesis, remodelling and metastasis [118] the aim should be to predominately induce necroptotic cancer cell death by RT [67, 68] and to concurrently target TAM, e.g. by pharmacologic blockade of chemokine (C-X-C Motif) receptor 4 (CXCR4) [123]. Massive necrosis should be induced to counteract the reduction of the immunogenicity of the necrotic cells by lactoferrin [124].

Interestingly, lactoferrin also functions as a "keep-out" signal to granulocytes. Since activated eosinophils were recently demonstrated to be essential for tumour rejection in the presence of tumour-specific CD8+ T cells and for an M1-like phenotype of macrophages [125], tumour promoting effects of apoptotic cells might also be connected to this. To summarise, apoptosis is central in conditioning the tumour microenvironment [126] (Figs. 7.1 and 7.3). This almost mandatorily demands that RT is combined with selected immune therapies to counteract the in part non-beneficial pro-tumourigenic effects of RT (Fig. 7.4). The same applies for possible selection of radioresistant cancer stem cells during and after RT [8, 127]. Mesenchymal stem cells are highly sensitive to small molecule receptor kinase inhibitors and combination treatments incorporating RT [128].

7.8 Conclusions

Even though approximately 60 % of patients with solid tumours are treated with RT, much fewer studies evaluating local therapies are published in high-impact oncology and medicine literature compared to systemic and targeted therapies [129].

Fortunately, a paradigm shift has been implemented during the last years: besides the local effects of RT on the DNA, also non-DNA targeted effects, the so-called systemic ones, do exist [130]. In former times it was predominantly publicised that only immune suppressive effects of RT exist. This has been questioned by many studies and it has become clear that a timely restricted radiation-induced decrease of immune cells does not automatically indicate that the immune system is functionally impaired [131]. The growing knowledge on the various forms of tumour cell death that can be induced by RT and/or CT has paved the way for combination of RT with IT [70, 132]. As it is common for the immune system that nearly every mechanism has wanted and unwanted effects independent of the existing state, also tumour cell death induction by RT can be beneficial for local and systemic tumour control (Fig. 7.2) and on the other hand even promote tumour cell proliferation and repopulation (Fig. 7.3). This highlights that a very sophisticated view on cell death induction by RT including the triggered cell death pathways and resulting cell death forms is mandatory [80, 133]. This becomes particularly important when further improving combination therapies consisting of RT, targeted therapies and immunotherapy (Fig. 7.4). Radiation-induced cell death is the mediator that broadens the modes of action of RT from a local level to a systemic one.

Acknowledgments This work is in part funded by the German Federal Ministry of Education and Research (BMBF; m4 Cluster, 16EX1021R and GREWIS, 02NUK017G) and the European Commission (DoReMi, European Atomic Energy Community's Seventh Framework Programme (FP7/2007–2011) under grant agreement no. 249689).

Conflict of Interest Statement All authors declare that they have no competing interests.

References

1. Lederman M. The early history of radiotherapy: 1895-1939. Int J Radiat Oncol Biol Phys. 1981;7:639–48.
2. Goodman LS, Wintrobe MM, Dameshek W, et al. Landmark article Sept. 21, 1946: nitrogen mustard therapy. Use of methyl-bis(beta-chloroethyl)amine hydrochloride and tris(beta-chloroethyl)amine hydrochloride for Hodgkin's disease, lymphosarcoma, leukemia and certain allied and miscellaneous disorders. By Louis S. Goodman, Maxwell M. Wintrobe, William Dameshek, Morton J. Goodman, Alfred Gilman and Margaret T. McLennan. JAMA. 1984;251:2255–61.
3. Kalbasi A, June CH, Haas N, et al. Radiation and immunotherapy: a synergistic combination. J Clin Invest. 2013;123:2756–63.
4. Aziz K, Nowsheen S, Pantelias G, et al. Targeting DNA damage and repair: embracing the pharmacological era for successful cancer therapy. Pharmacol Ther. 2012;133:334–50.
5. Ward JF. DNA damage produced by ionizing radiation in mammalian cells: identities, mechanisms of formation, and reparability. Prog Nucleic Acid Res Mol Biol. 1988;35:95–125.
6. Iliakis G. Backup pathways of NHEJ in cells of higher eukaryotes: cell cycle dependence. Radiother Oncol. 2009;92:310–5.
7. Bernier J, Hall EJ, Giaccia A. Radiation oncology: a century of achievements. Nat Rev Cancer. 2004;4:737–47.

8. Pajonk F, Vlashi E, McBride WH. Radiation resistance of cancer stem cells: the 4 R's of radiobiology revisited. Stem Cells. 2010;28:639–48.
9. Helbig L, Koi L, Bruchner K, et al. Hypoxia-inducible factor pathway inhibition resolves tumor hypoxia and improves local tumor control after single-dose irradiation. Int J Radiat Oncol Biol Phys. 2014;88:159–66.
10. Murata T, Akagi K, Kimura H, et al. Analysis of cell kinetics after irradiation by flow cytometry: proliferative ability of G2-blocked cells after cell sorting. Oncol Rep. 1998;5:385–8.
11. Maier P, Wenz F, Herskind C. Radioprotection of normal tissue cells. Strahlenther Onkol. 2014;190:745–52.
12. Travis EL, Parkins CS, Down JD, et al. Repair in mouse lung between multiple small doses of X rays. Radiat Res. 1983;94:326–39.
13. Williams MV, Denekamp J, Fowler JF. A review of alpha/beta ratios for experimental tumors: implications for clinical studies of altered fractionation. Int J Radiat Oncol Biol Phys. 1985;11:87–96.
14. Qi XS, White J, Li XA. Is alpha/beta for breast cancer really low? Radiother Oncol. 2011;100:282–8.
15. Durante M, Reppingen N, Held KD. Immunologically augmented cancer treatment using modern radiotherapy. Trends Mol Med. 2013;19:565–82.
16. Orth M, Lauber K, Niyazi M, et al. Current concepts in clinical radiation oncology. Radiat Environ Biophys. 2014;53:1–29.
17. Amornwichet N, Oike T, Shibata A, et al. Carbon-ion beam irradiation kills X-ray-resistant p53-null cancer cells by inducing mitotic catastrophe. PLoS One. 2014;9, e115121.
18. Fragkos M, Beard P. Mitotic catastrophe occurs in the absence of apoptosis in p53-null cells with a defective G1 checkpoint. PLoS One. 2011;6, e22946.
19. Castedo M, Perfettini JL, Roumier T, et al. Cell death by mitotic catastrophe: a molecular definition. Oncogene. 2004;23:2825–37.
20. Hayflick L. The limited in vitro lifetime of human diploid cell strains. Exp Cell Res. 1965;37:614–36.
21. Kurz DJ, Decary S, Hong Y, et al. Senescence-associated (beta)-galactosidase reflects an increase in lysosomal mass during replicative ageing of human endothelial cells. J Cell Sci. 2000;113(Pt 20):3613–22.
22. Jones KR, Elmore LW, Jackson-Cook C, et al. p53-Dependent accelerated senescence induced by ionizing radiation in breast tumour cells. Int J Radiat Biol. 2005;81:445–58.
23. Kotter B, Frey B, Winderl M, et al. The in vitro immunogenic potential of caspase-3 proficient breast cancer cells with basal low immunogenicity is increased by hypofractionated irradiation. Radiat Oncol. 2015;10:197.
24. Kroemer G, Galluzzi L, Vandenabeele P, et al. Classification of cell death: recommendations of the Nomenclature Committee on Cell Death 2009. Cell Death Differ. 2009;16:3–11.
25. Shelton DN, Chang E, Whittier PS, et al. Microarray analysis of replicative senescence. Curr Biol. 1999;9:939–45.
26. Coppe JP, Patil CK, Rodier F, et al. Senescence-associated secretory phenotypes reveal cell-nonautonomous functions of oncogenic RAS and the p53 tumor suppressor. PLoS Biol. 2008;6:2853–68.
27. Acosta JC, O'Loghlen A, Banito A, et al. Chemokine signaling via the CXCR2 receptor reinforces senescence. Cell. 2008;133:1006–18.
28. Kuilman T, Michaloglou C, Vredeveld LC, et al. Oncogene-induced senescence relayed by an interleukin-dependent inflammatory network. Cell. 2008;133:1019–31.
29. Glick D, Barth S, Macleod KF. Autophagy: cellular and molecular mechanisms. J Pathol. 2010;221:3–12.
30. Klionsky DJ, Abdalla FC, Abeliovich H, et al. Guidelines for the use and interpretation of assays for monitoring autophagy. Autophagy. 2012;8:445–544.
31. Dalby KN, Tekedereli I, Lopez-Berestein G, et al. Targeting the prodeath and prosurvival functions of autophagy as novel therapeutic strategies in cancer. Autophagy. 2010;6:322–9.

32. Gewirtz DA. The four faces of autophagy: implications for cancer therapy. Cancer Res. 2014;74:647–51.
33. Shimizu S, Kanaseki T, Mizushima N, et al. Role of Bcl-2 family proteins in a non-apoptotic programmed cell death dependent on autophagy genes. Nat Cell Biol. 2004;6:1221–8.
34. Yang Y, Yang Y, Yang X, et al. Autophagy and its function in radiosensitivity. Tumour Biol. 2015;36:4079–87.
35. Liang XH, Jackson S, Seaman M, et al. Induction of autophagy and inhibition of tumorigenesis by beclin 1. Nature. 1999;402:672–6.
36. Zhang D, Tang B, Xie X, et al. The interplay between DNA repair and autophagy in cancer therapy. Cancer Biol Ther. 2015;16:1005–13.
37. SenGupta T, Torgersen ML, Kassahun H, et al. Base excision repair AP endonucleases and mismatch repair act together to induce checkpoint-mediated autophagy. Nat Commun. 2013;4:2674.
38. Figueiredo N, Chora A, Raquel H, et al. Anthracyclines induce DNA damage response-mediated protection against severe sepsis. Immunity. 2013;39:874–84.
39. Medzhitov R. Septic shock: on the importance of being tolerant. Immunity. 2013;39:799–800.
40. Lin W, Yuan N, Wang Z, et al. Autophagy confers DNA damage repair pathways to protect the hematopoietic system from nuclear radiation injury. Sci Rep. 2015;5:12362.
41. Cottone L, Capobianco A, Gualteroni C, et al. 5-Fluorouracil causes leukocytes attraction in the peritoneal cavity by activating autophagy and HMGB1 release in colon carcinoma cells. Int J Cancer. 2015;136:1381–9.
42. Klug F, Prakash H, Huber PE, et al. Low-dose irradiation programs macrophage differentiation to an iNOS(+)/M1 phenotype that orchestrates effective T cell immunotherapy. Cancer Cell. 2013;24:589–602.
43. Kerr JF, Wyllie AH, Currie AR. Apoptosis: a basic biological phenomenon with wide-ranging implications in tissue kinetics. Br J Cancer. 1972;26:239–57.
44. Lauber K, Blumenthal SG, Waibel M, et al. Clearance of apoptotic cells: getting rid of the corpses. Mol Cell. 2004;14:277–87.
45. Voll RE, Herrmann M, Roth EA, et al. Immunosuppressive effects of apoptotic cells. Nature. 1997;390:350–1.
46. Lauber K, Munoz LE, Berens C, et al. Apoptosis induction and tumor cell repopulation: the yin and yang of radiotherapy. Radiat Oncol. 2011;6:176.
47. Belka C, Heinrich V, Marini P, et al. Ionizing radiation and the activation of caspase-8 in highly apoptosis-sensitive lymphoma cells. Int J Radiat Biol. 1999;75:1257–64.
48. Kerbel RS. Impact of multicellular resistance on the survival of solid tumors, including micrometastases. Invasion Metastasis. 1994;14:50–60.
49. Rafii A, Touboul C, Al Thani H, et al. Where cancer genomics should go next: a clinician's perspective. Hum Mol Genet. 2014;23:R69–75.
50. Jiang M, Axe T, Holgate R, et al. p53 binds the nuclear matrix in normal cells: binding involves the proline-rich domain of p53 and increases following genotoxic stress. Oncogene. 2001;20:5449–58.
51. Sengupta S, Harris CC. p53: traffic cop at the crossroads of DNA repair and recombination. Nat Rev Mol Cell Biol. 2005;6:44–55.
52. Baptiste N, Prives C. p53 in the cytoplasm: a question of overkill? Cell. 2004;116:487–9.
53. Sheikh MS, Burns TF, Huang Y, et al. p53-dependent and -independent regulation of the death receptor KILLER/DR5 gene expression in response to genotoxic stress and tumor necrosis factor alpha. Cancer Res. 1998;58:1593–8.
54. Yoon KW, Byun S, Kwon E, et al. Control of signaling-mediated clearance of apoptotic cells by the tumor suppressor p53. Science. 2015;349:1261669.
55. Wakatsuki M, Ohno T, Iwakawa M, et al. p73 protein expression correlates with radiation-induced apoptosis in the lack of p53 response to radiation therapy for cervical cancer. Int J Radiat Oncol Biol Phys. 2008;70:1189–94.
56. Kroemer G, Galluzzi L, Kepp O, et al. Immunogenic cell death in cancer therapy. Annu Rev Immunol. 2013;31:51–72.

57. de Bruyn M, Wiersma VR, Helfrich W, et al. The ever-expanding immunomodulatory role of calreticulin in cancer immunity. Front Oncol. 2015;5:35.
58. Tesniere A, Schlemmer F, Boige V, et al. Immunogenic death of colon cancer cells treated with oxaliplatin. Oncogene. 2010;29:482–91.
59. Zitvogel L, Kepp O, Senovilla L, et al. Immunogenic tumor cell death for optimal anticancer therapy: the calreticulin exposure pathway. Clin Cancer Res. 2010;16:3100–4.
60. Frey B, Rubner Y, Wunderlich R, et al. Induction of abscopal anti-tumor immunity and immunogenic tumor cell death by ionizing irradiation—implications for cancer therapies. Curr Med Chem. 2012;19:1751–64.
61. Gaipl US, Sheriff A, Franz S, et al. Inefficient clearance of dying cells and autoreactivity. Curr Top Microbiol Immunol. 2006;305:161–76.
62. Kocher M, Wittig A, Piroth MD, et al. Stereotactic radiosurgery for treatment of brain metastases. A report of the DEGRO Working Group on Stereotactic Radiotherapy. Strahlenther Onkol. 2014;190:521–32.
63. Krysko O, Love Aaes T, Bachert C, et al. Many faces of DAMPs in cancer therapy. Cell Death Dis. 2013;4, e631.
64. Holler N, Zaru R, Micheau O, et al. Fas triggers an alternative, caspase-8-independent cell death pathway using the kinase RIP as effector molecule. Nat Immunol. 2000;1:489–95.
65. Degterev A, Hitomi J, Germscheid M, et al. Identification of RIP1 kinase as a specific cellular target of necrostatins. Nat Chem Biol. 2008;4:313–21.
66. Pasparakis M, Vandenabeele P. Necroptosis and its role in inflammation. Nature. 2015;517:311–20.
67. Fulda S. Therapeutic exploitation of necroptosis for cancer therapy. Semin Cell Dev Biol. 2014;35:51–6.
68. Mantel F, Frey B, Haslinger S, et al. Combination of ionising irradiation and hyperthermia activates programmed apoptotic and necrotic cell death pathways in human colorectal carcinoma cells. Strahlenther Onkol. 2010;186:587–99.
69. Werthmoller N, Frey B, Wunderlich R, et al. Modulation of radiochemoimmunotherapy-induced B16 melanoma cell death by the pan-caspase inhibitor zVAD-fmk induces anti-tumor immunity in a HMGB1-, nucleotide- and T-cell-dependent manner. Cell Death Dis. 2015;6, e1761.
70. Kepp O, Senovilla L, Vitale I, et al. Consensus guidelines for the detection of immunogenic cell death. Oncoimmunology. 2014;3, e955691.
71. Stadler K, Frey B, Munoz LE, et al. Photopheresis with UV-A light and 8-methoxypsoralen leads to cell death and to release of blebs with anti-inflammatory phenotype in activated and non-activated lymphocytes. Biochem Biophys Res Commun. 2009;386:71–6.
72. Pineda JR, Daynac M, Chicheportiche A, et al. Vascular-derived TGF-beta increases in the stem cell niche and perturbs neurogenesis during aging and following irradiation in the adult mouse brain. EMBO Mol Med. 2013;5:548–62.
73. Xie Y, Bai O, Yuan J, et al. Tumor apoptotic bodies inhibit CTL responses and antitumor immunity via membrane-bound transforming growth factor-beta1 inducing CD8+ T-cell anergy and CD4+ Tr1 cell responses. Cancer Res. 2009;69:7756–66.
74. Derer A, Deloch L, Rubner Y, et al. Radio-immunotherapy-induced immunogenic cancer cells as basis for induction of systemic anti-tumor immune responses—pre-clinical evidence and ongoing clinical applications. Front Immunol. 2015;6:1–19.
75. Multhoff G, Pockley AG, Streffer C, et al. Dual role of heat shock proteins (HSPs) in anti-tumor immunity. Curr Mol Med. 2012;12:1174–82.
76. Galluzzi L, Senovilla L, Zitvogel L, et al. The secret ally: immunostimulation by anticancer drugs. Nat Rev Drug Discov. 2012;11:215–33.
77. Fucikova J, Moserova I, Truxova I, et al. High hydrostatic pressure induces immunogenic cell death in human tumor cells. Int J Cancer. 2014;135:1165–77.
78. Kang TW, Yevsa T, Woller N, et al. Senescence surveillance of pre-malignant hepatocytes limits liver cancer development. Nature. 2011;479:547–51.
79. Meng Y, Efimova EV, Hamzeh KW, et al. Radiation-inducible immunotherapy for cancer: senescent tumor cells as a cancer vaccine. Mol Ther. 2012;20:1046–55.

80. Lauber K, Ernst A, Orth M, et al. Dying cell clearance and its impact on the outcome of tumor radiotherapy. Front Oncol. 2012;2:116.
81. Rovere P, Vallinoto C, Bondanza A, et al. Bystander apoptosis triggers dendritic cell maturation and antigen-presenting function. J Immunol. 1998;161:4467–71.
82. Hanahan D, Weinberg RA. Hallmarks of cancer: the next generation. Cell. 2011;144:646–74.
83. Dunn GP, Bruce AT, Ikeda H, et al. Cancer immunoediting: from immunosurveillance to tumor escape. Nat Immunol. 2002;3:991–8.
84. Candeias SM, Gaipl US. The immune system in cancer prevention, development and therapy. Anticancer Agents Med Chem. 2015;16:101–7.
85. Frey B, Rubner Y, Kulzer L, et al. Antitumor immune responses induced by ionizing irradiation and further immune stimulation. Cancer Immunol Immunother. 2014;63:29–36.
86. Hodge JW, Ardiani A, Farsaci B, et al. The tipping point for combination therapy: cancer vaccines with radiation, chemotherapy, or targeted small molecule inhibitors. Semin Oncol. 2012;39:323–39.
87. Vanpouille-Box C, Diamond JM, Pilones KA, et al. TGFbeta is a master regulator of radiation therapy-induced antitumor immunity. Cancer Res. 2015;75:2232–42.
88. Demaria S, Ng B, Devitt ML, et al. Ionizing radiation inhibition of distant untreated tumors (abscopal effect) is immune mediated. Int J Radiat Oncol Biol Phys. 2004;58:862–70.
89. Mavragani IV, Laskaratou DA, Frey B, et al. Key mechanisms involved in ionizing radiation-induced systemic effects. A current review. Toxicol Res. 2016;5:12–33.
90. Prasanna A, Ahmed MM, Mohiuddin M, et al. Exploiting sensitization windows of opportunity in hyper and hypo-fractionated radiation therapy. J Thorac Dis. 2014;6:287–302.
91. Zeng J, See AP, Phallen J, et al. Anti-PD-1 blockade and stereotactic radiation produce long-term survival in mice with intracranial gliomas. Int J Radiat Oncol Biol Phys. 2013;86:343–9.
92. Dewan MZ, Galloway AE, Kawashima N, et al. Fractionated but not single-dose radiotherapy induces an immune-mediated abscopal effect when combined with anti-CTLA-4 antibody. Clin Cancer Res. 2009;15:5379–88.
93. Kulzer L, Rubner Y, Deloch L, et al. Norm- and hypo-fractionated radiotherapy is capable of activating human dendritic cells. J Immunotoxicol. 2014;11:328–36.
94. Burnette BC, Liang H, Lee Y, et al. The efficacy of radiotherapy relies upon induction of type i interferon-dependent innate and adaptive immunity. Cancer Res. 2011;71:2488–96.
95. Lee Y, Auh SL, Wang Y, et al. Therapeutic effects of ablative radiation on local tumor require CD8+ T cells: changing strategies for cancer treatment. Blood. 2009;114:589–95.
96. Ganss R, Ryschich E, Klar E, et al. Combination of T-cell therapy and trigger of inflammation induces remodeling of the vasculature and tumor eradication. Cancer Res. 2002;62:1462–70.
97. Hatfield P, Merrick A, Harrington K, et al. Radiation-induced cell death and dendritic cells: potential for cancer immunotherapy? Clin Oncol (R Coll Radiol). 2005;17:1–11.
98. Barcellos-Hoff MH, Derynck R, Tsang ML, et al. Transforming growth factor-beta activation in irradiated murine mammary gland. J Clin Invest. 1994;93:892–9.
99. Shahabi V, Postow MA, Tuck D, et al. Immune-priming of the tumor microenvironment by radiotherapy: rationale for combination with immunotherapy to improve anticancer efficacy. Am J Clin Oncol. 2015;38:90–7.
100. Ott PA, Hodi FS, Robert C. CTLA-4 and PD-1/PD-L1 blockade: new immunotherapeutic modalities with durable clinical benefit in melanoma patients. Clin Cancer Res. 2013;19:5300–9.
101. Schadendorf D, Hodi FS, Robert C, et al. Pooled analysis of long-term survival data from phase II and phase III trials of ipilimumab in unresectable or metastatic melanoma. J Clin Oncol. 2015;33:1889–94.
102. Alexandrov LB, Nik-Zainal S, Wedge DC, et al. Signatures of mutational processes in human cancer. Nature. 2013;500:415–21.
103. Persa E, Balogh A, Safrany G, et al. The effect of ionizing radiation on regulatory T cells in health and disease. Cancer Lett. 2015;368:252–61.

104. Bos PD, Plitas G, Rudra D, et al. Transient regulatory T cell ablation deters oncogene-driven breast cancer and enhances radiotherapy. J Exp Med. 2013;210:2435–66.
105. Durakovic N, Bezak KB, Skarica M, et al. Host-derived Langerhans cells persist after MHC-matched allografting independent of donor T cells and critically influence the alloresponses mediated by donor lymphocyte infusions. J Immunol. 2006;177:4414–25.
106. Price JG, Idoyaga J, Salmon H, et al. CDKN1A regulates Langerhans cell survival and promotes Treg cell generation upon exposure to ionizing irradiation. Nat Immunol. 2015;16:1060–8.
107. Xu J, Escamilla J, Mok S, et al. CSF1R signaling blockade stanches tumor-infiltrating myeloid cells and improves the efficacy of radiotherapy in prostate cancer. Cancer Res. 2013;73:2782–94.
108. Frey B, Gaipl US. Radio-immunotherapy: the focused beam expands. Lancet Oncol. 2015;16:742–3.
109. Zitvogel L, Kroemer G. Subversion of anticancer immunosurveillance by radiotherapy. Nat Immunol. 2015;16:1005–7.
110. Golden EB, Chhabra A, Chachoua A, et al. Local radiotherapy and granulocyte-macrophage colony-stimulating factor to generate abscopal responses in patients with metastatic solid tumours: a proof-of-principle trial. Lancet Oncol. 2015;16:795–803.
111. Lauber K, Herrmann M. Tumor biology: with a little help from my dying friends. Curr Biol. 2015;25:R198–201.
112. Ford CA, Petrova S, Pound JD, et al. Oncogenic properties of apoptotic tumor cells in aggressive B cell lymphoma. Curr Biol. 2015;25:577–88.
113. Biswas SK, Allavena P, Mantovani A. Tumor-associated macrophages: functional diversity, clinical significance, and open questions. Semin Immunopathol. 2013;35:585–600.
114. Weber M, Moebius P, Buttner-Herold M, et al. Macrophage polarisation changes within the time between diagnostic biopsy and tumour resection in oral squamous cell carcinomas—an immunohistochemical study. Br J Cancer. 2015;113:510–9.
115. Revesz L. Effect of tumour cells killed by x-rays upon the growth of admixed viable cells. Nature. 1956;178:1391–2.
116. van den Brenk HA, Crowe MC, Stone MG. Reactions of the tumour bed to lethally irradiated tumour cells, and the Revesz effect. Br J Cancer. 1977;36:94–104.
117. Chaurio R, Janko C, Schorn C, et al. UVB-irradiated apoptotic cells induce accelerated growth of co-implanted viable tumor cells in immune competent mice. Autoimmunity. 2013;46:317–22.
118. Willems JJ, Arnold BP, Gregory CD. Sinister self-sacrifice: the contribution of apoptosis to malignancy. Front Immunol. 2014;5:299.
119. Truman LA, Ford CA, Pasikowska M, et al. CX3CL1/fractalkine is released from apoptotic lymphocytes to stimulate macrophage chemotaxis. Blood. 2008;112:5026–36.
120. Tardaguila M, Mira E, Garcia-Cabezas MA, et al. CX3CL1 promotes breast cancer via transactivation of the EGF pathway. Cancer Res. 2013;73:4461–73.
121. Moschini I, Dell'Anna C, Losardo PL, et al. Radiotherapy of non-small-cell lung cancer in the era of EGFR gene mutations and EGF receptor tyrosine kinase inhibitors. Future Oncol. 2015;11:2329–42.
122. Huang Q, Li F, Liu X, et al. Caspase 3-mediated stimulation of tumor cell repopulation during cancer radiotherapy. Nat Med. 2011;17:860–6.
123. Hughes R, Qian BZ, Rowan C, et al. Perivascular M2 macrophages stimulate tumor relapse after chemotherapy. Cancer Res. 2015;75:3479–91.
124. Miles K, Clarke DJ, Lu W, et al. Dying and necrotic neutrophils are anti-inflammatory secondary to the release of alpha-defensins. J Immunol. 2009;183:2122–32.
125. Carretero R, Sektioglu IM, Garbi N, et al. Eosinophils orchestrate cancer rejection by normalizing tumor vessels and enhancing infiltration of CD8(+) T cells. Nat Immunol. 2015;16:609–17.
126. Gregory CD, Pound JD. Cell death in the neighbourhood: direct microenvironmental effects of apoptosis in normal and neoplastic tissues. J Pathol. 2011;223:177–94.

127. Oishi N, Wang XW. Novel therapeutic strategies for targeting liver cancer stem cells. Int J Biol Sci. 2011;7:517–35.
128. Nicolay NH, Sommer E, Perez RL, et al. Mesenchymal stem cells are sensitive to treatment with kinase inhibitors and ionizing radiation. Strahlenther Onkol. 2014;190:1037–45.
129. Holliday EB, Ahmed AA, Yoo SK, et al. Does cancer literature reflect multidisciplinary practice? A systematic review of oncology studies in the medical literature over a 20-year period. Int J Radiat Oncol Biol Phys. 2015;92:721–31.
130. Kadhim M, Salomaa S, Wright E, et al. Non-targeted effects of ionising radiation—implications for low dose risk. Mutat Res. 2013;752:84–98.
131. Belka C, Ottinger H, Kreuzfelder E, et al. Impact of localized radiotherapy on blood immune cells counts and function in humans. Radiother Oncol. 1999;50:199–204.
132. Demaria S, Golden EB, Formenti SC. Role of local radiation therapy in cancer immunotherapy. JAMA Oncol. 2015;1(9):1325–32.
133. Frey B, Hehlgans S, Rodel F, et al. Modulation of inflammation by low and high doses of ionizing radiation: implications for benign and malign diseases. Cancer Lett. 2015;368(2):230–7.

Chapter 8
Novel Approaches to Apoptosis-Inducing Therapies

Mike-Andrew Westhoff, Nicolas Marschall, and Klaus-Michael Debatin

Abstract Induction of apoptotic programmed cell death is one of the underlying principles of most current cancer therapies. In this review, we discuss the limitations and drawbacks of this approach and identify three distinct, but overlapping strategies to avoid these difficulties and further enhance the efficacy of apoptosis-inducing therapies. We postulate that the application of multi-targeted small molecule inhibitor cocktails will reduce the risk of the cancer cell populations developing resistance towards therapy. Following from these considerations regarding population genetics and ecology, we advocate the reconsideration of therapeutic end points to maximise the benefits, in terms of quantity and quality of life, for the patients. Finally, combining both previous points, we also suggest an altered focus on the cellular and molecular targets of therapy, i.e. targeting the (cancer cells') interaction with the tumour microenvironment.

Keywords Adaptive therapy • Intratumour heterogeneity • Combination therapy • Inducer & sensitiser • Microenvironment • Bcl-2 family • IAPs • PI3K signalling • ERK signalling

8.1 Introduction

Resistance to cellular suicide programmes, such as apoptosis, is not only a defining feature of cancer [1, 2], but it is also a development which must emerge at the very early stages of disease seeding. If one considers that the physiological role of apoptosis is the clearance of damaged or excessive cells [3], individual cancer cells have to devise hiding/escape mechanisms prior to the establishment of tumour bulk. Yet, of the three traditional pillars of cancer therapy—surgery, radio- and chemotherapy—two are effective based on their ability to induce cell death. In essence, cancer

M.-A. Westhoff • N. Marschall • K.-M. Debatin (✉)
Department of Pediatrics and Adolescent Medicine, University Medical Center Ulm, Eythstrasse 24, 89075 Ulm, Germany
e-mail: klaus-michael.debatin@uniklinik-ulm.de

C.D. Gregory (ed.), *Apoptosis in Cancer Pathogenesis and Anti-cancer Therapy*, Advances in Experimental Medicine and Biology 930,
DOI 10.1007/978-3-319-39406-0_8

therapy tries to induce a cellular response in a population that only exists due to their ability to avoid precisely that response. Newer, promising treatment approaches, such as immunotherapy or the use of oncolytic viruses, still work on the same fundamental principles (whether they work *better* remains to be seen). While the therapeutic arsenal at our disposal has improved tremendously over the last few decades—mortality for selected cancers has decreased by 30–60 % in the last 40 years [4]—the underlying strategy has remained unaltered since Emil Grubbe treated the first cancer patient with radiation in 1896 and mustard gas was introduced as chemotherapy in 1946 [5]. Ever since then our primary aim has remained unwavering and unaltered: Attempt to cure the patient by killing all cancer cells. And while now standing on the shoulders of giants we are closer to this goal than ever before, it is precisely why we should take a step back and reassess our strategy and our aims.

Treating a cancer patient is like a chess game, where we have just begun to learn the rules while our opponent has been playing since time immemorial. We have been able to penetrate the opponent's defence and are defining several interesting strategies, but with the patients' lives at stake (both in terms of *quality* and *quantity*) we must not get overconfident. Sometimes playing not to lose, can be better than trying to win, while the best option still seems to be to cheat and take away our opponent's best pieces.

8.2 Basic Rules: Why Do Cancer Cells Acquire Resistance to Apoptosis?

Cancer cells have evolved multiple mechanisms to escape apoptosis and other forms of cell death. We know that both epigenetic changes, e.g. microenvironmental cues from non-transformed cells or extracellular matrix proteins [6], and genetic changes, the mutational transformation of proto-oncogenes to activated oncogenes and the inactivation of tumour suppressor genes, can mediate the underlying signalling that leads to cell death resistance. While these alterations can be specific to a given tumour, a tumour type or even a minor subpopulation within a neoplasm, they broadly fall into two distinct categories:

1. The inactivation of death signalling cascades. Probably the most obvious example for this is the inactivation of caspase 8, a central mediator of the extrinsic death signalling cascade [7]. Caspase 8 is inactivated in several cancers, such as lung [8], gastric [9] and hepatocellular cancer [10], as well as childhood malignancies like neuroblastoma [11] and can be achieved via different routes, such as inhibition of gene expression via promoter methylation [8, 11], or mutations that yield a non-functional protein [9, 10].
2. The enhanced activation of survival signalling cascades which override the death signal. A good example of this is the PI3K/Akt/mTOR pathway, which consists of some of the most frequently mutated proteins in cancer [12].

Importantly, therapy failure which we ascribe (correctly) to increased apoptosis/cell death resistance is obviously not due to a pre-emptive attempt of cancer cells to circumvent treatment-induced cytotoxicity, but rather an essential part of the a priori escape and defence mechanism. Thus, cancer cells may be resistant or sensitive to many therapeutic approaches, at least until treatment leads to selective pressure to increase resistance, as a secondary side effect. The establishment of a malignant growth puts considerable stress upon the transformed cells, both intrinsically due to increased proliferation, lack of growth signals and nutrients and extrinsically due to an initially hostile microenvironment and an active immune system. The natural response to this type of stress is cell death [3, 13]. As therapies also induce stress, mainly via DNA damage [14, 15], the underlying resistance mechanisms are also relevant here. A classic example for this is the transformation of cells via overexpression of the *myc* gene, which leads to increased cell cycle progression and proliferation, but also increased apoptosis [16, 17]. Cells which evade cell death do so by, for example, enhanced autophagy, the unfolded protein response [18] or activation of PI3K-mediated survival signalling [19], all mechanisms that are also implicated in therapy resistance [20–22]. Similar results were obtained when chick embryo fibroblasts were transformed with v-Src oncoprotein; here, cells also entered into enhanced rounds of cell cycle progression and increased apoptosis was prevented by FAK phosphorylation [23]. If one ablated this phosphorylation, cells were still proliferating at an increased rate, but were concurrently driven into a higher rate of apoptosis [23]. FAK, which is currently clinically investigated as a potential therapeutic target [24], interacts with other mediators of therapy resistance, such as PI3K and MEK [24, 25]—demonstrating the intricate interconnectedness of what we generally tend to consider independent survival pathways. In summary, a priori therapy resistance is the by-product of stress resistance that cancer cells need to acquire early on during transformation and may be instrumental in cloaking the entirety of the tumour from immune surveillance [26]. As apoptosis indication is a common, but by no means unique response to stress, i.e. there are other forms of cell death that can be induced under these circumstances [27], we would postulate that a more promising approach to novel cancer therapies is to block also survival signals, rather than to enhance the classical apoptosis pathway.

Interestingly, targeting the most obvious targets in cancer, i.e. the most frequently inactivated proteins that enhance death signalling, or the most often activated signalling cascades that mediate survival, should theoretically be a promising approach, as they represent an obvious target in the largest fraction of malignancies, but practically this has not been a great success. If one looks at the mutational landscape of cancers, p53, so-called the guardian of the genome, presents itself as an ideal candidate, being mutated in ~50 % of all tumours [12]. Yet, while several promising small molecules have been developed that either block the inhibitor of p53, MDM2 or chaperone mutant p53 back into its active wild-type configuration [28] clinical trials have only shown limited promise so far [29]. The same holds true for therapies that target the PI3K/Akt/mTOR pathway, the most frequently activated survival cascade in tumours [12]. Its major negative regulator, PTEN, has even been dubbed 'the new guardian of the genome' [30]. However the evaluation of pre-clinically promising molecules that interfere with this signalling cascade has been rather sobering in a clinical setting. For example in glioblastoma, a particularly aggressive brain tumour, the PI3K/Akt/mTOR signalling cascade is hyperactivated in ~88 % of all tumours [31, 32]. However, despite intensive research and several clinical trials [33, 34], modulation of this pathway has led to either controversial [35] or disappointing results [33]. To date two mTOR (complex 1) inhibitors, Everolimus and Temsirolimus, have been approved for glioblastoma therapy [34], while emerging in vivo data suggest that mTOR is not a particularly promising target in glioblastoma monotherapy [36]. We have argued elsewhere that the apparent failure successfully to target the PI3K pathway in cancer therapy is not due to overestimating its importance, but due to underestimating its complexity [37]. The same also seems to hold true with regards to the p53 signalling network [38]. In Sect. 8.3, we will argue that these findings which explain the previous failure concurrently open up the opportunity of a novel treatment approach, that of sequential dosing, which we predict will in future drastically increase in importance.

But even on the rare occasions when an obvious therapy target makes a good target, as, for example, is the case for BCR-ABL fusion protein, there are still considerable drawbacks. Chronic myelogenous leukaemia (CML) accounts for 1–2 cases per 100,000 capita for approximately 15 % of all diagnosed leukaemias and ~90 % present with the so-called Philadelphia Chromosome which is indicative for BCR-ABL expression [39]. First studies with the novel tyrosine kinase inhibitor (TKI) imatinib mesylate, marketed by Novartis as Gleevec (USA) or Glivec (Europe) euphorically indicated that treatment led to equal life expectancies in CML patient to those found in the general population [40]. However, it soon became apparent that while patients' quantity and quality of life were greatly improved by imatinib, treatment-resistant cancer cells frequently emerged upon monotherapy, highlighting the potency and limitations of TKI therapy with the requirement of the consecutive use of novel TKIs and leading to the introduction of combination therapies including several TKIs [41, 42].

8.3 Strategic Thoughts: Different Treatment Options

In the clinic one often encounters a greater or equally great fear of chemotherapy-induced side effects compared to the actual disease. While the first generations of chemotherapy were efficient due to rather unspecific mechanisms that caused DNA damage via several different routes, e.g. topoisomerase inhibition, direct DNA integration or stabilisation of microtubules, new pharmacological drugs are designed with more specific targets or groups of targets in mind and therefore fewer side effects [43].

Interestingly, the large-scale expression profiling data made possible in recent years by rapid advancements in technology have mostly not revealed a new mutational landscape, i.e. did not identify any novel potential targets, but specified/adjusted frequencies of mutations (summarised in [12], examples: glioblastoma in [44], high-risk neuroblastoma in [45]). Even when studies yield surprising results, it is most commonly due to the involvement of an already known oncogene in an additional cancer entity and not due to the identification of a novel candidate gene. For example, of the nine novel, i.e. unexpected, driver mutations identified in breast cancer, only one was found in a protein not implicated in other cancers [46]. This particular gene, *tbx3*, however, had already been previously associated with breast tissue-specific hypo- or aplasia [47]. In summary, we would argue that the genetic and epigenetic mechanisms within the mutated cancer cells are rather well understood and—in terms of pathways, if not on the level of protein or mutation—delineated, so that it is unlikely that novel apoptosis therapies will focus on new targets. Two qualifiers that need to be added to the previous statement, as this does not hold true for (a) all subpopulations of mutant cancer cells found within a tumour (how that affects therapies end points will be discussed in Sect. 8.4) and (b) the microenvironment, i.e. the non-mutated components of a tumour (how that affects future therapies is the subject of Sect. 8.5). Monotherapies, even with promising and potent new drugs—as discussed earlier in the context of imatinib—frequently lead to the emergence of resistance, often necessitating the application of different chemotherapies, to which, depending on the underlying alteration that caused resistance to the first treatment, the majority of tumour cells are also already resistant. Until the recent past the protocol response to this vicious circle was increased dosage of chemotherapy until the maximal tolerated doses (MTDs) are reached, under the full knowledge that side effects of treatment might kill the patient before the cancer will. This approach was only challenged with the introduction of the metronomic chemotherapy protocol some 10 years ago, where the continuous or more frequent administration of lower therapeutic doses is postulated [48]. While large-scale randomised trials are still needed to verify the general superiority of metronomic treatment compared to MTD therapy, the data strongly suggest that this is the case, particularly with respect to toxicity [48]. The concept of combination therapy was introduced to avoid emergence of resistance to metronomic therapy which—by definition—would necessitate an increase in dosage until once again the MTD is reached and therefore it was suggested to combine the inducer of cell death, i.e. chemo- and radiotherapy, with a sensitiser which makes the cancer cells more amenable to treatment, ideally without also sensitising normal, non-cancerous cells for apoptosis.

Since the early 1990s, when it was shown that the newly discovered human Bcl-2 protein could prevent cell death in *C. elegans* [49], no other proteins, with the possible exception of p53, have been so closely associated with cancer and, in particular, apoptosis resistance. For example, while Hanahan and Weinberg initially discuss the Bcl-2 family in their seminal hallmarks paper from 2000, they choose IGF production as a mechanistical example for apoptosis evasion [1]; however, by the time of their 2011 update they have selected pro-apoptotic BH3 mimetics as a key therapeutic strategy to target the *resisting cell death* hallmark [2]. It is not surprising that therapies targeting this protein family have been clinically evaluated since the mid-1990s (Fig. 8.1a). The initial antisense approach showed very little clinical efficacy and many of the first-generation small molecule inhibitors showed toxic off-target effects [50], which led to a waning interest in exploring modulation of Bcl-2-mediating

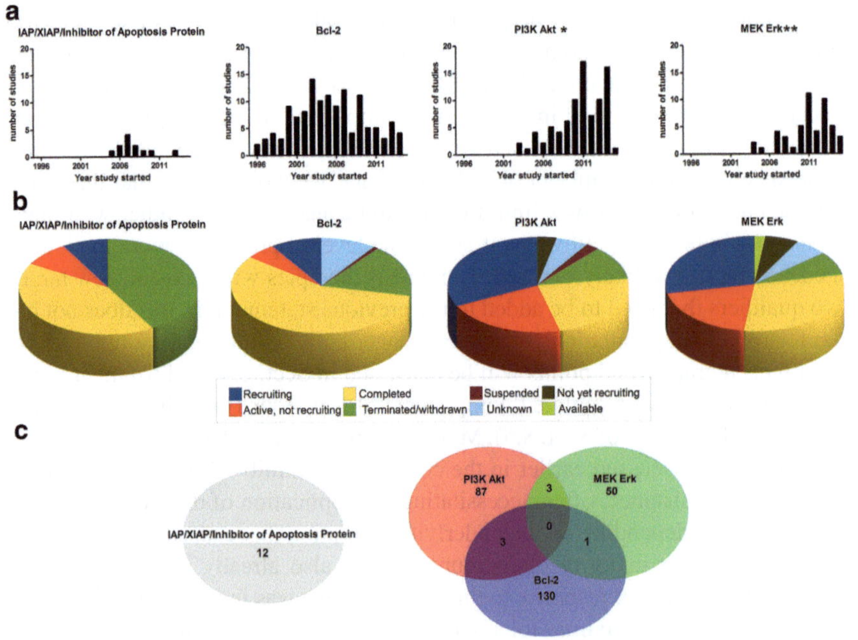

Fig. 8.1 Clinical Cancer Trials focusing on IAPs, Bcl-2 family members, PI3K- and MEK-mediated signalling. Utilising the searchable online database provided by the U.S. National Institutes of Health we analysed how many clinical cancer trials are currently focusing on one or more of the four signalling pathways we have identified as potentially interesting therapeutic targets. (**a**) Using search strings that either use a combination of 'IAP, XIAP or Inhibitor of Apoptosis Protein', 'BCL-2', 'PI3K, Akt' and 'MEK, Erk' we assessed how many clinical trials were initiated in the last two decades. (**b**) Using the same search criteria as earlier we looked at the current status of those trials identified. (**c**) Finally, we looked at how many trials focus on more than one of the four signalling pathways. As shown in the Venn diagram, even when including trials that just look at the signalling pathways as biomarkers and not potential therapeutic targets, those numbers are remarkably low. In conclusion, while a considerable set of data was obtained from trials focusing on BCL-2 family members, targeting either PI3K or MEK is a strategy that is currently still being developed and the clinical relevance of which needs to be further determined, as seen by the relatively high proportion of still recruiting and active studies. Surprisingly little work seems to focus on IAPs, which might be due to early difficulties as indicated by the high percentage of withdrawn/terminated studies. In this context it is important to note that early (less successful) trials concerning IAP and BCL-2 family inhibition were based on an anti-sense approach. Next generation inhibitors that target IAPs and proteins of the BCL-2 family by mimicking Smac and the BH3 domain, respectively, show a greater therapeutic potential. Far too little work is done on investigating possible combinations of inhibiting more than one of those four signalling cascades. *Asterisk* indicates that two studies had to be excluded from the analysis as they were withdrawn/terminated prior to being allocated an actual start date. *Double asterisk* indicates that one study had to be excluded from the analysis as it was classified as 'available' and thus does not have an official start date yet (*source*: https://clinicaltrials.gov/)

signalling as a therapeutic option (Fig. 8.1a, b). However, with recent improvements in rational design strategies and high-throughput screening facilities there is a recent renaissance in using Bcl-2 interacting molecules in treatment strategies, with Navitoclax and Obatoclax emerging as most promising clinical candidates [50].

Generally speaking, there are two alternative approaches, specific to the underlying cause of resistance as discussed earlier, aiming to increase efficacy of combination treatment. Both have their specific advantages and drawbacks: Either one aims to enhance the death signal or one can reduce the survival signalling. While conceptionally those two methodologies are fundamentally different—the former can be seen as an accelerant while the latter is akin to taking off the brakes—due to the intricate Byzantine nature of those signalling cascades involved, there often is a factual overlap between those two. For example, targeting XIAP, a negative regulator of the apoptosis cascade, via small molecule antagonists enhances the apoptotic signalling cascade, as does changing the balance of pro- and anti-apoptotic Bcl-2 proteins at the mitochondria [51]. Inhibiting the PI3K/Akt/mTOR survival cascade can have the same effect, as both XIAP and members of the Bcl-2 family can be regulated by Akt via phosphorylation [52–55]. Here, both enhancing apoptosis signalling and blocking survival signalling converge at the level of XIAP and Bcl-2 proteins. In addition, while XIAP can be regulated by the PI3K pathway, it is also upstream of MEK/ERK signalling [56], which in turn has been shown to regulate the Bcl-2 family of proteins [57]. A close look at four examples (summarised in Fig. 8.2) will elucidate their different strengths and weaknesses. Importantly, while targeting all four signalling pathways individually has shown pre-clinical promise that—at least, so far—has not translated into clinical success, we postulate that the success of novel treatment approaches will depend on the use of combination therapies.

8.3.1 *Targeting Apoptosis Signalling I: Inhibitor of Apoptosis Proteins*

The Inhibitor of Apoptosis Proteins (IAPs) are thus named as they were originally identified as preventing defensive apoptosis in insect cells infected with baculovirus [66]. Mirroring the discovery of the first viral oncogene/mammalian proto-oncogene Src, the viral protein was found to have both insect and vertebrates counterparts [66]. However, the eight mammalian IAPs and BIRPs (BIR-domain-containing proteins) which are defined by the presence of a BIR (baculoviral IAP repeat) have been shown to be involved in several additional aspects of cellular behaviour, such as the innate immune response and cell proliferation [67]. While the precise extent by which the various IAPs contribute to apoptosis resistance remains open to debate, the key contribution of XIAP remains unchallenged, as it binds and thus inhibits activity of caspases 3, 7 and 9 [56, 66]. The other two ubiquitously expressed IAPS, cIAP1 and cIAP2, share some functional overlap with XIAP, but they seem more predominately involved with RIP1-dependent necrosis or necroptosis [56]. Inhibition of IAPs mainly sensitises cells for, but does not induce, apoptosis, i.e. IAP inhibitors lend themselves to combination therapy, but not monotherapy [56]. Small molecule inhibitors of IAPs are generally modelled on the naturally occurring IAP antagonist Smac/DIABLO [68], but other strategies are also pursued [69]. The full potential of this approach was first demonstrated in 2006, when mice expressing

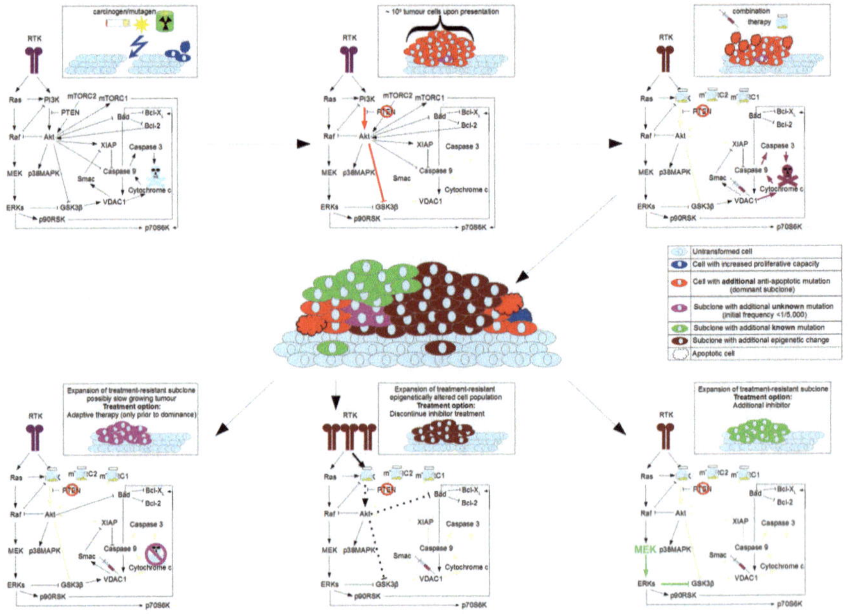

Fig. 8.2 The problem of modern (apoptosis-inducing) cancer therapies. Shown here is the (by no means comprehensive) interplay between the four signalling networks discussed in this review: the pro-survival cascades MEK/ERK and PI3K/Akt/mTOR and the apoptosis regulating IAPs and Bcl-2 family members. The *upper panel* shows the progression from the initial carcinogenic insult that induces increased proliferation, which only leads to tumour formation if additional cellular alterations protect cells from the enhanced stress (in this particular case, inactivation of PTEN that leads to enhanced PI3K-mediated signalling). Tumours upon clinical presentation often comprise more than 10^9 mutated cancer cells, but newer research indicates that therapy-resistant subclones may already be present at this stage. If the distribution of those clones is highly localised or their frequency is below the current detection limit of 1:5,000–10,000 cells, the clinician might not be able to consider them when devising a treatment strategy. Based on the available information a treatment is chosen, in the depicted case a combination of apoptosis-inducing chemotherapy and a pharmacological inhibitor of PI3K/mTOR. The *central image* depicts a recurrent tumour that has developed resistance to the chosen therapy. Note the highly mosaic nature of this tumour which potentially consists of multiple distinct subclones (this only considers the initial tumour site, distant metastatic loci further enhance complexity). The *lower panel* depicts the individual subclones and shows a possible treatment option for each population. The current challenge remains in finding an optimal strategy for combining these individual strategies. (Additional references used in this figure: [58–65])

an orthotopic human brain tumour were treated with Apo2L/TRAIL and a synthetic Smac peptide. While the Smac agonist alone had no discernible effect and TRAIL only marginally extended survival, the combination of both substances de facto cured the mice [70].

Another potential target within the IAP family is Survivin, which is expressed in foetal tissue and probably most cancers, but not in differentiated cells [71]. Unlike other IAPs Survivin expression is cell cycle-dependent [72] and plays a crucial role in cell division and genomic integrity [72, 73]. While this IAP has been also suggested to function as an inhibitor of caspase 9—interestingly, by associating with XIAP or another protein, the hepatitis B X-interacting protein [72]—its contribution to apoptosis resistance (at least, via caspase inhibition) seems to be dwarfed by its key roles in mitotic spindle checkpoint regulation, promotion of angiogenesis and preventing mitotic catastrophe, an alternative form of cell death [74–78]. In addition, Survivin together with XIAP has also been shown to up-regulate the production of extracellular matrix proteins and, thus, modulate the tumour microenvironment [68].

8.3.2 Targeting Apoptosis Signalling II: Bcl-2 Protein Family

The balance between the diverse members of the Bcl-2 protein family is the threshold that determines the mitochondrial commitment to apoptosis [79]. This protein family is defined by the presence of at least one Bcl-2 homology (BH) domain and can be divided into three subgroups: the BH3-only, the pro-survival and the pro-apoptotic proteins and thus their role in cell death has been described as a tripartite Bcl-2 apoptotic switch [79]: cellular stress stimulates BH3-only proteins that alter the balance between pro- and anti-apoptotic (or anti- and pro-survival) Bcl-2 family proteins leading to BAX and BAK forming pores within the outer membrane of the mitochondria and releasing Cytochrome c (essential for caspase 9 activation), as well as other proteins, such as Smac/DIABLO into the cytosol [79]. Although there are several different therapeutic approaches that modulate the Bcl-2 protein family [51], the most promising avenue seems to be the use of BH3 mimetics [80]. Basically, small molecules are designed to mimic the effect of BH3-only proteins and thus tip the balance between the various Bcl-2 family members towards apoptosis, i.e. treating cancers with these molecules can be sufficient to induce apoptosis even in the absence of additional stimuli, such as chemo- or radiotherapy [81]. Unfortunately high expression of the molecular targets of BH3 mimetics, often Bcl-2, Bcl-X_L, Mcl-1 or combinations thereof [80], is not sufficient to predict sensitivity towards treatment [82]. For a good sensitivity/strong response towards this treatment, some form of oncogenic addiction, whereby the intended targets of BH3 mimetics are already primed by naturally occurring BH3-only proteins, seems necessary [82]. In essence the cell population must have committed to countering pre-existing stress with the dependence on (up-regulating) Bcl-2 pro-survival proteins.

8.3.3 Targeting Survival Signalling I: PI3K/Akt/mTOR Pathway

In the context of tumour cells, the PI3K/Akt/mTOR signalling cascade is often considered a survival pathway, with its negative regulator PTEN, aforementioned 'new guardian of the genome' [30], being among the most frequently inactivated genes in cancer [83]. For example, in glioblastoma PI3K/Akt/mTOR signalling is elevated in ~88 % of all tumours [31, 84] and while there is a considerable body of promising pre-clinical data suggesting that modulation of this signalling cascade in glioblastoma sensitises these tumour cells for apoptosis (for example [85, 86]), clinical data has been less forthcoming [34]. We recently formulated three lines of reasoning which might explain this apparent discrepancy [37]: (1) Activation of PI3K signalling functions as a driver mutation for the cell of origin to reacquire stem cell characteristics/re-enter the cell cycle. (2) PI3K/Akt facilitates the invasive phenotype which is characteristic for GBM, both in terms of motility and survival under stress. (3) The central role of PI3K/Akt/mTOR in GBM biology has led us grossly to underestimate its importance.

The pre-clinical successes, as far as cell culture conditions and experimental animal models can be a good indicator, argue against (1) being the unique cause and the latter two explanations can be—should they be shown to be true—circumvented, as seen later. While combining a pharmacological inhibitor of PI3K/Akt/mTOR signalling, of which there are several [87], with an apoptosis-inducing signal is unlikely to have high therapeutic potential—both potentially due to subclones already present that are non-addicted to PI3K signalling, as well as the possibility of mutational escape—as part of a complex cocktail of inhibitors a substance targeting this signal cascade seems to be rather promising. In addition, inhibition of PI3K signalling can lead to cytostasis, for example in Hodgkin lymphoma [88], glioblastoma [89] and neuroblastoma [90], making any pharmacological inhibitor a potentially promising addition to a chronification regime.

8.3.4 Targeting Survival Signalling II: MEK/ERK Pathway

One of the major reasons that targeting a 'survival' cascade often does not fulfil its therapeutic potential is interconnectivity with other signalling cascades. In the case of PI3K/Akt/mTOR and MEK/ERK signalling, one can even wonder whether one would not be justified in treating those two pathways as one (for example [91]). With four MEK/MAP Kinase pathways and seven MEK proteins [92], there is great redundancy in the MEK/ERK signalling network, which has also been implicated in a large variety of cellular functions, such as learning, development, differentiation, as well as proliferation, survival and apoptosis [93]. While inhibition of signalling within the PI3K network has predominately focused on the PI3K and mTOR kinase function [94], within the MEK/ERK cascade MEK seems to be the preferred target [92, 95]. Small molecule inhibitors of B-RAF, an upstream activator of MEK, are already in clinical use and several MEK inhibitors currently being clinically evaluated have surprisingly little side effects [92]. Figure 8.1 gives an overview on the clinical evaluation of pharmacological inhibitors whose target molecules are discussed here.

8.3.5 Additional Considerations I: Timing

One additional aspect of multi-modular therapy that was highlighted by several recent studies is that it is not sufficient to identify a potent combination of inducer(s) and sensitiser(s), but also the sequential application needs to be considered. The maximal inhibition of a target—experimentally shown by the prolonged de-phosphorylation of a downstream molecule, while clinically deduced by maximal plasma levels—is not necessarily the best starting condition for the application of chemotherapy. Our own work in both neuroblastoma and glioblastoma indicates that a short inhibition of, in this particular case, PI3K signalling after prolonged exposure to chemotherapy can lead to increased apoptosis compared to blocking the signalling cascade prior to or concurrent with application of the death-inducing substance [96]. A similar effect was also observed when a pharmacological PI3K inhibitor was combined with radiotherapy [97]. These results are surprising only when, as alluded to already, we underestimate the complexity of the signalling cascade involved. Semantically we do ourselves no favour referring to certain pathways as 'survival signalling cascades' (as we ourselves do earlier). PI3K signalling regulates many aspects of cellular behaviour—such as proliferation, survival, metabolism and motility [6, 37]—and which aspect is most strongly affected is dependent on the molecular target(s) within the PI3K signalling network that is/are affected by the inhibitor, the cellular context and the length of inhibition. For example, in neuroblastoma maximal inhibition of the survival cascade leads to a measurable arrest in the G_1 phase of the cell cycle after 12 h [96]. This can increase the resistance of cancer cells towards chemo- and radiotherapy, as it is assumed that proliferating cells are most sensitive to apoptosis induction [98]: Resistance to apoptosis is lowered by inhibition of PI3K-mediated survival signalling and concurrently resistance to apoptosis is increased by inhibition of PI3K-mediated proliferation signalling. A similar argument can be made with regards to PI3K's role in regulating metabolism; inhibition of PI3K signalling can lead to the induction of autophagy [99], which given the precise context can either enhance [100] or inhibit [101] apoptosis induction. Importantly, the most efficiently timed combination in vitro is not necessarily more effective in vivo, as we were recently able to show using the RIST protocol, a complex combination therapy consisting of two sensitisers and two inducers of apoptosis [102]: Comparing the RIST (rapamycin, irinotecan, sunitinib, temozolomide) with the variant aRIST (alternative to rapamycin, the PI3K inhibitor GDC-0941), we found that the latter, which targets the PI3K cascade further upstream than the former, is more potent in vitro. However, this effect is not found in vivo, where the aRIST treatment barely differs from control. Interestingly, we obtained tantalising data that suggest the observed difference in efficacy is due to the aRIST variant being too potent. The tumour of intracranial xenografted, human glioma cells was less vascularised upon aRIST treatment, preventing future delivery of the chemotherapy and suggesting that sequential timing does not only need to be considered during the acute phase of combination therapy.

Furthermore, there are additional, recently identified mechanisms that indicate the importance of timing. Lee and colleagues showed that carefully timed inhibition of EGRR signalling can—through a process they termed "dynamic rewiring of oncogenic signalling pathways" (and which is independent of cell cycle progression/proliferation and initial DNA damage)—induce a less tumourigenic state in triple-negative breast cancer cells, that in essence makes them more receptive to doxorubicin-induced apoptosis [103]. Importantly, enhanced apoptosis could be achieved by 24 h pre-stimulation with erlotinib, but was lost when pre-stimulation was performed for 48 h [103].

Finally, another aspect of therapy that is affected by timing is the emergence of resistance. There are three common mechanisms by which resistance can emerge: de novo mutations, expansion of a pre-existing resistant subclone or epigenetic changes. The latter aspect has recently gained renewed interest [104–106] as it potentially is easy to revert. Stopping exposure to the inhibitor may lead to re-sensitisation of the cancer cell population [106], i.e. while high permanent concentrations of pharmacological inhibitors are often maintained to prevent mutational escape, treatment breaks upon the emergence of treatment resistance can reverse epigenetic escape. Interestingly, recent modelling work indicates that high phenotypic plasticity is only maintained by a reduction in average fitness [107], suggesting a potentially different treatment strategy depending on the dominant form of resistance that emerges within the tumour population upon treatment.

8.3.6 Additional Considerations II: Specificity

Perusing lists of current clinical trials (for example at clinicaltrials.gov) there seems to be an over-abundance of evaluating TKIs either as monotherapy or in combination with chemotherapeutic agents, the implicit hope presumably being to find the 'next imatinib', i.e. a substance which potently and specifically only targets tumour cells. We would argue that this is both unlikely and undesirable. While certain cancers—at least initially—respond to targeted therapies due to oncogenic addiction [108], this is by no means necessarily a universal feature of malignancies. The alternative model, referred to as 'Nile Distributary Problem', postulates that some cancers, for example glioblastoma, utilise "multiple cross-covering growth enhancing pathways to grow and avoid cytotoxic interventions" [109]. Under these circumstances a cocktail covering several signalling cascades appears a promising option; in the case of glioblastoma, for example, successes have been reported using the CUSP9 and the RIST treatment protocols ([109] and [102], respectively). Such a drug cocktail would also potentially reduce the second problem associated with TKI use: the emergence of secondary resistance [108]. Blocking an essential signal in a, per definition, highly genetically unstable population creates high pressure to mutationally escape the valley of the newly created fitness landscape. Indeed, the use of a single TKI as a mode to eliminate cancer would only be feasible in a monoclonal disease (i.e. all cancer cells are equally addicted to the TKI's target). As this is not the case, we will next discuss the consequences this has for future therapeutic advances.

8.3.7 *Additional Considerations III: Alternative, Alternate and Additional Treatment Options*

A mistake we often seem to make is falling back into binary thought patterns, supporting only a particular type of treatment (naturally most often the therapy one is most familiar with), thus not considering the full spectrum of potential combinations, such as:

- Alternative approaches once the primary treatment option loses its potency
- Alternate approaches that might lower the side effects for the patients without allowing the tumour cells to recover
- Additional treatments that enhance the potency of our chosen primary strategy.

In the last two decades, tumour immune therapy has increasingly presented itself as a potent and promising therapeutic approach [110, 111] and while not considered standard of care in most cancers, immune therapy should be considered next to surgery, radiotherapy and chemotherapy, as a key therapeutic intervention. While the use of monoclonal antibodies or indeed bone marrow transplantations for haematologic malignancies [110] are among the more obvious examples of its potential, the clinical evaluation of immune therapy is well on its way in a wide range of cancers, for example in lung cancer [112], metastatic melanoma [113] or glioblastoma [114]. Although there is some excellent literature discussing immune therapy (for example [115–117]), we would like to draw the reader's attention to the potential use of immune therapy combined with radiation [118] or targeted therapy [119], in essence suggesting that this approach should be an integral part of the above described combination therapy. Unmasking the tumour hiding from the immune system, for example by using dendritic cell vaccines or blocking inhibitory signals aimed at activated T cells, can elicit potent and long-lasting therapeutic effects by inducing an anti-tumour immune memory, often with negligible side effects for the patient [119]. Therefore, one could easily envision future therapeutic approaches that first reduce tumour burden by combination/targeted therapy allowing the patient's health to recover sufficiently for the immune system to continue eradicating the tumour and—more importantly—to initiate tumour surveillance in order to control minimal residual disease and prevent potential relapses.

8.4 Winning Versus Stalemate Draw: Therapeutic End Points

With the heightened sensitivity of our diagnostic arsenal the heterogeneous composition of tumours has become increasingly apparent. Currently, the detection limit for subclones is somewhere between 1:5,000 [120] and 1:10,000 [121], with multiregional sequencing on a single cell level remaining the only option to get a comprehensive picture of tumour heterogeneity [122]. Importantly, the individual populations within a tumour do not need to be in direct competition and even potentially co-operate [123], so that a minor subclone can be the driving force of tumour growth [124]. Yet, treatment decisions are predominately based on histological or molecular data which focuses on the most common, i.e. (perceived as) dominant subclonal population in a tumour.

It has been suggested that the very nature of cancer, the high genetic instability of the mutant cancer cell, combined with the microenvironmental forces that further enhance this instability, is so conducive to somatic mutations that the introduction of a strong selective force, such as therapy, almost inevitably leads to the mutational escape of a tumour cell subpopulation and thus treatment failure [125]. In addition, treatment can also lead to the expansion of pre-existing subclones, if their genotypes are associated with a phenotype that leads to better survival under therapy [122]. The rapid clearance of the dominant tumour population by effective therapy can create an ecological niche which is then filled by these slow growing, but resistant subclones. We previously used the meteor strike that wiped out the dinosaurs and thus created a niche into which mammals could expand as a well-known example to illustrate the process under discussion [6]. This has been well documented in both haematologic malignancies and solid tumours; for example in leukaemia it was shown that relapse can be mediated by a pre-existing drug-resistant subclone [126], while the dominance of subclones in colorectal cancer was greatly affected by chemotherapy [127].

One potential strategy to prevent the emergence of therapy resistance forces us to consider the therapeutic end point which is to be achieved by the treatment. While undoubtedly the improvement of the patient's health should (and is) always at the forefront of a good clinician's mind, it is a logical fallacy to assume attempting to eradicate a malignancy is synonymous with the maximal extension of quantity and quality of a patient's life.

The relapse risks and the unwanted consequences of the death-inducing effects of current therapies are well known; therefore, it must be legitimate to question whether novel approaches to apoptosis-inducing therapies should just focus on new molecular targets or application methods/schedules or whether a critical re-evaluation of the therapeutic end point is also an appropriate topic.

An alternative protocol, dubbed 'adaptive therapy' has been postulated and its mathematical and experimental foundation were worked out in 2009 [128]. Here the stabilisation of tumour size is the focus of medical intervention, the patient is closely monitored and treatment is only initiated when this stability is compromised, i.e. upon tumour expansion, and discontinued upon size stabilisation, implicitly acknowledging the need to prevent sub-clonal outgrowth [5, 128]. While such an approach that explicitly aims at chronifying a disease and—*in extremis*—eschews any attempt at healing the patients (in order to maximise quantity and quality of life!) is clinically not yet feasible for most tumours, we expect most difficulties not to arise from the translational aspect associated with this line of therapeutic research, but from ethical and monetary considerations. A treatment schedule would almost certainly be based on an apoptosis/cell death-inducing stimulus at a low metronomic dose to prevent tumour expansion, paired with a cocktail of sensitisers to reduce side effects, while concurrently preventing mutational escape by not focussing on a single signalling pathway. At least some sensitisers should ideally have anti-proliferative properties of their own, further aiding the stabilisation of the tumour population. Importantly, dose and/or schedule must be selected so as not to induce a too high selective force onto the tumour environment. Probably depending on type and state of the malignancy an additional focus should be on preventing local invasion and metastasis, as discussed in the following section.

But when to apply a chronifying schedule and when attempt to cure? While ultimately a decision can only ever be reached with the active involvement of the patient who should be the ultimate arbitrator of the future direction to be taken, we must ensure all relevant information is present and presented. Therefore, one must contemplate what a good guideline would be. Hypothetically, assuming epidemiological data suggesting a cancer has a 50:50 chance of being cured, what additional considerations are needed? Surely, if we want to compare cure with chronification a 5-year survival rate as definition of cured seems insufficient. How does 20 years cancer-free survival compare to 20 years chronic therapy? How to evaluate the loss of quality of life by having to undergo regular treatment and check-ups? Is epidemiological data sufficient to base a decision on it? How to take the personal background of the patient into account? Age will be an important factor, 5 years of cancer-free survival for a 70-year old would bring him/her towards normal life expectancy, so does it make sense to attempt chronification here, when maybe 20 years could be reached, but the patient is likely to die of unrelated causes much earlier? Importantly, while we understand cancer as a disease of old age, it is still among the leading causes of death in children and adolescents [129]. We have previously suggested that indicators towards answering these questions can be gleaned from other chronic diseases, such as (Type 1) insulin-dependent diabetes [5]. Here, chronification of the disease, i.e. the regular treatment with insulin, has led to an increase in average life expectancy of 15 years, when comparing subjects born between 1950 and 1964 to those born between 1965 and 1980 [130]. Indeed, a diabetic can—on average—lead a normal life with only few (although by no means minor) constrictions. Efforts in curing this disease rather than improving its management are minimal and there are no serious suggestions that a potential risky cure is preferable to the high quality and quantity of life that can be achieved by regular insulin treatment. There is no a priori reason why such an approach would not also work with some/many cancers. There is, however, no doubt that such an approach would entail considerable costs. For any form of adaptive therapy to be effective, close and regular monitoring of the tumour is absolutely essential. While novel approaches, such as liquid biopsies [131, 132], might prove viable alternatives in the future, currently expensive imaging techniques, such as MRI and PET scan, are the best options to get a comprehensive picture of the tumour state. Here, efforts to lower costs and reduce the health burden those imaging techniques exert on the patient need to be made.

We recently published our own experience with a variant of the adaptive therapy, where we describe the treatment of adolescent patients with relapsed, therapy-resistant glioblastoma/astrocytoma grade 4 [102]. Brain tumours are relatively rare in children, affecting between 1.7 and 4.1 per 100,000 children, but unfortunately around 4 % of these are high-grade astrocytoma [133]. While the survival probability with a 5-year survival of 24 % is slightly better than in adults [133], it is in absolute potential years of life lost one of the most devastating diseases. It is also incredibly difficult to treat, like adult astrocytoma grade 4, paediatric glioblastoma is also extremely aggressive and highly invasive, making localised treatment particularly hard [134]. However, as they are genetically distinct from their adult counterpart [135], many of the treatment options available for adult glioblastoma show little effect in paediatric brain tumours and due to the low number of incidences large clinical trials, even circumventing the difficulties associated with under-age patients, are difficult to establish [134]. In our house we have established in the context of a compassionate use setting the RIST protocol, which is a multi-modal, metronomic therapy approach combining two small molecule pharmacological inhibitors with low doses of chemotherapy [102]. Considering that this protocol is only used in adolescent glioblastoma patients where conventional therapy has already failed, i.e. has presumably led to a tumour that is highly resistant to treatment, it is remarkable how effective this approach can be in prolonging quantity and quality of life, the two patients whose clinical history we discussed were treated for 34 and more than 60 months, respectively, with the RIST therapy [102]. It is also gratifying to note the positive effect this prolonged survival with relatively little side effects has on the clinicians and nursing staff who accompanied the patients on their journey, not to mention the patients themselves and their families.

8.5 Redefining the Rules: Expanding the Cancer Definitions

A further consequence of applying evolutionary and ecological considerations to cancer therapies and highlighting the intratumour heterogeneity is to understand that a cancerous malignancy is not just a three-dimensional cluster of mutated tumour cells, but, in essence, a new tissue and, thus, a complex ecosystem [136]. Tumours consist of, as we discussed earlier, competing and cooperating populations of mutant cancer cells in a unique, specialised habitat made up from genetically unaltered (but often epigenetically distinct) supporting cells (as illustrated in [129]) and extracellular proteins, which—taken together—are referred to as the microenvironment.

The key role of the microenvironment in tumour progression is already highlighted in one of the oldest hypotheses of modern oncology, the ‘Soil and Seed’ hypothesis [137] and it has been long known that a tightly controlled and regulated microenvironment can prevent tumour formation even in the presence of transformed cells: Dolberg and Bissell showed that freshly hatched chicks develop palpable sarcomas within 1 week of infection with the Rous Sarcoma Virus, while avian embryos exposed to the virus do not form tumours. However, when infected embryonic cells are placed in cell culture they rapidly develop a transformed phenotype [138]. Here, the environment provided by the embryo is obviously (a) restrictive and (b) different from that encountered in a neonatal chick. Other examples from the literature also demonstrate a permissive role of the microenvironment, e.g. the contribution of proteolytic proteins by tumour-associated macrophages is essential for breast cancer cells to metastasise to the brain [139]. Indeed the contribution of the microenvironment seems to be a crucial determinant as to which site cancer cells can and will metastasise [139], which is of particular importance with respect to the—presumably still best described as speculative—premetastatic niche [140]. It has been proposed that non-tumour cells prepare a "nest" for malignant metastatic cells and that these alterations in the microenvironment precede and are essential for the actual establishment of a metastatic tumour [141]. This has been demonstrated for VEGFR1-expressing haematopoietic precursor cells that home in on specific sites within the bone marrow alter the existing structure and make it more amenable to migrating tumour cells [142]. Importantly, targeting the microenvironment altered by the presence of the mutant cancer cells and restoring its normal phenotype can be sufficient to prevent tumour progression [143]. This is particularly tempting as the microenvironment provides a genetically much more stable therapeutic target, making mutation escape and thus treatment resistance a more unlikely potential outcome. The disadvantage of targeting non-tumourigenic cells is that due to their lack of a distinguished genetic identity they are potentially a more difficult target, i.e. their lack of uniqueness might lead to increased side effects. One way to have the best of both worlds might be to target the points of interaction between the genetically unstable, but distinct tumour cells and their surroundings, be that extracellular components, such as fibronectin or collagen, genetically normal, but epigenetically altered cells, such as tumour-associated stroma, or other tumour cells. However, in the latter case, there is again the increased risk of mutational escape.

Already in 1972, the same year the term *apoptosis* was introduced to the wider scientific community [144], Durand and Sutherland showed in a series of experiments that the sensitivity of cells towards treatment is dependent on their interaction with their microenvironment, in this case predominately cell–cell contacts with other tumour cells [145, 146]. Over the last several decades this mechanism has been identified in a wide variety of cancers, such as multiple myeloma, various forms of lymphoma and leukaemia, as well as breast cancer, hepatocellular carcinoma, small cell lung cancer and glioblastoma ([147], updated in [6]), and we have come to refer to it as AMAR, or adhesion-mediated apoptosis resistance. AMAR can be further subdivided into three distinct, but overlapping categories: CAM-DR, cell-adhesion mediated drug resistance (for example [148–152]), CAM-RR, cell-adhesion mediated radiation resistance (for example [153, 154]) and anoikis resistance [155].

Importantly, many of the signalling cascades which have been identified as promising targets of small molecule inhibitors have also been implicated in mediating AMAR, for example the Bcl-2 family [156–158], IAPs [159], PI3K-mediated signalling [160–162], or—yet again emphasising the interconnectivity of these pathways—combinations thereof, e.g. the adhesion of acute myeloid leukaemia to extracellular matrix has been shown to activate both Bcl-2 and XIAP, in a PI3K-dependent manner ([163] and [164], respectively). The MAP Kinase signalling has also been implicated in mediating CAM-DR [165] and anoikis resistance [166].

This further emphasises the importance of these pathways, as cancer cells either have to activate them via different modes of adhesion [147] or mutationally/epigenetically activate them to avoid the risk of anoikis [166]. Is it, therefore, not sufficient to use small molecule inhibitors to block Bcl-2-, XIAP-, MAP Kinase- and PI3K-mediating signalling? After all, loss of cell polarity and cell-cell adhesion is also a prerequisite of invasion and metastasis during the epithelial-mesenchymal transition (EMT) [167, 168]. Potentially, targeting the interaction between tumour cells and microenvironment might make the tumour both more sensitive to apoptosis and more invasive. This very much depends on the choice of target, as dynamic adhesion is also essential for motility, and targeting these points can also block invasion, as we have recently shown for glioblastoma [169]. Furthermore, recent data suggest that circulating tumour cells exhibit a higher metastatic potential when they are present in (rare) circulating clusters compared to the more frequent individual cells [170].

While we suggest that the interaction of tumour cell and environment is a promising target to block, there is also an alternative strategy of activating the microenvironment. Here, not the destruction of the supporting structure for the mutated tumour cells is the focus of the intervention but the blocked immune response. For example, the MPDL3280A antibody, which blocks PD-L1-PD-1 interaction has been shown to enhance anti-cancer immunity in a clinical trial setting [171]. Interestingly, while the antibody's target is expressed on mutant tumour cells as well as on immune cells that reside in the microenvironment, among them macrophages, it is the interaction of the antibody on the latter which seems predominately to mediate the therapeutic effect [171]. Indeed the position of antigen-specific $CD8^+$ T cells within the tumour microenvironment seems to be the crucial determinant as to whether therapy is successful [172].

8.6 Conclusions

While we wish we could confidently predict that novel efficient approaches in apoptosis-inducing therapies will be sheer elegance in their simplicity, it is more likely that changes that will benefit the patients will come from reassessing molecular/cellular targets and therapeutic end points of existing therapies. Determining the aim of the therapy is certainly not as trivial as it sounds: While from a purely scientific standpoint the study of the diverse mechanisms evolved in cancer cell populations is fascinating and devising strategies to overcome these resistant mechanism is akin to an intellectually satisfying game of chess, from a clinical (and ultimately and most importantly from a humanitarian) point of view the maximal well-being of our patients must be our *summum bonum*. Importantly, this must not be equated to curing the malignancy. Therefore, the first step towards an open and honest debate in this regard is to acknowledge that the attempt of chronifying malignancies must—under certain circumstances—not be understood as the clinician's failure to cure the patient, but as following the highest Hippocratic ideals of *first, do no harm*.

The individual use of apoptosis-inducing therapies and the targeted outcome of the chosen treatment will/should most certainly be devised in an open and frank dialogue between patient and clinician, but generally speaking we identified three novel approaches which will become increasingly important for future treatment strategies:

1. There is an old maxim which postulates that a treatment without side effects can only be achieved by a treatment without effects. While this might seem a rather flippant or even callous way of putting it, with regards to cancer, which—compared to a disease-causing microorganism or virus—is genetically almost identical to the patient, this is unfortunately very true. Hence we see the common side effects with chemo- and radiotherapy, where the DNA damage that induces apoptosis in cancer cells also causes the death of other rapidly dividing cell populations, such as hair follicle stem cells and cells of the intestinal lining. While molecular targeted therapies are set to reduce the side effects by targeting specific, but generally speaking not unique molecules within the tumour, this also drastically enhances the selective pressure on these genetically highly unstable populations, leading to the emergence of resistance. There are compelling data suggesting that combination regimens are the most promising future option (for example [102], also reviewed in [173]). Here, a balance has to be struck between broad-spectrum efficacy, i.e. general TKIs, and increased specificity, i.e. ideally inhibitors which target tumour-specific proteins, such as Bcr-Abl [40], or tumour-specific mutations, for example the T790M mutation in EGFR [174]. Increased specificity leads to reduced off-target effects but will also frequently induce relapse due to the selection of therapy-resistant subclones within the tumour (for Bcr-Abl see [41, 42], for EGFR-driven tumours see [175]). In essence, here, treatment creates more aggressive tumours while only allowing a short reprieve for the patients. TKIs can cause some serious side effects [176], but with reduced specificity can not only block a signalling cascade to which tumour cells are addicted, but also inhibit potential escape routes which the malignant population could reach via mutation or dynamic rewiring of oncogenic signalling. On the downside, relatively unspecific TKIs could potentially also inhibit kinases that function as tumour suppressors and thus enhance tumour progression [177]. While only a few mutations are needed to initiate tumour progression, many malignancies harbour an excess of 10,000 mutations [178] and the total number of receptor tyrosine kinases expressed, i.e. potential targets, not only varies significantly between malignancies, but they are also often associated with different risk factors, such as a role in tumourigenesis, tumour survival and inhibition or enhancement of metastasis [177]. Finding a potent inhibitor combination for different categories of given malignancies, for different tumours and for different subclones within a tumour will be a formidable task. Experimentally, this will be challenging. We already described how misleading in vitro data might appear without in vivo confirmation [102], but also the reliability of animal models needs to be reassessed [179], in particular in terms of pharmacokinetics, drug delivery and interaction of drugs with the immune system.

2. "Chronification" is a term not found in the standard dictionaries of the English language, yet it has increasingly become an important term in medical science (for example [180–183]). With respect to cancer, the possibility of using chronification as a treatment strategy arises from rigorously applying population genetic and ecological models to the cancer ecosystem. While it has been suggested [136] that the foundation for the evolutionary theory of cancer was laid as far back as 1976 [184], it is only in the last decade that those theories have been used to devise new treatment strategies [128, 136, 185–189]. While this is doubtless—at least in part—due to improved experimental techniques that allow us meaningfully to address these questions, e.g. cancer genomics, next generation sequencing and utilising reverse phase protein array [190–192], we like to think that closer cooperation and appreciation between the different disciplines involved, e.g. the various branches of biology and medicine, as well as computer sciences, have contributed to the recent advances. The most pressing scientific question, as opposed to ethical and economic considerations, is when to use apoptosis-inducing therapies to cure the patients and when to reduce tumour burden and stabilise the disease. We suspect that critical for the decision-making process will be our ability to predict which mutational escape route will be taken by the treatment-induced arising dominant subclone, i.e. do we need to know the complete tumour heterogeneity to predict the most likely fitness peak, or can we predict the next likely fitness peak, i.e. will there be a preferred resistance mechanism arising, as seen—for example—upon treatment with the EGFR inhibitors erlotinib and gefitinib, where ~60 % of the resistances arising are mediated by the T790M mutation [174], and will we be able therapeutically to react to the evolved and selected resistance mechanism.
3. We feel that one of the most unappreciated aspects of cancer is the interaction of the mutated cancer cell with its microenvironment, i.e. the non-transformed cells and the extracellular matrix. The microenvironment is so altered by the presence of mutant cells and—in turn—such a potent influence on those cells, that we have argued that we need to re-evaluate our definition of cancer to include it as an integral part of the malignancy and therefore also as a potential target [129]. While the updated discussion on the hallmarks of cancer further emphasises the role of the microenvironment/the non-transformed cells within or surrounding the tumour compared to the original work from 2000 [1, 2], this expanded perspective has not yet reached the laboratory benches. A considerable amount of research and probably most large-scale screening assays are still performed on two-dimensional cultures of cells seeded on plastic. In addition, often established cell lines are used, the protein and gene expression profiles of which barely resemble that of the tumour entities they were derived from (for example glioblastoma versus glioblastoma cell line [193, 194]). In vivo models, mice in particular, are also limited in mimicking the complexity of human cancers [195], as they consist of highly inbred strains, often with little or no immune response and no gut flora. The importance of the latter points has recently been further highlighted by several pieces of data elucidating the role of macrophages, particularly in a bacteria-dependent immune response [196–198]. Macrophages

also serve as a potent reminder that we are still at the beginning of understanding the role of the microenvironment and how best to use it in future therapies [5]: Traditionally, the M1 phenotype has been associated with antitumourigenic properties of macrophages [199, 200], while the M2 phenotype has been connected to tumour-promoting TAMs (Tumour-associated Macrophages) [201]. However, this distinction seems far from universal, for example macrophages induce epithelial-mesenchymal transition in colorectal carcinoma cells independently of their phenotype [202]. Targeting the microenvironment, either to overcome AMAR [6, 147], or to modulate the immune checkpoint response in immunotherapy [115], is an active and very promising area of research; e.g. while writing this chapter two new studies came out highlighting the role of exosomes in mediating interactions between tumour cells and their microenvironment [203, 204].

Taken together, we are confident that apoptosis-inducing therapy will remain one of our primary tools in fighting cancer. However, we believe there is still potential greatly to enhance its potency by combining individual targeting strategies, both on a molecular and a cellular level, and by reassessing the intended therapeutic endpoint.

Acknowledgements We thank S. E. Barry for critically reading early drafts of the manuscript and B. Welz as well as S. Glatzel for their many years of expert secretarial assistance.

References

1. Hanahan D, Weinberg RA. The hallmarks of cancer. Cell. 2000;100(1):57–70.
2. Hanahan D, Weinberg RA. Hallmarks of cancer: the next generation. Cell. 2011; 144(5):646–74.
3. Renehan AG, Booth C, Potten CS. What is apoptosis, and why is it important? BMJ. 2001;322(7301):1536–8.
4. Esserman LJ, Thompson Jr IM, Reid B. Overdiagnosis and overtreatment in cancer: an opportunity for improvement. JAMA. 2013;310(8):797–8.
5. Stroh S, Debatin K-M, Westhoff M-A. Darwinian principles in cancer therapy. Eur Oncol Haematol. 2014;10(2):116–120.
6. Westhoff MA, Brühl O, Nonnenmacher L, Karpel-Massler G, Debatin KM. Killing me softly-future challenges in apoptosis research. Int J Mol Sci. 2014;15(3):3746–67.
7. Fulda S, Debatin KM. Modulation of apoptosis signaling for cancer therapy. Arch Immunol Ther Exp (Warsz). 2006;54(3):173–5.
8. Shivapurkar N, Toyooka S, Eby MT, Huang CX, Sathyanarayana UG, Cunningham HT, Reddy JL, Brambilla E, Takahashi T, Minna JD, Chaudhary PM, Gazdar AF. Differential inactivation of caspase-8 in lung cancers. Cancer Biol Ther. 2002;1(1):65–9.
9. Soung YH, Lee JW, Kim SY, Jang J, Park YG, Park WS, Nam SW, Lee JY, Yoo NJ, Lee SH. CASPASE-8 gene is inactivated by somatic mutations in gastric carcinomas. Cancer Res. 2005;65(3):815–21.
10. Soung YH, Lee JW, Kim SY, Sung YJ, Park WS, Nam SW, Kim SH, Lee JY, Yoo NJ, Lee SH. Caspase-8 gene is frequently inactivated by the frameshift somatic mutation 1225_1226delTG in hepatocellular carcinomas. Oncogene. 2005;24(1):141–7.

11. Casciano I, De Ambrosis A, Croce M, Pagnan G, Di Vinci A, Allemanni G, Banelli B, Ponzoni M, Romani M, Ferrini S. Expression of the caspase-8 gene in neuroblastoma cells is regulated through an essential interferon-sensitive response element (ISRE). Cell Death Differ. 2004;11(1):131–4.
12. Kandoth C, McLellan MD, Vandin F, Ye K, Niu B, Lu C, Xie M, Zhang Q, McMichael JF, Wyczalkowski MA, Leiserson MD, Miller CA, Welch JS, Walter MJ, Wendl MC, Ley TJ, Wilson RK, Raphael BJ, Ding L. Mutational landscape and significance across 12 major cancer types. Nature. 2013;502(7471):333–9.
13. Fulda S, Gorman AM, Hori O, Samali A. Cellular stress responses: cell survival and cell death. Int J Cell Biol. 2010;2010:214074.
14. Woods D, Turchi JJ. Chemotherapy induced DNA damage response: convergence of drugs and pathways. Cancer Biol Ther. 2013;14(5):379–89.
15. Cheung-Ong K, Giaever G, Nislow C. DNA-damaging agents in cancer chemotherapy: serendipity and chemical biology. Chem Biol. 2013;20(5):648–59.
16. Facchini LM, Penn LZ. The molecular role of Myc in growth and transformation: recent discoveries lead to new insights. FASEB J. 1998;12(9):633–51.
17. Pelengaris S, Khan M, Evan G. c-MYC: more than just a matter of life and death. Nat Rev Cancer. 2002;2(10):764–76.
18. Hart LS, Cunningham JT, Datta T, Dey S, Tameire F, Lehman SL, Qiu B, Zhang H, Cerniglia G, Bi M, Li Y, Gao Y, Liu H, Li C, Maity A, Thomas-Tikhonenko A, Perl AE, Koong A, Fuchs SY, Diehl JA, Mills IG, Ruggero D, Koumenis C. ER stress-mediated autophagy promotes Myc-dependent transformation and tumor growth. J Clin Invest. 2012;122(12):4621–34.
19. Chesler L, Schlieve C, Goldenberg DD, Kenney A, Kim G, McMillan A, Matthay KK, Rowitch D, Weiss WA. Inhibition of phosphatidylinositol 3-kinase destabilizes Mycn protein and blocks malignant progression in neuroblastoma. Cancer Res. 2006;66(16):8139–46.
20. Sui X, Chen R, Wang Z, Huang Z, Kong N, Zhang M, Han W, Lou F, Yang J, Zhang Q, Wang X, He C, Pan H. Autophagy and chemotherapy resistance: a promising therapeutic target for cancer treatment. Cell Death Dis. 2013;4, e838.
21. Rouschop KM, Dubois LJ, Keulers TG, van den Beucken T, Lambin P, Bussink J, van der Kogel AJ, Koritzinsky M, Wouters BG. PERK/eIF2alpha signaling protects therapy resistant hypoxic cells through induction of glutathione synthesis and protection against ROS. Proc Natl Acad Sci U S A. 2013;110(12):4622–7.
22. Burris III HA. Overcoming acquired resistance to anticancer therapy: focus on the PI3K/AKT/mTOR pathway. Cancer Chemother Pharmacol. 2013;71(4):829–42.
23. Westhoff MA, Serrels B, Fincham VJ, Frame MC, Carragher NO. SRC-mediated phosphorylation of focal adhesion kinase couples actin and adhesion dynamics to survival signaling. Mol Cell Biol. 2004;24(18):8113–33.
24. Golubovskaya VM. Targeting FAK in human cancer: from finding to first clinical trials. Front Biosci. 2014;19:687–706.
25. Waters AM, Beierle EA. The interaction between FAK, MYCN, p53 and Mdm2 in neuroblastoma. Anticancer Agents Med Chem. 2014;14(1):46–51. pii: ACAMC-EPUB-55926.
26. Dunn GP, Old LJ, Schreiber RD. The immunobiology of cancer immunosurveillance and immunoediting. Immunity. 2004;21(2):137–48.
27. Galluzzi L, Vitale I, Abrams JM, Alnemri ES, Baehrecke EH, Blagosklonny MV, Dawson TM, Dawson VL, El-Deiry WS, Fulda S, Gottlieb E, Green DR, Hengartner MO, Kepp O, Knight RA, Kumar S, Lipton SA, Lu X, Madeo F, Malorni W, Mehlen P, Nunez G, Peter ME, Piacentini M, Rubinsztein DC, Shi Y, Simon HU, Vandenabeele P, White E, Yuan J, Zhivotovsky B, Melino G, Kroemer G. Molecular definitions of cell death subroutines: recommendations of the Nomenclature Committee on Cell Death 2012. Cell Death Differ. 2012;19(1):107–20.
28. Saha MN, Qiu L, Chang H. Targeting p53 by small molecules in hematological malignancies. J Hematol Oncol. 2013;6:23.
29. Khoo KH, Verma CS, Lane DP. Drugging the p53 pathway: understanding the route to clinical efficacy. Nat Rev Drug Discov. 2014;13(3):217–36.

30. Yin Y, Shen WH. PTEN: a new guardian of the genome. Oncogene. 2008;27(41):5443–53.
31. Fan QW, Weiss WA. Targeting the RTK-PI3K-mTOR axis in malignant glioma: overcoming resistance. Curr Top Microbiol Immunol. 2010;347:279–96.
32. Cancer-Genome-Atlas-Research-Network. Comprehensive genomic characterization defines human glioblastoma genes and core pathways. Nature. 2008;455(7216):1061–8.
33. Akhavan D, Cloughesy TF, Mischel PS. mTOR signaling in glioblastoma: lessons learned from bench to bedside. Neuro Oncol. 2010;12(8):882–9.
34. Wen PY, Lee EQ, Reardon DA, Ligon KL, Alfred Yung WK. Current clinical development of PI3K pathway inhibitors in glioblastoma. Neuro Oncol. 2012;14(7):819–29.
35. de Groot JF, Fuller G, Kumar AJ, Piao Y, Eterovic K, Ji Y, Conrad CA. Tumor invasion after treatment of glioblastoma with bevacizumab: radiographic and pathologic correlation in humans and mice. Neuro Oncol. 2010;12(3):233–42.
36. Mendiburu-Elicabe M, Gil-Ranedo J, Izquierdo M. Efficacy of rapamycin against glioblastoma cancer stem cells. Clin Transl Oncol. 2014;16(5):495–502.
37. Westhoff M-A, Karpel-Massler G, Bruhl O, Enzenmuller S, La Ferla-Bruhl K, Siegelin M, Nonnenmacher L, Debatin K-M. A critical evaluation of PI3K inhibition in Glioblastoma and Neuroblastoma therapy. Mol Cell Ther. 2014;2(1):32.
38. Bieging KT, Mello SS, Attardi LD. Unravelling mechanisms of p53-mediated tumour suppression. Nat Rev Cancer. 2014;14(5):359–70.
39. Jabbour E, Kantarjian H. Chronic myeloid leukemia: 2014 update on diagnosis, monitoring, and management. Am J Hematol. 2014;89(5):547–56.
40. Gambacorti-Passerini C, Antolini L, Mahon FX, Guilhot F, Deininger M, Fava C, Nagler A, Della Casa CM, Morra E, Abruzzese E, D'Emilio A, Stagno F, le Coutre P, Hurtado-Monroy R, Santini V, Martino B, Pane F, Piccin A, Giraldo P, Assouline S, Durosinmi MA, Leeksma O, Pogliani EM, Puttini M, Jang E, Reiffers J, Valsecchi MG, Kim DW. Multicenter independent assessment of outcomes in chronic myeloid leukemia patients treated with imatinib. J Natl Cancer Inst. 2011;103(7):553–61.
41. An X, Tiwari AK, Sun Y, Ding PR, Ashby Jr CR, Chen ZS. BCR-ABL tyrosine kinase inhibitors in the treatment of Philadelphia chromosome positive chronic myeloid leukemia: a review. Leuk Res. 2010;34(10):1255–68.
42. Mughal A, Aslam HM, Khan AM, Saleem S, Umah R, Saleem M. Bcr-Abl tyrosine kinase inhibitors- current status. Infect Agent Cancer. 2013;8(1):23.
43. DeVita Jr VT, Chu E. A history of cancer chemotherapy. Cancer Res. 2008;68(21):8643–53.
44. Brennan CW, Verhaak RG, McKenna A, Campos B, Noushmehr H, Salama SR, Zheng S, Chakravarty D, Sanborn JZ, Berman SH, Beroukhim R, Bernard B, Wu CJ, Genovese G, Shmulevich I, Barnholtz-Sloan J, Zou L, Vegesna R, Shukla SA, Ciriello G, Yung WK, Zhang W, Sougnez C, Mikkelsen T, Aldape K, Bigner DD, Van Meir EG, Prados M, Sloan A, Black KL, Eschbacher J, Finocchiaro G, Friedman W, Andrews DW, Guha A, Iacocca M, O'Neill BP, Foltz G, Myers J, Weisenberger DJ, Penny R, Kucherlapati R, Perou CM, Hayes DN, Gibbs R, Marra M, Mills GB, Lander E, Spellman P, Wilson R, Sander C, Weinstein J, Meyerson M, Gabriel S, Laird PW, Haussler D, Getz G, Chin L. The somatic genomic landscape of glioblastoma. Cell. 2013;155(2):462–77.
45. Pugh TJ, Morozova O, Attiyeh EF, Asgharzadeh S, Wei JS, Auclair D, Carter SL, Cibulskis K, Hanna M, Kiezun A, Kim J, Lawrence MS, Lichenstein L, McKenna A, Pedamallu CS, Ramos AH, Shefler E, Sivachenko A, Sougnez C, Stewart C, Ally A, Birol I, Chiu R, Corbett RD, Hirst M, Jackman SD, Kamoh B, Khodabakshi AH, Krzywinski M, Lo A, Moore RA, Mungall KL, Qian J, Tam A, Thiessen N, Zhao Y, Cole KA, Diamond M, Diskin SJ, Mosse YP, Wood AC, Ji L, Sposto R, Badgett T, London WB, Moyer Y, Gastier-Foster JM, Smith MA, Guidry Auvil JM, Gerhard DS, Hogarty MD, Jones SJ, Lander ES, Gabriel SB, Getz G, Seeger RC, Khan J, Marra MA, Meyerson M, Maris JM. The genetic landscape of high-risk neuroblastoma. Nat Genet. 2013;45(3):279–84.
46. Stephens PJ, Tarpey PS, Davies H, Van Loo P, Greenman C, Wedge DC, Nik-Zainal S, Martin S, Varela I, Bignell GR, Yates LR, Papaemmanuil E, Beare D, Butler A, Cheverton A, Gamble J, Hinton J, Jia M, Jayakumar A, Jones D, Latimer C, Lau KW, McLaren S, McBride DJ,

Menzies A, Mudie L, Raine K, Rad R, Chapman MS, Teague J, Easton D, Langerod A, Lee MT, Shen CY, Tee BT, Huimin BW, Broeks A, Vargas AC, Turashvili G, Martens J, Fatima A, Miron P, Chin SF, Thomas G, Boyault S, Mariani O, Lakhani SR, van de Vijver M, van't Veer L, Foekens J, Desmedt C, Sotiriou C, Tutt A, Caldas C, Reis-Filho JS, Aparicio SA, Salomon AV, Borresen-Dale AL, Richardson AL, Campbell PJ, Futreal PA, Stratton MR. The landscape of cancer genes and mutational processes in breast cancer. Nature. 2012;486(7403):400–4.

47. Klopocki E, Neumann LM, Tonnies H, Ropers HH, Mundlos S, Ullmann R. Ulnar-mammary syndrome with dysmorphic facies and mental retardation caused by a novel 1.28 Mb deletion encompassing the TBX3 gene. Eur J Hum Genet. 2006;14(12):1274–9.
48. Romiti A, Cox MC, Sarcina I, Di Rocco R, D'Antonio C, Barucca V, Marchetti P. Metronomic chemotherapy for cancer treatment: a decade of clinical studies. Cancer Chemother Pharmacol. 2013;72(1):13–33.
49. Hengartner MO, Horvitz HR. C. elegans cell survival gene ced-9 encodes a functional homolog of the mammalian proto-oncogene bcl-2. Cell. 1994;76(4):665–76.
50. Leber B, Geng F, Kale J, Andrews DW. Drugs targeting Bcl-2 family members as an emerging strategy in cancer. Expert Rev Mol Med. 2010;12, e28.
51. Fischer U, Schulze-Osthoff K. Apoptosis-based therapies and drug targets. Cell Death Differ. 2005;12 Suppl 1:942–61.
52. Dan HC, Sun M, Kaneko S, Feldman RI, Nicosia SV, Wang HG, Tsang BK, Cheng JQ. Akt phosphorylation and stabilization of X-linked inhibitor of apoptosis protein (XIAP). J Biol Chem. 2004;279(7):5405–12.
53. Datta SR, Dudek H, Tao X, Masters S, Fu H, Gotoh Y, Greenberg ME. Akt phosphorylation of BAD couples survival signals to the cell-intrinsic death machinery. Cell. 1997;91(2):231–41.
54. Gardai SJ, Hildeman DA, Frankel SK, Whitlock BB, Frasch SC, Borregaard N, Marrack P, Bratton DL, Henson PM. Phosphorylation of Bax Ser184 by Akt regulates its activity and apoptosis in neutrophils. J Biol Chem. 2004;279(20):21085–95.
55. Qi XJ, Wildey GM, Howe PH. Evidence that Ser87 of BimEL is phosphorylated by Akt and regulates BimEL apoptotic function. J Biol Chem. 2006;281(2):813–23.
56. Berthelet J, Dubrez L. Regulation of apoptosis by inhibitors of apoptosis (IAPs). Cells. 2013;2(1):163–87.
57. Boucher MJ, Morisset J, Vachon PH, Reed JC, Laine J, Rivard N. MEK/ERK signaling pathway regulates the expression of Bcl-2, Bcl-X(L), and Mcl-1 and promotes survival of human pancreatic cancer cells. J Cell Biochem. 2000;79(3):355–69.
58. Meier F, Schittek B, Busch S, Garbe C, Smalley K, Satyamoorthy K, Li G, Herlyn M. The RAS/RAF/MEK/ERK and PI3K/AKT signaling pathways present molecular targets for the effective treatment of advanced melanoma. Front Biosci. 2005;10:2986–3001.
59. Unni E, Koul D, Yung WK, Sinha R. Se-methylselenocysteine inhibits phosphatidylinositol 3-kinase activity of mouse mammary epithelial tumor cells in vitro. Breast Cancer Res. 2005;7(5):R699–707.
60. Mortenson MM, Galante JG, Gilad O, Schlieman MG, Virudachalam S, Kung HJ, Bold RJ. BCL-2 functions as an activator of the AKT signaling pathway in pancreatic cancer. J Cell Biochem. 2007;102(5):1171–9.
61. Babchia N, Calipel A, Mouriaux F, Faussat AM, Mascarelli F. The PI3K/Akt and mTOR/P70S6K signaling pathways in human uveal melanoma cells: interaction with B-Raf/ERK. Invest Ophthalmol Vis Sci. 2010;51(1):421–9.
62. Shen J, Wu Y, Xu JY, Zhang J, Sinclair SH, Yanoff M, Xu G, Li W, Xu GT. ERK- and Akt-dependent neuroprotection by erythropoietin (EPO) against glyoxal-AGEs via modulation of Bcl-xL, Bax, and BAD. Invest Ophthalmol Vis Sci. 2010;51(1):35–46.
63. Greger JG, Eastman SD, Zhang V, Bleam MR, Hughes AM, Smitheman KN, Dickerson SH, Laquerre SG, Liu L, Gilmer TM. Combinations of BRAF, MEK, and PI3K/mTOR inhibitors overcome acquired resistance to the BRAF inhibitor GSK2118436 dabrafenib, mediated by NRAS or MEK mutations. Mol Cancer Ther. 2012;11(4):909–20.

64. Yamaguchi H, Hsu JL, Hung MC. Regulation of ubiquitination-mediated protein degradation by survival kinases in cancer. Front Oncol. 2012;2:15.
65. Suh DH, Kim MK, Kim HS, Chung HH, Song YS. Mitochondrial permeability transition pore as a selective target for anti-cancer therapy. Front Oncol. 2013;3:41.
66. Verhagen AM, Coulson EJ, Vaux DL. Inhibitor of apoptosis proteins and their relatives: IAPs and other BIRPs. Genome Biol. 2001;2(7):REVIEWS3009.
67. Gyrd-Hansen M, Meier P. IAPs: from caspase inhibitors to modulators of NF-kappaB, inflammation and cancer. Nat Rev Cancer. 2010;10(8):561–74.
68. Owens TW, Gilmore AP, Streuli CH, Foster FM. Inhibitor of apoptosis proteins: promising targets for cancer therapy. J Carcinog Mutagen Suppl. 2013;14:S14.
69. Mannhold R, Fulda S, Carosati E. IAP antagonists: promising candidates for cancer therapy. Drug Discov Today. 2010;15(5-6):210–9.
70. Fulda S, Wick W, Weller M, Debatin KM. Smac agonists sensitize for Apo2L/TRAIL- or anticancer drug-induced apoptosis and induce regression of malignant glioma in vivo. Nat Med. 2002;8(8):808–15.
71. Sah NK, Khan Z, Khan GJ, Bisen PS. Structural, functional and therapeutic biology of survivin. Cancer Lett. 2006;244(2):164–71.
72. Mita AC, Mita MM, Nawrocki ST, Giles FJ. Survivin: key regulator of mitosis and apoptosis and novel target for cancer therapeutics. Clin Cancer Res. 2008;14(16):5000–5.
73. Castedo M, Perfettini JL, Roumier T, Andreau K, Medema R, Kroemer G. Cell death by mitotic catastrophe: a molecular definition. Oncogene. 2004;23(16):2825–37.
74. Li F, Ambrosini G, Chu EY, Plescia J, Tognin S, Marchisio PC, Altieri DC. Control of apoptosis and mitotic spindle checkpoint by survivin. Nature. 1998;396(6711):580–4.
75. Fernandez JG, Rodriguez DA, Valenzuela M, Calderon C, Urzua U, Munroe D, Rosas C, Lemus D, Diaz N, Wright MC, Leyton L, Tapia JC, Quest AF. Survivin expression promotes VEGF-induced tumor angiogenesis via PI3K/Akt enhanced beta-catenin/Tcf-Lef dependent transcription. Mol Cancer. 2014;13:209.
76. Carvalho A, Carmena M, Sambade C, Earnshaw WC, Wheatley SP. Survivin is required for stable checkpoint activation in taxol-treated HeLa cells. J Cell Sci. 2003;116(Pt 14):2987–98.
77. Lens SM, Wolthuis RM, Klompmaker R, Kauw J, Agami R, Brummelkamp T, Kops G, Medema RH. Survivin is required for a sustained spindle checkpoint arrest in response to lack of tension. EMBO J. 2003;22(12):2934–47.
78. Maskey D, Yousefi S, Schmid I, Zlobec I, Perren A, Friis R, Simon HU. ATG5 is induced by DNA-damaging agents and promotes mitotic catastrophe independent of autophagy. Nat Commun. 2013;4:2130.
79. Czabotar PE, Lessene G, Strasser A, Adams JM. Control of apoptosis by the BCL-2 protein family: implications for physiology and therapy. Nat Rev Mol Cell Biol. 2014;15(1):49–63.
80. Billard C. BH3 mimetics: status of the field and new developments. Mol Cancer Ther. 2013;12(9):1691–700.
81. Ni Chonghaile T, Letai A. Mimicking the BH3 domain to kill cancer cells. Oncogene. 2008;27 Suppl 1:S149–57.
82. Stamelos VA, Redman CW, Richardson A. Understanding sensitivity to BH3 mimetics: ABT-737 as a case study to foresee the complexities of personalized medicine. J Mol Signal. 2012;7(1):12.
83. Carico C, Nuno M, Mukherjee D, Elramsisy A, Dantis J, Hu J, Rudnick J, Yu JS, Black KL, Bannykh SI, Patil CG. Loss of PTEN is not associated with poor survival in newly diagnosed glioblastoma patients of the temozolomide era. PLoS One. 2012;7(3), e33684.
84. Shi Y, Paluch BE, Wang X, Jiang X. PTEN at a glance. J Cell Sci. 2012;125(Pt 20):4687–92.
85. Opel D, Westhoff MA, Bender A, Braun V, Debatin KM, Fulda S. Phosphatidylinositol 3-kinase inhibition broadly sensitizes glioblastoma cells to death receptor- and drug-induced apoptosis. Cancer Res. 2008;68(15):6271–80.
86. Westhoff MA, Kandenwein JA, Karl S, Vellanki SH, Braun V, Eramo A, Antoniadis G, Debatin KM, Fulda S. The pyridinylfuranopyrimidine inhibitor, PI-103, chemosensitizes

glioblastoma cells for apoptosis by inhibiting DNA repair. Oncogene. 2009;28(40): 3586–96.
87. Akinleye A, Avvaru P, Furqan M, Song Y, Liu D. Phosphatidylinositol 3-kinase (PI3K) inhibitors as cancer therapeutics. J Hematol Oncol. 2013;6(1):88.
88. Georgakis GV, Li Y, Rassidakis GZ, Medeiros LJ, Mills GB, Younes A. Inhibition of the phosphatidylinositol-3 kinase/Akt promotes G1 cell cycle arrest and apoptosis in Hodgkin lymphoma. Br J Haematol. 2006;132(4):503–11.
89. Zhu Y, Shah K. Multiple lesions in receptor tyrosine kinase pathway determine glioblastoma response to pan-ERBB inhibitor PF-00299804 and PI3K/mTOR dual inhibitor PF-05212384. Cancer Biol Ther. 2014;15(6):815–22.
90. Ho R, Minturn JE, Hishiki T, Zhao H, Wang Q, Cnaan A, Maris J, Evans AE, Brodeur GM. Proliferation of human neuroblastomas mediated by the epidermal growth factor receptor. Cancer Res. 2005;65(21):9868–75.
91. Will M, Qin AC, Toy W, Yao Z, Rodrik-Outmezguine V, Schneider C, Huang X, Monian P, Jiang X, de Stanchina E, Baselga J, Liu N, Chandarlapaty S, Rosen N. Rapid induction of apoptosis by PI3K inhibitors is dependent upon their transient inhibition of RAS-ERK signaling. Cancer Discov. 2014;4(3):334–47.
92. Akinleye A, Furqan M, Mukhi N, Ravella P, Liu D. MEK and the inhibitors: from bench to bedside. J Hematol Oncol. 2013;6:27.
93. Shaul YD, Seger R. The MEK/ERK cascade: from signaling specificity to diverse functions. Biochim Biophys Acta. 2007;1773(8):1213–26.
94. Porta C, Paglino C, Mosca A. Targeting PI3K/Akt/mTOR signaling in cancer. Front Oncol. 2014;4:64.
95. Zhao Y, Adjei AA. The clinical development of MEK inhibitors. Nat Rev Clin Oncol. 2014;11(7):385–400.
96. Westhoff MA, Faham N, Marx D, Nonnenmacher L, Jennewein C, Enzenmuller S, Gonzalez P, Fulda S, Debatin KM. Sequential dosing in chemosensitization: targeting the PI3K/Akt/mTOR pathway in neuroblastoma. PLoS One. 2013;8(12), e83128.
97. Kuger S, Graus D, Brendtke R, Gunther N, Katzer A, Lutyj P, Polat B, Chatterjee M, Sukhorukov VL, Flentje M, Djuzenova CS. Radiosensitization of glioblastoma cell lines by the dual PI3K and mTOR inhibitor NVP-BEZ235 depends on drug-irradiation schedule. Transl Oncol. 2013;6(2):169–79.
98. Mitchison TJ. The proliferation rate paradox in antimitotic chemotherapy. Mol Biol Cell. 2012;23(1):1–6.
99. Das G, Shravage BV, Baehrecke EH. Regulation and function of autophagy during cell survival and cell death. Cold Spring Harb Perspect Biol. 2012;4(6):a008813.
100. Saiki S, Sasazawa Y, Imamichi Y, Kawajiri S, Fujimaki T, Tanida I, Kobayashi H, Sato F, Sato S, Ishikawa K, Imoto M, Hattori N. Caffeine induces apoptosis by enhancement of autophagy via PI3K/Akt/mTOR/p70S6K inhibition. Autophagy. 2011;7(2):176–87.
101. Castino R, Bellio N, Follo C, Murphy D, Isidoro C. Inhibition of PI3k class III-dependent autophagy prevents apoptosis and necrosis by oxidative stress in dopaminergic neuroblastoma cells. Toxicol Sci. 2010;117(1):152–62.
102. Nonnenmacher L, Westhoff MA, Fulda S, Karpel-Massler G, Halatsch ME, Engelke J, Simmet T, Corbacioglu S, Debatin KM. RIST: A potent new combination therapy for glioblastoma. Int J Cancer. 2015;136(4):E173–87.
103. Lee MJ, Ye AS, Gardino AK, Heijink AM, Sorger PK, MacBeath G, Yaffe MB. Sequential application of anticancer drugs enhances cell death by rewiring apoptotic signaling networks. Cell. 2012;149(4):780–94.
104. Zhou J, Bi C, Janakakumara JV, Liu SC, Chng WJ, Tay KG, Poon LF, Xie Z, Palaniyandi S, Yu H, Glaser KB, Albert DH, Davidsen SK, Chen CS. Enhanced activation of STAT pathways and overexpression of survivin confer resistance to FLT3 inhibitors and could be therapeutic targets in AML. Blood. 2009;113(17):4052–62.

105. Stolzel F, Steudel C, Oelschlagel U, Mohr B, Koch S, Ehninger G, Thiede C. Mechanisms of resistance against PKC412 in resistant FLT3-ITD positive human acute myeloid leukemia cells. Ann Hematol. 2010;89(7):653–62.
106. Herrmann MD, Lennerz JK, Bullinger L, Bartholomae S, Holzmann K, Westhoff MA, Corbacioglu S, Debatin KM. Transitory dasatinib-resistant states in KITmut t(8;21) acute myeloid leukemia cells correlate with altered KIT expression. Exp Hematol. 2014; 42:90–100.
107. Saito N, Ishihara S, Kaneko K. Baldwin effect under multipeaked fitness landscapes: phenotypic fluctuation accelerates evolutionary rate. Phys Rev E Stat Nonlin Soft Matter Phys. 2013;87(5):052701.
108. Torti D, Trusolino L. Oncogene addiction as a foundational rationale for targeted anti-cancer therapy: promises and perils. EMBO Mol Med. 2011;3(11):623–36.
109. Kast RE, Karpel-Massler G, Halatsch ME. CUSP9* treatment protocol for recurrent glioblastoma: aprepitant, artesunate, auranofin, captopril, celecoxib, disulfiram, itraconazole, ritonavir, sertraline augmenting continuous low dose temozolomide. Oncotarget. 2014;5(18):8052–82.
110. Dougan M, Dranoff G. Immune therapy for cancer. Annu Rev Immunol. 2009;27:83–117.
111. Harris TJ, Drake CG. Primer on tumor immunology and cancer immunotherapy. J Immunother Cancer. 2013;1:12.
112. Hall RD, Gray JE, Chiappori AA. Beyond the standard of care: a review of novel immunotherapy trials for the treatment of lung cancer. Cancer Control. 2013;20(1):22–31.
113. Weber JS. Current perspectives on immunotherapy. Semin Oncol. 2014;41 Suppl 5:S14–29.
114. Okonogi N, Shirai K, Oike T, Murata K, Noda SE, Suzuki Y, Nakano T. Topics in chemotherapy, molecular-targeted therapy, and immunotherapy for newly-diagnosed glioblastoma multiforme. Anticancer Res. 2015;35(3):1229–35.
115. Pardoll DM. The blockade of immune checkpoints in cancer immunotherapy. Nat Rev Cancer. 2012;12(4):252–64.
116. Chen DS, Mellman I. Oncology meets immunology: the cancer-immunity cycle. Immunity. 2013;39(1):1–10.
117. Raval RR, Sharabi AB, Walker AJ, Drake CG, Sharma P. Tumor immunology and cancer immunotherapy: summary of the 2013 SITC primer. J Immunother Cancer. 2014;2:14.
118. Vatner RE, Cooper BT, Vanpouille-Box C, Demaria S, Formenti SC. Combinations of immunotherapy and radiation in cancer therapy. Front Oncol. 2014;4:325.
119. Vanneman M, Dranoff G. Combining immunotherapy and targeted therapies in cancer treatment. Nat Rev Cancer. 2012;12(4):237–51.
120. Campbell PJ, Pleasance ED, Stephens PJ, Dicks E, Rance R, Goodhead I, Follows GA, Green AR, Futreal PA, Stratton MR. Subclonal phylogenetic structures in cancer revealed by ultra-deep sequencing. Proc Natl Acad Sci U S A. 2008;105(35):13081–6.
121. Gerecke C, Mascher C, Gottschalk U, Kleuser B, Scholtka B. Ultrasensitive detection of unknown colon cancer-initiating mutations using the example of the Adenomatous polyposis coli gene. Cancer Prev Res (Phila). 2013;6(9):898–907.
122. Hiley C, de Bruin EC, McGranahan N, Swanton C. Deciphering intratumor heterogeneity and temporal acquisition of driver events to refine precision medicine. Genome Biol. 2014;15(8):453.
123. Cleary AS, Leonard TL, Gestl SA, Gunther EJ. Tumour cell heterogeneity maintained by cooperating subclones in Wnt-driven mammary cancers. Nature. 2014;508(7494):113–7.
124. Marusyk A, Tabassum DP, Altrock PM, Almendro V, Michor F, Polyak K. Non-cell-autonomous driving of tumour growth supports sub-clonal heterogeneity. Nature. 2014;514(7520):54–8.
125. Gillies RJ, Verduzco D, Gatenby RA. Evolutionary dynamics of carcinogenesis and why targeted therapy does not work. Nat Rev Cancer. 2012;12(7):487–93.
126. Choi S, Henderson MJ, Kwan E, Beesley AH, Sutton R, Bahar AY, Giles J, Venn NC, Pozza LD, Baker DL, Marshall GM, Kees UR, Haber M, Norris MD. Relapse in children with acute lymphoblastic leukemia involving selection of a preexisting drug-resistant subclone. Blood. 2007;110(2):632–9.

127. Kreso A, O'Brien CA, van Galen P, Gan OI, Notta F, Brown AM, Ng K, Ma J, Wienholds E, Dunant C, Pollett A, Gallinger S, McPherson J, Mullighan CG, Shibata D, Dick JE. Variable clonal repopulation dynamics influence chemotherapy response in colorectal cancer. Science. 2013;339(6119):543–8.
128. Gatenby RA, Brown J, Vincent T. Lessons from applied ecology: cancer control using an evolutionary double bind. Cancer Res. 2009;69(19):7499–502.
129. Westhoff MA, Bruhl O, Debatin KM. Cancer therapy: know your enemy? Mol Cell Pediatr. 2014;1(1):10.
130. Miller RG, Secrest AM, Sharma RK, Songer TJ, Orchard TJ. Improvements in the life expectancy of type 1 diabetes: the Pittsburgh Epidemiology of Diabetes Complications study cohort. Diabetes. 2012;61(11):2987–92.
131. Crowley E, Di Nicolantonio F, Loupakis F, Bardelli A. Liquid biopsy: monitoring cancer-genetics in the blood. Nat Rev Clin Oncol. 2013;10(8):472–84.
132. Yoshioka Y, Kosaka N, Konishi Y, Ohta H, Okamoto H, Sonoda H, Nonaka R, Yamamoto H, Ishii H, Mori M, Furuta K, Nakajima T, Hayashi H, Sugisaki H, Higashimoto H, Kato T, Takeshita F, Ochiya T. Ultra-sensitive liquid biopsy of circulating extracellular vesicles using ExoScreen. Nat Commun. 2014;5:3591.
133. Kaatsch P, Rickert CH, Kuhl J, Schuz J, Michaelis J. Population-based epidemiologic data on brain tumors in German children. Cancer. 2001;92(12):3155–64.
134. MacDonald TJ, Aguilera D, Kramm CM. Treatment of high-grade glioma in children and adolescents. Neuro Oncol. 2011;13(10):1049–58.
135. Karsy M, Huang T, Kleinman G, Karpel-Massler G. Molecular, histopathological, and genomic variants of glioblastoma. Front Biosci. 2014;19:1065–87.
136. Greaves M, Maley CC. Clonal evolution in cancer. Nature. 2012;481(7381):306–13.
137. Paget S. The distribution of secondary growths in cancer of the breast. 1889. Cancer Metastasis Rev. 1989;8(2):98–101.
138. Dolberg DS, Bissell MJ. Inability of Rous sarcoma virus to cause sarcomas in the avian embryo. Nature. 1984;309(5968):552–6.
139. Sevenich L, Bowman RL, Mason SD, Quail DF, Rapaport F, Elie BT, Brogi E, Brastianos PK, Hahn WC, Holsinger LJ, Massague J, Leslie CS, Joyce JA. Analysis of tumour- and stroma-supplied proteolytic networks reveals a brain-metastasis-promoting role for cathepsin S. Nat Cell Biol. 2014;16(9):876–88.
140. Sceneay J, Smyth MJ, Moller A. The pre-metastatic niche: finding common ground. Cancer Metastasis Rev. 2013;32(3-4):449–64.
141. Steeg PS. Cancer biology: emissaries set up new sites. Nature. 2005;438(7069):750–1.
142. Kaplan RN, Riba RD, Zacharoulis S, Bramley AH, Vincent L, Costa C, MacDonald DD, Jin DK, Shido K, Kerns SA, Zhu Z, Hicklin D, Wu Y, Port JL, Altorki N, Port ER, Ruggero D, Shmelkov SV, Jensen KK, Rafii S, Lyden D. VEGFR1-positive haematopoietic bone marrow progenitors initiate the pre-metastatic niche. Nature. 2005;438(7069):820–7.
143. Bissell MJ, Radisky D. Putting tumours in context. Nat Rev Cancer. 2001;1(1):46–54. doi:10.1038/35094059.
144. Kerr JF, Wyllie AH, Currie AR. Apoptosis: a basic biological phenomenon with wide-ranging implications in tissue kinetics. Br J Cancer. 1972;26(4):239–57.
145. Sutherland RM, Durand RE. Cell contact as a possible contribution to radiation resistance of some tumours. Br J Radiol. 1972;45(538):788–9.
146. Durand RE, Sutherland RM. Radiation-resistant tumor cells may be more sensitive in vitro. Cancer Res. 1972;32(11):2587–8.
147. Westhoff MA, Fulda S. Adhesion-mediated apoptosis resistance in cancer. Drug Resist Updat. 2009;12(4-5):127–36.
148. Damiano JS, Cress AE, Hazlehurst LA, Shtil AA, Dalton WS. Cell adhesion mediated drug resistance (CAM-DR): role of integrins and resistance to apoptosis in human myeloma cell lines. Blood. 1999;93(5):1658–67.

149. Hazlehurst LA, Argilagos RF, Emmons M, Boulware D, Beam CA, Sullivan DM, Dalton WS. Cell adhesion to fibronectin (CAM-DR) influences acquired mitoxantrone resistance in U937 cells. Cancer Res. 2006;66(4):2338–45.
150. Westhoff MA, Zhou S, Bachem MG, Debatin KM, Fulda S. Identification of a novel switch in the dominant forms of cell adhesion-mediated drug resistance in glioblastoma cells. Oncogene. 2008;27(39):5169–81.
151. Zhu B, Zhao L, Zhu L, Wang H, Sha Y, Yao J, Li Z, You Q, Guo Q. Oroxylin A reverses CAM-DR of HepG2 cells by suppressing Integrinbeta1 and its related pathway. Toxicol Appl Pharmacol. 2012;259(3):387–94.
152. Kiziltepe T, Ashley JD, Stefanick JF, Qi YM, Alves NJ, Handlogten MW, Suckow MA, Navari RM, Bilgicer B. Rationally engineered nanoparticles target multiple myeloma cells, overcome cell-adhesion-mediated drug resistance, and show enhanced efficacy in vivo. Blood Cancer J. 2012;2(4), e64.
153. Cordes N, Meineke V. Cell adhesion-mediated radioresistance (CAM-RR). Extracellular matrix-dependent improvement of cell survival in human tumor and normal cells in vitro. Strahlenther Onkol. 2003;179(5):337–44.
154. Sandfort V, Koch U, Cordes N. Cell adhesion-mediated radioresistance revisited. Int J Radiat Biol. 2007;83(11-12):727–32.
155. Chiarugi P, Giannoni E. Anoikis: a necessary death program for anchorage-dependent cells. Biochem Pharmacol. 2008;76(11):1352–64.
156. Uhm JH, Dooley NP, Kyritsis AP, Rao JS, Gladson CL. Vitronectin, a glioma-derived extracellular matrix protein, protects tumor cells from apoptotic death. Clin Cancer Res. 1999;5(6):1587–94.
157. Taylor ST, Hickman JA, Dive C. Epigenetic determinants of resistance to etoposide regulation of Bcl-X(L) and Bax by tumor microenvironmental factors. J Natl Cancer Inst. 2000;92(1):18–23.
158. Wang L, Li Z, Wang C, Yang Y, Sun L, Yao W, Cai X, Wu G, Zhou F, Zha X. E-cadherin decreased human breast cancer cells sensitivity to staurosporine by up-regulating Bcl-2 expression. Arch Biochem Biophys. 2009;481(1):116–22.
159. Lwin T, Hazlehurst LA, Li Z, Dessureault S, Sotomayor E, Moscinski LC, Dalton WS, Tao J. Bone marrow stromal cells prevent apoptosis of lymphoma cells by upregulation of anti-apoptotic proteins associated with activation of NF-kappaB (RelB/p52) in non-Hodgkin's lymphoma cells. Leukemia. 2007;21(7):1521–31.
160. Aoudjit F, Vuori K. Integrin signaling inhibits paclitaxel-induced apoptosis in breast cancer cells. Oncogene. 2001;20(36):4995–5004.
161. Niedermeier M, Hennessy BT, Knight ZA, Henneberg M, Hu J, Kurtova AV, Wierda WG, Keating MJ, Shokat KM, Burger JA. Isoform-selective phosphoinositide 3′-kinase inhibitors inhibit CXCR4 signaling and overcome stromal cell-mediated drug resistance in chronic lymphocytic leukemia: a novel therapeutic approach. Blood. 2009;113(22):5549–57.
162. Bourogaa E, Bertrand J, Despeaux M, Jarraya R, Fabre N, Payrastre L, Demur C, Fournie JJ, Damak M, Feki AE, Racaud-Sultan C. Hammada scoparia flavonoids and rutin kill adherent and chemoresistant leukemic cells. Leuk Res. 2011;35(8):1093–101.
163. Matsunaga T, Fukai F, Miura S, Nakane Y, Owaki T, Kodama H, Tanaka M, Nagaya T, Takimoto R, Takayama T, Niitsu Y. Combination therapy of an anticancer drug with the FNIII14 peptide of fibronectin effectively overcomes cell adhesion-mediated drug resistance of acute myelogenous leukemia. Leukemia. 2008;22(2):353–60.
164. Wang X, Wang C, Qin YW, Yan SK, Gao YR. The association of up-regulation of X-linked inhibitor of apoptosis protein with cell adhesion-mediated drug resistance in U937 cells. Hematol Oncol. 2008;26(1):21–6.
165. Liu L, Zhang H, Sun L, Gao Y, Jin H, Liang S, Wang Y, Dong M, Shi Y, Li Z, Fan D. ERK/MAPK activation involves hypoxia-induced MGr1-Ag/37LRP expression and contributes to apoptosis resistance in gastric cancer. Int J Cancer. 2010;127(4):820–9.
166. Guadamillas MC, Cerezo A, Del Pozo MA. Overcoming anoikis—pathways to anchorage-independent growth in cancer. J Cell Sci. 2011;124(Pt 19):3189–97.

167. Yang J, Weinberg RA. Epithelial-mesenchymal transition: at the crossroads of development and tumor metastasis. Dev Cell. 2008;14(6):818–29.
168. Kalluri R, Weinberg RA. The basics of epithelial-mesenchymal transition. J Clin Invest. 2009;119(6):1420–8.
169. Westhoff MA, Zhou S, Nonnenmacher L, Karpel-Massler G, Jennewein C, Schneider M, Halatsch ME, Carragher NO, Baumann B, Krause A, Simmet T, Bachem MG, Wirtz CR, Debatin KM. Inhibition of NF-kappaB signaling ablates the invasive phenotype of glioblastoma. Mol Cancer Res. 2013;11(12):1611–23.
170. Aceto N, Bardia A, Miyamoto DT, Donaldson MC, Wittner BS, Spencer JA, Yu M, Pely A, Engstrom A, Zhu H, Brannigan BW, Kapur R, Stott SL, Shioda T, Ramaswamy S, Ting DT, Lin CP, Toner M, Haber DA, Maheswaran S. Circulating tumor cell clusters are oligoclonal precursors of breast cancer metastasis. Cell. 2014;158(5):1110–22.
171. Herbst RS, Soria JC, Kowanetz M, Fine GD, Hamid O, Gordon MS, Sosman JA, McDermott DF, Powderly JD, Gettinger SN, Kohrt HE, Horn L, Lawrence DP, Rost S, Leabman M, Xiao Y, Mokatrin A, Koeppen H, Hegde PS, Mellman I, Chen DS, Hodi FS. Predictive correlates of response to the anti-PD-L1 antibody MPDL3280A in cancer patients. Nature. 2014;515(7528):563–7.
172. Tumeh PC, Harview CL, Yearley JH, Shintaku IP, Taylor EJ, Robert L, Chmielowski B, Spasic M, Henry G, Ciobanu V, West AN, Carmona M, Kivork C, Seja E, Cherry G, Gutierrez AJ, Grogan TR, Mateus C, Tomasic G, Glaspy JA, Emerson RO, Robins H, Pierce RH, Elashoff DA, Robert C, Ribas A. PD-1 blockade induces responses by inhibiting adaptive immune resistance. Nature. 2014;515(7528):568–71.
173. Lau D, Magill ST, Aghi MK. Molecularly targeted therapies for recurrent glioblastoma: current and future targets. Neurosurg Focus. 2014;37(6), E15.
174. Tjin Tham Sjin R, Lee K, Walter AO, Dubrovskiy A, Sheets M, Martin TS, Labenski MT, Zhu Z, Tester R, Karp R, Medikonda A, Chaturvedi P, Ren Y, Haringsma H, Etter J, Raponi M, Simmons AD, Harding TC, Niu D, Nacht M, Westlin WF, Petter RC, Allen A, Singh J. In vitro and in vivo characterization of irreversible mutant-selective EGFR inhibitors that are wild-type sparing. Mol Cancer Ther. 2014;13(6):1468–79.
175. Yun CH, Mengwasser KE, Toms AV, Woo MS, Greulich H, Wong KK, Meyerson M, Eck MJ. The T790M mutation in EGFR kinase causes drug resistance by increasing the affinity for ATP. Proc Natl Acad Sci U S A. 2008;105(6):2070–5.
176. Eckstein N, Roper L, Haas B, Potthast H, Hermes U, Unkrig C, Naumann-Winter F, Enzmann H. Clinical pharmacology of tyrosine kinase inhibitors becoming generic drugs: the regulatory perspective. J Exp Clin Cancer Res. 2014;33:15.
177. Broekman F, Giovannetti E, Peters GJ. Tyrosine kinase inhibitors: multi-targeted or single-targeted? World J Clin Oncol. 2011;2(2):80–93.
178. Tomlinson I, Sasieni P, Bodmer W. How many mutations in a cancer? Am J Pathol. 2002;160(3):755–8.
179. Seok J, Warren HS, Cuenca AG, Mindrinos MN, Baker HV, Xu W, Richards DR, McDonald-Smith GP, Gao H, Hennessy L, Finnerty CC, Lopez CM, Honari S, Moore EE, Minei JP, Cuschieri J, Bankey PE, Johnson JL, Sperry J, Nathens AB, Billiar TR, West MA, Jeschke MG, Klein MB, Gamelli RL, Gibran NS, Brownstein BH, Miller-Graziano C, Calvano SE, Mason PH, Cobb JP, Rahme LG, Lowry SF, Maier RV, Moldawer LL, Herndon DN, Davis RW, Xiao W, Tompkins RG. Genomic responses in mouse models poorly mimic human inflammatory diseases. Proc Natl Acad Sci U S A. 2013;110(9):3507–12.
180. Hashmi JA, Baliki MN, Huang L, Baria AT, Torbey S, Hermann KM, Schnitzer TJ, Apkarian AV. Shape shifting pain: chronification of back pain shifts brain representation from nociceptive to emotional circuits. Brain. 2013;136(Pt 9):2751–68.
181. Marinelli S, Nazio F, Tinari A, Ciarlo L, D'Amelio M, Pieroni L, Vacca V, Urbani A, Cecconi F, Malorni W, Pavone F. Schwann cell autophagy counteracts the onset and chronification of neuropathic pain. Pain. 2014;155(1):93–107.
182. Ossipov MH, Morimura K, Porreca F. Descending pain modulation and chronification of pain. Curr Opin Support Palliat Care. 2014;8(2):143–51.

183. Miranda J, Lamana SM, Dias EV, Athie M, Parada CA, Tambeli CH. Effect of pain chronification and chronic pain on an endogenous pain modulation circuit in rats. Neuroscience. 2015;286:37–44.
184. Nowell PC. The clonal evolution of tumor cell populations. Science. 1976;194(4260):23–8.
185. Brock A, Chang H, Huang S. Non-genetic heterogeneity—a mutation-independent driving force for the somatic evolution of tumours. Nat Rev Genet. 2009;10(5):336–42.
186. Denison TA, Bae YH. Tumor heterogeneity and its implication for drug delivery. J Control Release. 2012;164(2):187–91.
187. Huang S. Genetic and non-genetic instability in tumor progression: link between the fitness landscape and the epigenetic landscape of cancer cells. Cancer Metastasis Rev. 2013;32(3-4):423–48.
188. Foo J, Leder K, Mumenthaler SM. Cancer as a moving target: understanding the composition and rebound growth kinetics of recurrent tumors. Evol Appl. 2013;6(1):54–69.
189. Durrett R. Population genetics of neutral mutations in exponentially growing cancer cell populations. Ann Appl Probab. 2013;23(1):230–50.
190. Lee SH, Sim SH, Kim JY, Cha S, Song A. Application of cancer genomics to solve unmet clinical needs. Genomics Inform. 2013;11(4):174–9.
191. Kim M, Lee KH, Yoon SW, Kim BS, Chun J, Yi H. Analytical tools and databases for metagenomics in the next-generation sequencing era. Genomics Inform. 2013;11(3):102–13.
192. van Oostrum J, Calonder C, Rechsteiner D, Ehrat M, Mestan J, Fabbro D, Voshol H. Tracing pathway activities with kinase inhibitors and reverse phase protein arrays. Proteomics Clin Appl. 2009;3(4):412–22.
193. De Witt Hamer PC, Van Tilborg AA, Eijk PP, Sminia P, Troost D, Van Noorden CJ, Ylstra B, Leenstra S. The genomic profile of human malignant glioma is altered early in primary cell culture and preserved in spheroids. Oncogene. 2008;27(14):2091–6.
194. Li A, Walling J, Kotliarov Y, Center A, Steed ME, Ahn SJ, Rosenblum M, Mikkelsen T, Zenklusen JC, Fine HA. Genomic changes and gene expression profiles reveal that established glioma cell lines are poorly representative of primary human gliomas. Mol Cancer Res. 2008;6(1):21–30.
195. Mak IW, Evaniew N, Ghert M. Lost in translation: animal models and clinical trials in cancer treatment. Am J Transl Res. 2014;6(2):114–8.
196. Pennisi E. Biomedicine. Cancer therapies use a little help from microbial friends. Science. 2013;342(6161):921.
197. Iida N, Dzutsev A, Stewart CA, Smith L, Bouladoux N, Weingarten RA, Molina DA, Salcedo R, Back T, Cramer S, Dai RM, Kiu H, Cardone M, Naik S, Patri AK, Wang E, Marincola FM, Frank KM, Belkaid Y, Trinchieri G, Goldszmid RS. Commensal bacteria control cancer response to therapy by modulating the tumor microenvironment. Science. 2013;342(6161):967–70.
198. Viaud S, Saccheri F, Mignot G, Yamazaki T, Daillere R, Hannani D, Enot DP, Pfirschke C, Engblom C, Pittet MJ, Schlitzer A, Ginhoux F, Apetoh L, Chachaty E, Woerther PL, Eberl G, Berard M, Ecobichon C, Clermont D, Bizet C, Gaboriau-Routhiau V, Cerf-Bensussan N, Opolon P, Yessaad N, Vivier E, Ryffel B, Elson CO, Dore J, Kroemer G, Lepage P, Boneca IG, Ghiringhelli F, Zitvogel L. The intestinal microbiota modulates the anticancer immune effects of cyclophosphamide. Science. 2013;342(6161):971–6.
199. Keller R, Keist R, Wechsler A, Leist TP, van der Meide PH. Mechanisms of macrophage-mediated tumor cell killing: a comparative analysis of the roles of reactive nitrogen intermediates and tumor necrosis factor. Int J Cancer. 1990;46(4):682–6.
200. Jadus MR, Williams CC, Avina MD, Ly M, Kim S, Liu Y, Narasaki R, Lowell CA, Wepsic HT. Macrophages kill T9 glioma tumor cells bearing the membrane isoform of macrophage colony stimulating factor through a phagocytosis-dependent pathway. J Immunol. 1998;160(1):361–8.
201. Mills CD, Kincaid K, Alt JM, Heilman MJ, Hill AM. M-1/M-2 macrophages and the Th1/Th2 paradigm. J Immunol. 2000;164(12):6166–73.

202. Helm O, Held-Feindt J, Grage-Griebenow E, Reiling N, Ungefroren H, Vogel I, Kruger U, Becker T, Ebsen M, Rocken C, Kabelitz D, Schafer H, Sebens S. Tumor-associated macrophages exhibit pro- and anti-inflammatory properties by which they impact on pancreatic tumorigenesis. Int J Cancer. 2014;135(4):843–61.
203. Boelens MC, Wu TJ, Nabet BY, Xu B, Qiu Y, Yoon T, Azzam DJ, Twyman-Saint Victor C, Wiemann BZ, Ishwaran H, Ter Brugge PJ, Jonkers J, Slingerland J, Minn AJ. Exosome transfer from stromal to breast cancer cells regulates therapy resistance pathways. Cell. 2014;159(3):499–513.
204. Melo SA, Sugimoto H, O'Connell JT, Kato N, Villanueva A, Vidal A, Qiu L, Vitkin E, Perelman LT, Melo CA, Lucci A, Ivan C, Calin GA, Kalluri R. Cancer exosomes perform cell-independent MicroRNA biogenesis and promote tumorigenesis. Cancer Cell. 2014;26(5):707–21.

Chapter 9
Killing Is Not Enough: How Apoptosis Hijacks Tumor-Associated Macrophages to Promote Cancer Progression

Andreas Weigert, Javier Mora, Divya Sekar, Shahzad Syed, and Bernhard Brüne

Abstract Macrophages are a group of heterogeneous cells of the innate immune system that are crucial to the initiation, progression, and resolution of inflammation. Moreover, they control tissue homeostasis in healthy tissue and command a broad sensory arsenal to detect disturbances in tissue integrity. Macrophages possess a remarkable functional plasticity to respond to irregularities and to initiate programs that allow overcoming them in order to return back to normal. Thus, macrophages kill malignant or transformed cells, rearrange extracellular matrix, take up and recycle cellular as well as molecular debris, initiate cellular growth cascades, and favor directed migration of cells. As an example, apoptotic death of bystander cells is sensed by macrophages, initiating functional responses that support all hallmarks of cancer. In this chapter, we describe how tumor cell apoptosis hijacks tumor-associated macrophages to promote tumor growth. We propose that tumor therapy should not only kill malignant cells but also target the interaction of the host with apoptotic cancer cells, as this might be efficient to limit the protumor action of apoptotic cells and boost the antitumor potential of macrophages. Leaving the apoptotic cell/macrophage interaction untouched might also limit the benefit of conventional tumor cell apoptosis-focused therapy since surviving tumor cells might receive overwhelming support by the wound healing response that apoptotic tumor cells will trigger in local macrophages, thereby enhancing tumor recurrence.

Keywords Tumor-associated macrophages • Macrophage polarization • Apoptotic cells • Tumor immunosuppression • Hallmarks of cancer • Ontogeny of tumor macrophages • Tumor stroma interaction • Tumor therapy directed against tumor-associated macrophages • Role of tumor-associated macrophages

---

A. Weigert • J. Mora • D. Sekar • S. Syed • B. Brüne (✉)
Faculty of Medicine, Institute of Biochemistry I, Goethe-University Frankfurt, Theodor-Stern-Kai 7, 60590 Frankfurt, Germany
e-mail: b.bruene@biochem.uni-frankfurt.de

C.D. Gregory (ed.), *Apoptosis in Cancer Pathogenesis and Anti-cancer Therapy*, Advances in Experimental Medicine and Biology 930,
DOI 10.1007/978-3-319-39406-0_9

9.1 Provenance and Plasticity of Macrophages

Macrophages are a group of heterogeneous cells of the innate immune system that are crucial to the initiation, progression, and resolution of inflammation [1, 2]. However, the skills of macrophages extend far beyond the regulation of inflammation. They are guardians of tissue homeostasis even in the absence of inflammation [3, 4]. Any disturbance of macrophage-dependent tissue homeostasis may result in disease and different pathologies develop as a result of overshooting specific macrophage responses that, under physiological circumstances, would be part of the program to restore homeostasis [2, 4]. Several features enable macrophages to fulfill their function as homeostatic cells. They command a broad sensory arsenal to detect disturbances in tissue integrity and possess a remarkable functional plasticity to respond appropriately to such disorders in order to overcome them. Thus, macrophages kill malignant or transformed cells, rearrange extracellular matrix, take up and recycle cellular as well as molecular debris, initiate cellular growth cascades, and favor directed migration of cells. As an example, apoptotic death of bystander cells is sensed by macrophages, initiating a functional program that promotes the recruitment of new blood vessels and epithelial cell proliferation to induce tissue regeneration and healing [5, 6]. Thus, sensing damage directly triggers a healing response. In this chapter, we propose the hypothesis that tumor cell apoptosis hijacks this homeostatic function of (tumor-associated) macrophages to promote tumor growth.

9.1.1 Tissue Macrophage Origin

Macrophages are present in all tissues of the adult organism. Due to historical reasons, specific names have been assigned to distinct tissue macrophage populations indicating the heterogeneity of these cells. Among them are brain macrophages (microglia), liver macrophages (Kupffer cells), skin macrophages (Langerhans cells and dermal macrophages), spleen macrophages (marginal-zone macrophages, red-pulp macrophages, subcapsular sinus, medullary macrophages, metallophilic macrophages), lung macrophages (alveolar macrophages), and bone macrophages (osteoclasts). These macrophage populations fulfill their homeostatic function that is specific to the environmental niche. However, not only in the adult, but already in the embryo, macrophages play important roles in determining tissue architecture. They are required, among others, for branching morphogenesis, bone morphogenesis, generation of adipose tissue, and vascular patterning (both lymph and blood) [2, 7]. Consequently, macrophages appear early during embryonic development starting from embryonic day 8 (E8) in the mouse and maintain their presence throughout the entire life span of the organism. Two key mechanisms ensure the temporal persistence of macrophages in each organ during steady state as well as under inflammatory conditions. First, macrophage progenitors can be recruited to differentiate into tissue macrophages that meet the functional requirements of the organ-specific microenvironment. Second, tissue-resident macrophages can undergo in situ proliferation. Tissue macrophage ontogeny has been a matter of intense debate virtually since their original description by Elie Metchnikoff in 1887 [8–10]. Experiments conducted in the 1960s instilled the notion that tissue macrophages are generally terminally differentiated, nonproliferating, short-lived cells originating from monocytes that are produced in the bone marrow [9, 11]. However, this view was challenged very early afterward by findings that macrophages are produced in the yolk sac before the appearance of monocytes, and that tissue-resident macrophages can be long-lived cells and proliferate under certain experimental conditions [12]. Recently, a number of elegant studies using, among others, fate mapping and parabiosis approaches have shed new light onto this fundamental question. We are now rapidly approaching a unifying concept of tissue macrophage provenance and maintenance that gives merit to both theories mentioned earlier. According to this concept, the body is populated with macrophages in different waves during development. Starting in the embryo, the earliest macrophages develop as a result of primitive hematopoiesis from mesenchymal progenitors in the extra-embryonic yolk sac from where they populate the embryo as soon as a functional vasculature is established (starting from E8.5) [13, 14]. Simultaneously, early yolk sac-derived erythro-myeloid progenitors migrate to the fetal liver, where they give rise to fetal monocytes and macrophages [15]. Definite, hematopoietic stem cell (HSC)-dependent, hematopoiesis starts in the aorta-gonads-mesonephros region of the embryo (from E10) and is later (around E11) commenced in the fetal liver [1, 13]. Thus, fetal monocytes of embryonic as well as extra-embryonic progenitors are produced in the fetal liver, which colonize the embryo to differentiate into tissue macrophages. Bone marrow hematopoiesis starting after birth finally gives rise to a third wave of macrophage progenitors, bone marrow-derived monocytes. These different progenitors contribute to the tissue macrophage pool in the adult in an organ-specific manner, as illustrated by the following examples. Yolk-sac macrophages constitute the vast majority of microglia in the CNS. These cells renew via local proliferation, with minimal to no contribution of fetal or adult monocytes during steady state [14, 16, 17]. Apart from the brain, yolk sac macrophages are more or less replaced by fetal or adult monocyte-derived macrophages during development. Langerhans cells are a mixture of yolk sac macrophages and fetal monocytes from potentially both embryonic and extra-embryonic sources that proliferate locally [3, 15], whereas dermal macrophages are constantly replenished by adult monocytes [18]. Thus, the skin is populated by macrophages of all sources so far identified. At the other end of the spectrum, intestinal macrophages are strictly adult monocyte-derived cells that are constantly replaced and do not proliferate in situ [19]. For information concerning the origin of other tissue macrophage populations, we would like to direct your attention to recent excellent reviews [3, 13].

The situation gets more complex when the steady state is challenged during inflammation. Upon an inflammatory insult the tissue macrophage pool can undergo significant alterations due to the profound numbers of inflammatory monocytes that are recruited to the site of inflammation from the circulation and subsequently differentiate into macrophages. It is currently unclear to what extent elicited monocyte-derived macrophages integrate into the resident tissue macrophage pool in different organs in a self-sustaining manner. However, the picture emerges that the degree to which resident tissue macrophages are depleted in response to the inflammatory insult determines if bone marrow-derived progenitors stably contribute to the tissue macrophage pool once inflammation is terminated. A notable exception is the brain, where monocytes under each circumstance only contribute transiently to the macrophage pool [3, 13].

9.1.2 Ontogeny Versus Microenvironment in Tissue-Specific Macrophage Functionality

Another remaining question is whether macrophages of different origin in the same tissue differ in their functional capabilities, i.e., to what degree ontogeny or the local microenvironment determines macrophage phenotypes. Most tissue macrophage populations and all adult monocytes depend on colony stimulating factor-1 (CSF1) receptor and its ligands CSF1 or interleukin-34 (IL-34) and the downstream lineage-determining transcription factor PU.1 [20, 21]. Master transcriptional regulators such as PU.1 provide the basis of transcriptional availability, i.e., regions of open chromatin, in a cell-type-specific manner, on top of which other specific transcription factors can be installed to shape tissue-specific cell function. A few of such specific transcription factors for tissue macrophage subtypes have been identified recently, including myocyte-specific enhancer factor 2c (MEF2c) in microglia, liver x receptor (LXRα) in Kupffer cells and selected splenic macrophages, peroxisome proliferator-activated receptor γ (PPARγ) in splenic red pulp and alveolar macrophages, PU.1-related factor (SPI-C) in splenic red pulp macrophages and bone marrow macrophages, Runt-related transcription factor 3 (RUNX3) in intestinal macrophages, and GATA-binding protein 6 (GATA6) in peritoneal macrophages [22]. Moreover, for some of these transcription factors the tissue-specific cues that determine their expression/activation have been suggested. Heme induced SPI-C expression in progenitors of red pulp and bone marrow macrophage development (which are professional iron-recycling cells) and in monocytes to induce their differentiation to iron-recycling cells [23]. Maintenance of peritoneal macrophages depends on retinoic acid, which stimulates GATA6 expression [24, 25]. Granulocyte macrophage colony-stimulating factor (GM-CSF) induces PPARγ expression during alveolar macrophage development [26] and transforming growth factor-β (TGF-β) regulates microglia transcriptional programs through expression of SMAD transcription factors, which work cooperatively with MEF2C [27, 28]. Thus, the tissue microenvironment largely dictates the genetic signature of its resident macrophages. This notion is supported by a recent report demonstrating that transplantation of fully differentiated tissue macrophages into an alternate tissue was sufficient to alter their transcriptional program to integrate functionally into the recipient tissue macrophage pool [22]. Thus, it seems also likely that monocyte-derived macrophages can successfully replace fetal macrophage populations, e.g., after an inflammatory insult, as described in the heart [3, 13]. However, as mentioned earlier, distinct macrophage subpopulations with different functions are found virtually side by side in some tissues (e.g., skin, spleen, or peritoneum). Although the transcriptional profile of these distinct macrophage populations is similar enough to suggest exposure to common tissue-specific cues, the remaining differences might allow for an impact of ontogeny [28].

9.1.3 Macrophage Polarization

Independent of genetic imprinting due to ontogeny or differentiation in a specific microenvironment, macrophages must retain a high plasticity in their functional repertoire, enabling them to respond to inflammatory stimuli of varying nature [4, 29] (Fig. 9.1). Indeed, macrophages show such a degree of versatility that discrete macrophage phenotypes can not be easily correlated to a defined functional response or a defined stimulus. Considerable efforts have been undertaken to identify core signatures of human monocyte-derived macrophages exposed to distinct stimuli [30, 31]. These studies confirmed the antithetic nature of long-known extreme macrophage phenotypes, i.e., interferon (IFN)γ-stimulated versus IL-4-stimulated macrophages (often designated as M1 or M2 macrophages, respectively [32]), with their characteristic transcriptional repertoire. IFNγ-stimulated macrophages show a transcriptional signature defined by activation of signal transducer and activator of transcription 1 (STAT1), as well as interferon regulatory factor 5 (IRF5) [31, 33]. Such macrophages produce proinflammatory cytokines such IL-12 and IL-23 and show a high bactericidal as well as tumoricidal capacity. IL-4-stimulated or alternatively activated macrophages produce an alternative set of cytokines and

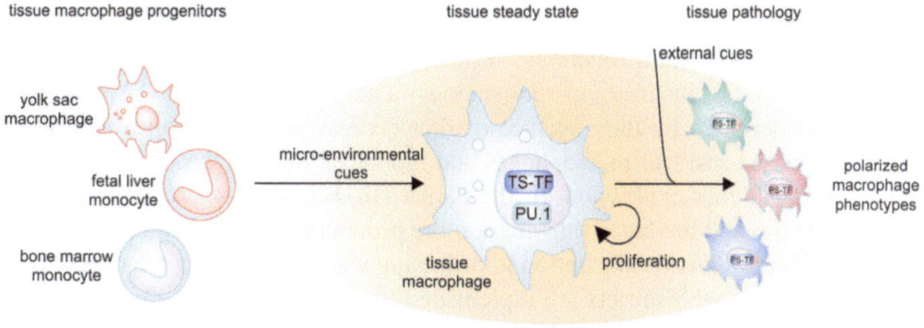

Fig. 9.1 Ontogeny and the microenvironment shape macrophage function. During development, distinct tissue macrophage progenitors (embryonic progenitors marked with *red border*) are recruited to distinct tissues. There, they express tissue-specific transcription factors (TS-TF) under the influence of cues present in their local environment, which together with lineage-determining transcription factors such as PU.1 determine their function. If tissues are challenged by external signals, their remaining remarkable plasticity enables macrophages to adopt multiple phenotypes shaped by polarization-specific transcription factors (PS-TF) to overcome and resolve the challenge

chemokines opposing the repertoire of classically activated macrophages. Moreover, they express specific phagocytic receptors such as CD206, produce extracellular matrix (ECM), and growth factors to promote tissue remodeling and to combat extracellular parasites [29]. Their transcriptional repertoire is characterized among others by STAT6 and IRF4 [30, 31]. Next to these two extreme phenotypes exists a plethora of hybrid and completely unrelated functional macrophage phenotypes that are poorly understood.

9.1.4 Ontogeny and Polarization of Tumor-Associated Macrophages

Summarizing the last paragraphs, the recent decade has seen considerable efforts that have advanced our understanding in macrophage origin and functional identity. However, most of the data obtained so far are only relevant in the steady state or very defined/polarized pathological conditions. Much less is known concerning complex pathological situations such as tumor development. When speculating about macrophage origin in tumors, which can be considered as 'new' organs developing in the adult, they will likely be populated by adult monocyte-derived macrophages. However, embryonic macrophages or adult monocyte-derived 'original' tissue macrophages might well constitute a significant part of the total macrophage pool in tumors when considering their strong proliferative capacity that depends upon factors such as CSF-1 or IL-4, which are abundant in tumors [34]. Indeed the few studies that investigated the origin of macrophages in endogenous tumors in brain [35] and mammary gland [36] provided evidence for a side-by-side existence of recruited inflammatory monocyte-derived macrophages and the 'locals.' Moreover, local proliferation of resident as well as elicited macrophages seems to contribute to the maintenance of the tumor-associated macrophage (TAM) pool [36, 37]. Interestingly, certain functional differences in resident versus elicited TAMs were proposed. For instance, in spontaneous mammary tumors two TAM populations were identified by their surface marker expression profile corresponding to mammary tissue-resident macrophages and monocyte-derived TAMs with the former showing higher expression of alternative (IL-4-driven) macrophage markers such as CD206 and the latter showing a more pronounced proinflammatory profile [36, 38]. This phenotypic separation is surprising when considering the observation that both macrophage subsets are replenished by adult monocytes [36]. Since the origin of mammary tissue macrophages in the adult organism is unclear, the phenotypic separation of two distinct macrophage populations in breast tumors might also be explained by their functional polarization in distinct microenvironmental niches characterized by varying cytokine, growth factor, and oxygen content, as suggested before [39]. Moreover, distinct microenvironmental niches might contain tumor cells in various stages of life and death, which drastically influences macrophage responses [40, 41]. Thus, the old question of nature versus nurture, in our case ontogeny versus microenvironment, is also unresolved when asking how TAM function is established. Likely their origin as well as the specific microenvironmental niche in which they reside generates cells with a multitude of different functional properties. Understanding this diversity will be instrumental to identify TAM subsets to target and to spare in cancer therapy.

9.2 TAM Function and Significance

A healthy organism is protected by a fine-tuned immune system reacting to invading pathogens and internally derived danger-associated molecular patterns to maintain tissue homeostasis. As a result of acute infection, circulating innate immune cells are recruited and the cross-talk between innate and adaptive immunity overcomes the acute inflammatory response and progresses toward its complete resolution. In contrast, cancer is characterized by a perturbed tissue homeostasis, with constant low-grade 'smoldering inflammation' that does never resolve. This type of inflammation can drive tumor progression, promoting basically all hallmarks of cancer. Tumors are multicellular ecosystems and innate immune cells are highly represented, with macrophages being the most abundant ones. Substantial experimental evidence in mice and clinical data in man substantiate that in the majority of cases TAMs enhance tumor progression to malignancy. A strict correlation between increased numbers and/or density of TAMs and poor prognosis has been demonstrated in various malignancies, including mammary carcinoma, prostate cancer, bladder cancer, glioma, and lymphoma [42–48]. TAMs may contribute to cancer initiation and promotion but also support angiogenesis as well as lymphangiogenesis, promote tumor cell invasion, migration, intravasation, and facilitate immunosuppression [34]. Here we briefly summarize the diverse tumor-promoting characteristics and functions of TAMs, which are reflected by dynamic changes in the TAM phenotypes and distinct TAM subpopulations. Moreover, we summarize current knowledge of how apoptotic cells add to the phenotype change and functional polarization of macrophages. The emerging wide spectrum of distinct TAM functional phenotypes is the result of signals in the tumor microenvironment, which may be unique among distinct tumors and depends on the fitness of the immune system. As outlined earlier, TAMs do not comprise a homogeneous cell population, rather representing cells of different ontogeny that dynamically change in their developing microenvironmental niche. Consequently, subpopulations of TAMs should be defined by their biological activity rather than nomenclatures that follow the expression of distinct marker profiles or their response to defined cytokines [7].

9.2.1 TAMs in Cancer Initiation and Promotion

Smoldering inflammation is now accepted as a hallmark of cancer [49]. Activated macrophages are central to perpetuate this low-grade inflammation and they have been postulated to produce a mutagenic environment by generating reactive nitrogen and oxygen species [50]. Inflammatory cytokines produced by macrophages

(IFNγ, tumor necrosis factor-α (TNF-α), IL-1β, IL-6) sustain chronic inflammation that seems to be required for tumor initiation and promotion [51]. Genetic ablation of the anti-inflammatory transcription factor STAT3 in macrophages resulted in chronic inflammation and caused invasive adenocarcinoma, substantiating the link between inflammation and cancer [52]. In addition, Langerhans cells can cause skin carcinogenesis by metabolic conversion of precarcinogens such as polycyclic aromatic hydrocarbons (PAHs) to their activated mutagenic state. Our own observations suggest a role of hypoxia inducible factors (HIF-1α) in this process, as conditional myeloid-specific HIF-1α knockout mice fail to express cytochrome P450 1A1 (CYP1A1), which is required to metabolically activate the carcinogen 3-methylcholanthrene (3-MCA). However, besides low-grade inflammation it appears that acute and full blown activation of macrophages can kill aberrant cells or at least contribute to this process [53]. Besides causing smoldering inflammation or contributing to chemical carcinogen activation, TAMs have trophic and immune regulatory roles in established tumors, thereby contributing to every hallmark of cancer progression.

9.2.2 Role of TAMs in Angiogenesis, Invasion, and Metastasis

The biological function of these macrophages is based on polarization programs that are different from classical activation (e.g., lipopolysaccharide (LPS)/IFNγ, pathogen-associated molecular patterns) and may occur during the transition from benign tumor growth to an invasive cancer. Thus, macrophages become tumor educated. The process is initiated by a number of factors such as a Th2-type immune environment, immune complexes, the formation of an apoptotic synapse with dying tumor cells, a vast number of protein and lipid soluble factors released from apoptotic cells, and hypoxia. Many publications that are more recent refer to this still growing list of molecules [29, 42, 54]. It becomes an even more exhausting endeavor to list all those factors released from TAMs to fulfill their protumor function. Major biological activities summarized in Fig. 9.2 seem to be directed at facilitating (1) tumor growth, (2) invasion and metastasis, (3) angiogenesis and lymphangiogenesis, and (4) immunosuppression. In addition, features of alternatively polarized macrophages comprise a shift to anti-inflammatory cytokine formation and chemokine production that attract regulatory T cells (Tregs), Th2 cells, eosinophils and basophils, upregulation of scavenger receptors (e.g., CD206, CD163, Lyve1, SR), a $iNOS^{lo}$ (inducible NO synthase)/$arginase1^{hi}$ expression profile, an altered lipid profile that is characteristic of wound healing, iron release, and an altered expression profile of amino acid metabolizing enzymes (e.g., $arginase^{hi}$, indoleamine 2,3-dioxygenase^{hi} (IDO)) (references in: [55–57]).

More specifically, growth factors (GF) such as epidermal GF (EGF), basic fibroblast GF (bFGF), platelet-derived GF (PDGF), or vascular endothelial GF (VEGF) may promote tumor growth, while a number of proteases such as plasmin, urokinase-type plasminogen activator, cathepsin B, or various matrix metalloproteases remodel the extracellular matrix to favor invasion and metastasis. Compounds such as TGF-β,

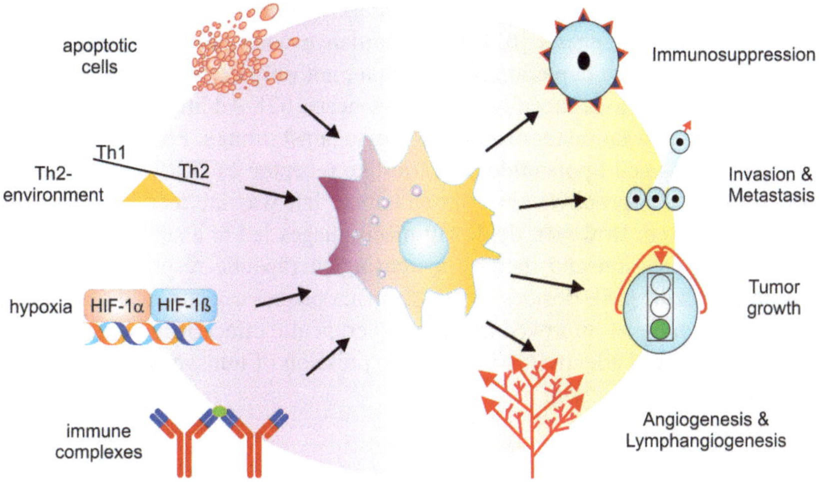

Fig. 9.2 Incoming and outgoing signals of tumor-associated macrophages. Unique properties of tumor associated macrophages with 'input signals' (*left side*) being apoptotic cells, a Th2-cytokine balance, hypoxia, and immune complexes. These signals program TAMs (output signals, *right side*) to support an immune-suppressive environment, to cause tumor cell invasion and metastasis, to enhance tumor cell proliferation, and to foster angiogenesis/lymphangiogenesis

VEGF-A, VEGF-C, PDGF, matrix metalloproteases (MMPs), or chemokines (e.g., CXCL8/IL-8) can control the process of angiogenesis/lymphangiogenesis, while the expression of TGF-β, IL-10, arginase1, and IDO enhance immunosuppression. In addition, TAMs upregulate the ligands for inhibitory coreceptors such as programmed cell death-1 (PD-1) or cytotoxic T-lymphocyte-associated protein 4 (CTLA-4) to block the activity of T cells, B cells, and NK cells, which adds to characteristics of a low CD8+ activity but enhanced Treg numbers and function in the tumor microenvironment. Despite our knowledge on cell surface receptor expression, secretion of cytokines, chemokines, or lipids in recruiting and activating Tregs and suppressing T effector cells, the dominant immunosuppressive function of TAMs in vivo still is largely unexplored.

9.3 TAM Depletion as an Antitumor Therapeutic Strategy

Given the evidence of TAM function summarized earlier, it seems rational to consider TAM depletion as a therapeutic strategy. Such an approach may serve to lower tumor burden by decreasing macrophage-dependent proliferation or survival effects on tumor cells, by decreasing TAM-derived proangiogenic factors that aid in vascularization of the tumor with blood as well as lymph vessels, and by relieving immune suppression, while allowing the inherent antitumor immune responses to resume. Experimental macrophage depletion has been used to elucidate their role in several clinical pathologies including neurodegeneration [58], colitis [59], diabetes [60], but also tumorigenesis [61]. In concordance with the multiplicity of TAM functions, macrophage depletion, for example, not only decreased lymphoma cell growth, but also led to reduced lymphangiogenesis [62]. Additionally, macrophage depletion improved immune- or gene therapy against tumors. For example, di-palmitic acid conjugated lipopeptide with toll-like receptor 2 (TLR2) agonist activity was proposed as a prophylactic tumor vaccine, but was ineffective against fully established tumors. However, depleting macrophages led to a subsequent removal of immunosuppression and thereby increasing therapeutic responses against the lipopeptide vaccine [63]. Besides, depleting macrophages can augment gene therapy. This was shown by an improved antitumor therapeutic effect of IFNγ gene transfer [64] or an improved adenovirus-mediated expression of transgene in the liver [65] upon macrophage depletion.

9.3.1 Methods of TAM Depletion

Several methods have been investigated to deplete macrophages systemically or from the tumor microenvironment. Initially, the ability of macrophages to ingest or phagocytize liposomes was exploited as a strategy to deplete them by loading the liposomes with cell death inducing drugs such as clodronate, which was effective in eradicating tumors in animal models. However, the therapeutic use and clinical applications of drug-loaded liposomes are limited due to high in vivo toxicity, shorter stability, and high production costs [66]. Hence, as an alternative to liposomes, drugs encapsulated in red blood cells to deplete macrophages [67] or *Shigella flexneri*-induced macrophage apoptosis are being currently investigated [68]. However, their applicability to deplete TAMs needs to be determined. The above-mentioned methods have a drawback of being nonspecific as they deplete not only TAMs but also other subtypes of macrophages or monocytes as well as some lymphocytes. Indeed TAMs in tumor stroma or tumor-draining lymph nodes are found scattered amid a heterogeneous mixture of blood capillaries and immune cells. The latter include cytotoxic, tumoricidal macrophages, or macrophages with the capacity to activate Th1/Tc cells [69] that could potentially mount antitumor responses in the absence of TAM. Therefore, it is essential to specifically target TAMs by strategies that leave such inflammatory macrophages unharmed. Thus, cytotoxic drug carriers coated with proteins or monoclonal antibodies that can specifically recognize TAMs, while leaving other immune cells and blood vessels intact are indispensable.

9.3.2 Capitalizing on TAM Markers for Specific Drug Delivery

Folate receptor-β (FR-β) is specifically overexpressed in TAMs residing in the tumor tissue. Indeed, delivering zoledronic acid entrapped in folate-linked liposome could result in in vitro cytotoxicity, specifically toward FR-β expressing RAW264.7

macrophages [70]. Alternatively, TAMs express high levels of the macrophage mannose receptor CD206 as mentioned earlier. Indeed, mannose-containing polymers such as glucomannan polysaccharide (from the herb *Bletilla Striata*) carrying alendronate (bisphosphonate) were preferentially taken up by RAW 264.7 macrophages as compared to endothelial or cancer cells. Moreover, F4/80$^+$ cells were remarkably reduced in tumor-bearing animals after administration of alendronate–glucomannan conjugate, which subsequently reduced the tumor burden [71]. Also, TAMs express high levels of macrophage galactose-type lectin (Mgl). Hence a galactosylated dextran was successfully used to deliver oligonucleotides [72] opening the possibility to target Mgl with a monoclonal antibody in order to selectively deplete TAMs. On the same line, legumain is an acidic cysteine endopeptidase that is overexpressed in tumor tissues and in F4/80$^+$ CD206$^+$ TAMs under tumor stress conditions such as hypoxia. Hence, a DNA vaccine encoding legumain was administered before tumor inoculation. This immunization strategy specifically depleted TAMs by means of a CTL (cytotoxic T cells) response that was mounted against legumain-expressing TAMs [73]. In conclusion, unique functions orchestrated by TAMs in tumor immunity often comprise the presence of specific surface molecular signatures. Hence depleting TAMs by exploiting these signatures might be highly beneficial in reducing tumor growth, angiogenesis, and improving antitumor immunity. However, in light of recent observations that TAMs in tumors come in different flavors [36, 38], subset-specific molecular surface signatures need to be explored in order to target the desired TAM subset, while sparing putatively tumoricidal cells.

9.3.3 Targeting TAM Progenitors

In addition to depleting TAMs by delivering cytolytic drugs/peptides to TAM-specific surface receptors, preventing their infiltration into the tumor stroma has been proposed as a promising TAM-depletion strategy. Tumor-secreted chemokines attract monocytes to the tumor site and eventually promote their differentiation into TAMs, among them being CSF-1, CCL2 (MCP1), VEGF, and angiopoietin-2 [42, 43, 74–76]. Accordingly, mice displaying a null mutation of CSF1, therefore lacking mature macrophages, for instance in the mammary gland, and displaying severely reduced monocyte infiltration in oncogene-driven breast tumors, showed decreased angiogenesis and tumor progression to malignancy [77]. Also, the in vivo infiltration of F4/80$^+$ (mouse) or CSF-1R$^+$ CD163$^+$ (human) TAMs into tumor tissues was inhibited using a monoclonal antibody (RG7155, Phase I clinical trial) designed to inhibit CSF-1R. This was associated with a concomitant increase in lymphocyte infiltration and an increased CD8/CD4 ratio, which then translated to an increased clinical benefit in several of the solid malignancies studied [78]. In another study, an antibody against CSF-1R prevented F4/80$^+$ CD163$^+$ TAM accumulation at the tumor site and inhibition of tumor outgrowth [79]. As CSF-1R is not only required for monocyte to macrophage differentiation, but is also a critical mediator of macrophage proliferation and survival, targeting this molecule seems reasonable. However, again, the question of specificity over other tissue macrophage populations in the body, especially those requiring constant monocyte influx, such as in the intestines, should not be neglected to predict and monitor unwanted side effects.

Next to CSF-1 or CSF-1R, CCL2 emerges as an alternative target to block monocyte recruitment to tumors. CCL2 is the ligand of the chemokine receptor CCR2, which is highly expressed on monocytes. CCL2 secreting tumors showed higher vascularization and accumulation of alternatively activated macrophages at the tumor site. Indeed, pharmacological inhibitors against CCL2 [61] reduced tumor growth and the associated vascular network formation. Similar to CCL2, IL-16 can be an attractive target to prevent monocyte recruitment to tumors. IL-16 is a pleiotropic cytokine shown to be crucial for tumor development by recruiting monocytes to tumor spheroids [80]. Interestingly, inflammatory, potentially tumoricidal versus anti-inflammatory macrophages show differences in chemokine receptor expression. Thus, recruiting the former over the latter might be beneficial to overcome TAM function. This principle has already been exploited in experimental animal models. Administering the CXCL16-expressing metastatic colon cancer cells promoted infiltration of antitumor macrophages to the tumor site that prevented metastasis formation [81]. Next to chemokine networks, the cytotoxic drug trabectedin has been shown to induce selective caspase-8-dependent apoptosis in monocytes, but not other immune cells, thereby reducing TAM density and improving survival in cancer [82]. In conclusion, these studies highlight the importance of preventing monocyte recruitment to the tumor site as a means to deplete TAMs and indicate putative targets, some of which are already approaching the clinical setting.

9.3.4 Targeting TAM Differentiation and Renewal

A third strategy to address depletion of TAMs is by preventing their differentiation from monocytes. MicroRNA (miR)-142-3p negatively controls the expression of IL6ST (receptor for IL6 family of cytokines), which is involved in promoting macrophage differentiation. Indeed, tumor-associated $CD11b^+$ macrophages show decreased expression of this miR. Interestingly, enforced expression of miR-142-3p in the bone marrow impaired differentiation of these macrophages, favoring control of tumor growth through the restoration of lymphocyte cytolytic activity as a consequence of inhibiting macrophage differentiation [83]. Molecular signatures of monocyte to macrophage differentiation might reveal new structures to interfere with this process in the context of the tumor microenvironment.

Finally, a way to decrease the TAM content in tumors might be targeting their proliferation. A subset of $CD68^+$ $PCNA^+$ macrophages called 'promacs' (proliferating macrophages) were identified in tumors and were associated with poor breast cancer prognosis and clinical outcome [84]. Since proliferating macrophages constitute a rather small subset of the TAM pool, such strategies will likely be not forefront. Moreover, targeting molecules such as CSF-1 or CSF-1R might lower TAM proliferation with the benefit of hindering the arrival on new macrophage progenitors. It is also unclear whether conventional chemotherapeutic drugs targeting proliferating cells might not already interfere with TAM proliferation in vivo.

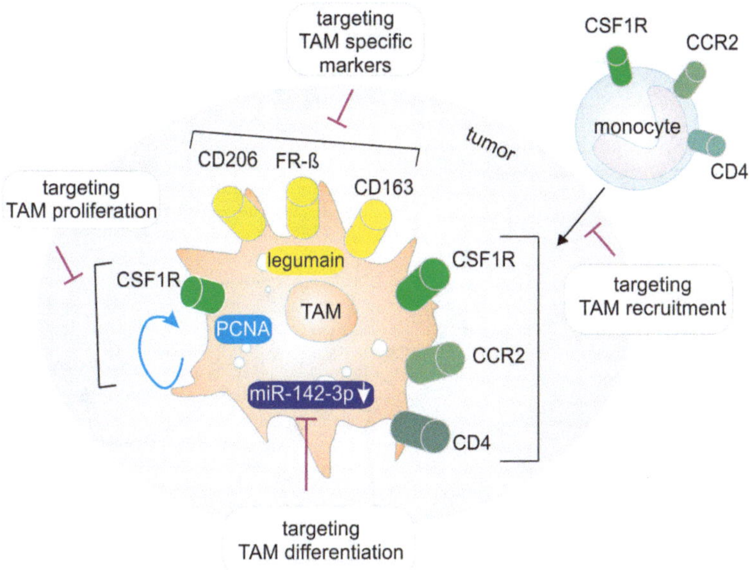

Fig. 9.3 Depleting TAMs by various strategies. TAMs can be depleted by targeting monocyte recruitment to the tumor stroma by employing monoclonal antibodies against CSF-1R, CCR2, or CD4. Once monocytes arrive at the tumor site, their differentiation to TAM can be limited, e.g., via antagonizing microRNA (miR) function. Once established, TAMs have the potential to renew by local proliferation, which might be abolished again by targeting CSF1R. Finally, TAMs can be directly depleted by cytotoxic drugs that are delivered via markers specifically expressed or enriched in TAMs such as CD206, CD163, or folate receptor β (FR-β)

The different approaches to reduce TAM numbers in tumors, as summarized in Fig. 9.3, will likely improve the therapeutic response of tumors to conventional therapy [85], especially when targeting protumor TAMs, while leaving potentially antitumor activities intact. Another way to achieve the goal of switching the TAM balance from pro- to antitumor is reprogramming of TAMs in situ, which will be discussed next.

9.3.5 TAM Reprogramming as an Antitumor Therapeutic Strategy

In mice and man, TAMs, as indicated earlier, are not only deceived by a 'wound-healing'-type tumor micromilieu but also orchestrate immune escape from other sentinels of adaptive immunity by escalating an immunosuppressive microenvironment. Interestingly, the majority of registered clinical trials regarding cancer are designed to chemotherapeutically target rapidly dividing tumor cells and/or reactivate antitumor adaptive immunity, whereas TAMs, in the background, discreetly perform the rebuilding tasks for tumor growth and metastasis, often provoking chemo-resistance. However, TAMs have the inherent potential to activate antitumor immunity and to

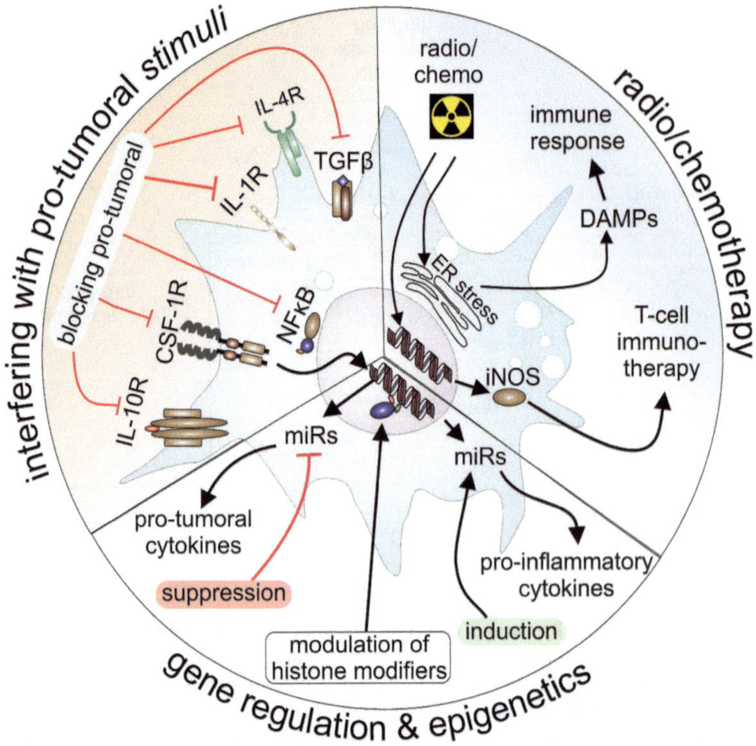

Fig. 9.4 Principles to target the protumor TAM phenotype. The TAM phenotype can be antagonized by interfering with the IL-10, CSF-1, IL-1, IL-4, or TGF-β receptors as well as antagonizing NF-kB signaling. In addition, radio/chemotherapy effectively programs TAMs by inducing ER stress and/or iNOS, thereby activating antitumor immunity. Gene regulation either by microRNA (miR)-mediated silencing of target genes or epigenetic changes alters the TAM phenotype. miRs (e.g., miR-155, miR-19a-3p, etc.) activate several proinflammatory signaling cascades to induce proinflammatory mediators. Thus, their expression is of benefit to antagonize protumor TAM functions. In contrast, suppression of protumor miRs (e.g., miR-21, miR-29a, etc.) might be equally effective

directly kill tumor cells. Recent studies suggest that this plasticity of TAMs can be exploited as a novel therapeutic strategy. This section addresses the mechanisms of TAM reprogramming in conjunction with chemotherapy as an effective antitumor therapy and highlights the importance of various *stimuli*, genetic and epigenetic regulations in the process (Fig. 9.4).

9.3.6 Modulating the Activation/Inhibition Stimuli

Interfering with the activation of tumor-promoting macrophages or boosting the activation of antitumor macrophages have been effective means to reprogram TAM phenotypes for antitumor therapy. Often, information regarding the reprogramming of TAMs as an antitumor therapy comes from in vivo murine studies. For instance, in a murine model of skin cancer, blockade of IL-4 signaling was sufficient to reprogram TAMs away from an alternatively activated phenotype and to inhibit tumor angiogenesis and growth [86]. Likewise, in the model of pancreatic ductal adenocarcinoma, inhibiting CSF-1R signaling not only reduced the number of TAMs, but also functionally reprogrammed remaining TAMs to induce antitumor T-cell activation by enhancing antigen presentation [87]. In a mouse xenograft model of human renal cell carcinoma (RCC), blocking the proinflammatory and protumor axis of IL-1-IL-1R by IL-1R antagonist decreased tumor growth and down-regulated protumor genes (*TNFA*, *IL6*, *VEGFA*, *PTGS2* (*prostaglandin-endoperoxide synthase* 2), and *MMP*) in $CD11b^{+}Ly6C^{hi/lo}F4/80^{hi}$ TAMs [88]. Furthermore, TAMs from IL-1RA-treated tumor-bearing mice increased *IL12B* and iNOS expression, accompanied by decreased *IL10* and *arginase1* expression, which suggests reprogramming toward an antitumor phenotype [89]. Likewise, CCL16 attracted macrophages and DCs in mammary and colon carcinomas, upon treatment with the Toll-like receptor 9 ligand CpG and α-IL-10 receptor antibody, resulted in reprogramming of TAMs, hemorrhagic tumor necrosis, activation of DCs as well as cytotoxic T cells, and clearance of tumor remnants [90]. Reprogramming of the suppressive TAM phenotype toward a proimmunogenic one could be achieved in a mouse model, as shown by Chatterjee et al. using nontoxic copper chelate that suppresses TGF-β production and elevates IL-12 production by inducing reactive oxygen species (ROS) generation, which triggers p38 MAPK and intercellular glutathione to elicit antitumor Th1 responses [91, 92].

Skewing signaling or the activity of nuclear factor-kappaB (NF-kB), the master regulator of cancer related signaling, has been a popular strategy for TAM reprogramming. Monocytes derived from RCC patients and healthy donor monocytes cocultured with RCC tumor cell lines displayed NF-kB activation and modulation of protumor genes [88]. Protumor TAMs attained a cytotoxic phenotype toward tumors when NF-kB signaling was inhibited, which is characterized by $IL\text{-}12^{hi}MHC\text{-}II^{hi}IL\text{-}10^{low}arginase\text{-}1^{low}$ expression. Pharmacological inhibition of NF-kB in TAMs also promoted regression of advanced tumors in vivo by induction of macrophage tumoricidal activity and activating antitumor activity through IL-12-dependent NK cell recruitment [93]. Taken together, pharmacological targeting of the activation stimuli or downstream signaling cascade in TAMs for reprogramming may be a useful addendum in combination with tumor-directed chemotherapy, especially in scenarios of high-risk chemo-resistance. However, directed delivery methods need to be developed to specifically target TAMs [94] in order to reduce toxicity and side effects in combinatorial therapy.

9.3.7 Radio/Chemotherapy

Recent advances in understanding the role of TAMs in tumor growth and metastasis provide a compelling argument for direct targeting of TAMs by radio/chemotherapy either for reprogramming to an antitumor/classically activated phenotype or to eliminate them from the tumor microenvironment (reviewed in [95]). Chemotherapies such as doxorubicin or oxaliplatin not only induce cell death by focused effects on DNA synthesis but also induce collateral effects via endoplasmic reticulum (ER) stress that causes the release of danger signals such as damage-associated molecular patterns (DAMPS), including ATP, high-mobility group B1 (HMGB1), thereby activating mononuclear phagocytes, enhancing their antigen-presentation capacity, and promoting T-cell responses against immunogenic tumors [96] (reviewed in [97]). This immunogenic cell death (ICD) reprograms TAMs to a proinflammatory phenotype. Reprogramming of TAMs in the tumor microenvironment could also be brought about by the taxane docetaxel. In a 4T1-Neu mammary tumor implant model, docetaxel depleted immunosuppressive TAMs, while concomitantly activating and expanding antitumor monocytes/MDSCs to enhance tumor-specific, cytotoxic T-cell responses [98]. However, in this study it was not clear, owing to the plasticity of TAM phenotypes in the tumor microenvironment, whether expansion of antitumor TAMs was due to predominantly reprogramming or due to the deletion of protumor TAMs. TAM reprogramming is also an effective means to increase the payload of chemotherapy at the tumor site by altering the tumor vascular architecture. Rolny et al. demonstrated that histidine-rich glycoproteins reprogram protumor TAMs to a tumor-inhibiting type by downregulating placental growth factor (PlGF), thereby normalizing tumor vessels, promoting antitumor immunity, inhibiting tumor growth and metastasis, while improving chemotherapy [99]. Interestingly, low-dose gamma irradiation of tumors also reprogrammed TAMs to an activated state characterized by iNOS induction that orchestrates T-cell immunotherapy [100]. Likewise, some forms of immunotherapy may also depend on effectively reprogramming TAMs toward an antitumor phenotype as demonstrated by Luo and Knudson by intravesical instillation of *Mycobacterium bovis* bacillus Calmette-Guérin (BCG), which induced macrophage cytotoxicity toward bladder cancer cells in both human and mouse. This macrophage-mediated killing of bladder cancer cells depended on both direct effector-target cell contact and release of soluble cytotoxic factors, such as TNF-α, IFNγ, and nitric oxide (NO) from macrophages [101]. Therefore, during radio/chemotherapy, targeting TAMs by ICD inducers is a win–win strategy as it not only serves as a contingency plan for chemo-resistance but would also make tumor-directed chemotherapy more effective by enabling optimal delivery of chemo-agents to tumor cells.

9.3.8 MicroRNA-Mediated Gene Regulation and Epigenetic Programming

Next-generation tumor therapy includes modulation of gene regulation that defines a protumoral TAM phenotype. In the last few years, posttranscriptional gene regulation by small noncoding RNA such as miR in TAM reprogramming has been demonstrated in experimental models [102] and promises potential for diagnosis and tumor therapy [103]. Direct targeting of miRs or targeting signaling pathways upstream of miRs in TAMs could support tumor therapy, especially in case of chemo-resistance. Using a myeloid cell-specific loss of function approach, it was proposed that miR-155 is required for the activation of $CD11c^{+}$ TAMs, and that this TAM subset, in turn, actively mediates antitumor immunity during early stages of breast carcinogenesis. However, it is unclear whether a specific antitumor subset in the $CD11c^{+}$ TAM population was targeted, since these cells were also connected to promoting tumor growth [36] or whether a phenotypic change involved the loss of CD11c expression. Stable knockdown of miR-155 in myeloid cells attenuated AKT protein kinase signaling, reprogramming TAMs to a protumor phenotype and accelerated tumor growth in a spontaneous breast cancer model [104]. miR-155 also promoted classical macrophage activation by not only downregulating inhibitors of the proinflammatory response, such as suppressor of cytokine signaling-1 [105] and B-cell lymphoma-6 protein [106] but also targeting the IL-13 receptor [107], which promotes alternative macrophage activation. Furthermore, a recent study showed that miR-155 delivery in alternatively activated macrophages was sufficient to reprogram these cells toward a more proinflammatory phenotype [108], increasing TNF-α production by stabilizing the TNF-α transcript, and downregulating genes associated with alternative activation such as arginase-1, chitinase 3-like 3, and CCAAT/enhancer binding protein β in macrophages [109].

miRs can directly target and suppress hallmark cancer genes [110] such as those involved in angiogenesis (e.g., *Col4a2* (collagen, type IV, alpha 2), *pry1* (pathogenesis-related protein 1), and *Timp3* (metalloproteinase inhibitor 3)) and establishing TAM function, thereby directly affecting tumor growth and metastasis. Mathsyaraja et al. recently published that the transcription factor ETS2, whose expression and phosphorylation is regulated by the CSF1-ERK pathway, regulates the expression of miR-21, miR-29a, miR-142-3p, and miR-223 in TAMs in a spontaneous mouse model of mammary carcinoma. Depletion of the miR-processing enzyme Dicer in TAMs blocked angiogenesis and metastatic tumor growth. Furthermore, expression of miR-21 and miR-29a in $CD115^{+}CD14^{lo}CD16^{hi}$ blood cells correlated with metastatic burden of breast cancer patients [102]. Similarly, miR-19a-3p downregulated the *Fra-1* proto-oncogene in TAMs, suppressed the protumoral phenotype and inhibited breast cancer progression [111]. Interestingly, the role of miRs not always correlates with the established phenotype of classically or alternatively activated TAMs. CD206 expression, which is often associated with a protumor TAM phenotype [89] and is required for tissue remodeling [112], has been implicated in antitumor gene regulation by virtue of the intronic expression of miRs. MiR-511-3p, an intronic miR encoded by the *Mrc1* gene (encodes for CD206) in mouse and human has been shown to provide a negative feedback on protumor genetic programs by directly targeting Rho-dependent kinase-2 [113]. Overexpression of miR-511-3p inhibited tumor growth, altered tumor blood vessel morphology, and tuned down the protumor gene signature of $CD206^{+}$ TAMs, thereby demonstrating an extra layer of gene expression control, which relies on an endogenous molecular switch and the plasticity of macrophage phenotypes in the tumor microenvironment.

Thus, interfering with miR activity and/or gene regulation for reprogramming TAMs is an attractive pharmacological target for tumor therapy. Recent preclinical studies successfully demonstrate the systemic use of nanoparticles loaded with miR-155 mimic to target CD11c+ TAMs/DCs [114]. Mannosylated polymer nanoparticles (MnNP) that are capable of escaping the endosomal compartment are used to deliver siRNA to TAMs in vitro and in vivo [115]. Targeting of miR-223 [116], miR-511-3p [113], or silencing miRs in mice and nonhuman primates [117, 118] may pave a way to use in human patients in the near future.

Genetic reprogramming in TAMs is not always associated with hard-wired genetic changes due to the fact that in vivo macrophages exhibit mixed and transient phenotypes as they are exposed to several potentially opposing polarizing factors, and transitions from protumor to antitumor phenotype (or vice versa) are quite rapid [119, 120]. A role of epigenetic modulators in macrophage reprogramming has been described in the literature. Alternative activation of macrophages, in vivo, is mediated by the histone demethylase JMJD3, which removes negative H3K27me3 marks at the *Irf4* locus, thereby facilitating expression of the key transcription factor IRF4 [121]. Furthermore, induction of *Ifnb* and the downstream IFN response, which is an important component of antitumor macrophages, is strongly dependent on histone deacetylase 3 (HDAC3) [122] and HDAC inhibitors suppress the induction of various inflammatory and IFN target genes [123]. Understanding the epigenetic changes during macrophage polarization and modulating the epigenetic regulators for combinatorial tumor therapy is a niche area and several clinical trials have been registered to address these aspects (ClinicalTrials.gov Identifier: NCT00262834, NCT01486277, NCT01738815).

9.4 The Impact of Apoptotic Cells on the TAM Phenotype and Function

Tumor growth is defined by the interplay of a variety of intrinsic and extrinsic factors. Among these, a high proliferative capacity and relatively low rates of cell death are conditions expected to favor tumor growth. Nevertheless, sustained and prominent apoptosis is a characteristic, inherent feature of different types of growing tumors, being described already before the concept of apoptosis was first described in 1972 [124]. The notion that tumor cell apoptosis, at least at low levels, contrary to conventional assumptions, promotes tumor growth and progression can be appreciated when looking at the physiological role of apoptosis, which surpasses the silent elimination of unwanted cells. Apoptosis under conditions of tissue stress such as in wounds supports healing and regeneration by providing specific signals directly to resident tissue cells [125] or by altering macrophage phenotypes [126]. Thus, during tissue stress the process initiated by apoptosis of cells culminates in a phenotype switch of macrophages to restore tissue homeostasis, e.g., by reducing tissue mononuclear and lymphoid cell numbers to a physiological level and function and by delivering proliferation and recruitment signals to epithelial and endothelial cells to promote regeneration [5, 6, 127]. In this auto-regulatory system, cell death programs the reconstitution of a new physiological environment. When considering tumors as "wounds that do not heal" [128], apoptotic tumor cells might well support tumor growth via similar mechanisms as in wounds. Apoptotic cell death as an inherent feature of growing tumors might provide a number of signals that force macrophages to acquire a tumor-supportive phenotype affecting all hallmarks of cancer such as tumor cell proliferation, angiogenesis, and immune suppression. Thus, the perturbed tissue homeostasis in tumors hijacks physiological regeneration systems to support its own growth and to limit its elimination. Intuitively presumed only to inhibit cell growth and limit malignancies, apoptotic cells may in this manner actually promote net tumor growth by releasing different molecules, which act on neighboring tumor cells to support tumor growth [129] and by educating macrophages to a protumor phenotype. Tumorigenic effects of apoptotic cells have been noticed already back in 1956 by Revesz [130], showing that coinjection of irradiated tumor cells with living tumor cells enhances tumor occurrence in experimental animals. This was not the case when tumor cells were killed by heat (necrosis) and coinjected with living cells, indicating the requirement of actively produced signals [131]. More recent studies have shown that radiotherapy, a commonly used anticancer therapy, can enhance tumor cell repopulation in vivo, through the induction of apoptosis [132]. This process termed 'sinister self-sacrifice' needs consideration when targeting malignant cells [133]. Although the consequences of this self-sacrifice can be explained by a feeder effect of dying cells toward living tumor cells, an interaction of apoptotic cells with macrophages might provide a similar physiological outcome. Of note, a study employing adoptive transfer of conditioned macrophages demonstrated that macrophage priming with apoptotic cells speeds up growth of implanted living tumors, whereas macrophage priming with necrotic cells protected against tumor growth [134].

There are striking similarities not only between the function of macrophages ingesting apoptotic cells and TAMs but also regarding the underlying molecular signature. In both cells expression of proinflammatory cytokines such as IL-12 is low, whereas anti-inflammatory cytokines like IL-10 are highly expressed. They express growth factors such as VEGF, TGF-β, and EGF to promote epithelial and endothelial proliferation and survival, matrix remodeling enzymes such as MMPs and display a low capacity to produce immunoregulatory gaseous molecules such as ROS and NO [41–44, 135]. On the following pages we will summarize the current evidence linking macrophage—apoptotic cell interaction to TAM generation and discuss the applicability of targeting this interaction for cancer therapy. For this, we will follow different steps of apoptotic cell/phagocyte interaction sequentially; phagocyte recruitment through find-me signals, apoptotic cell recognition via eat-me signals, and the consequences derived from these modes of interaction that affect the immunological phagocyte response (Fig. 9.5).

9.4.1 Apoptotic Cell-Derived Soluble Mediators in TAM Recruitment and Polarization

Pivotal for apoptotic cell removal is the presence of phagocytes. Thus, resident and recently recruited macrophages, as well as blood monocytes require a directed migratory signal to reach the location where apoptotic cells need to be removed.

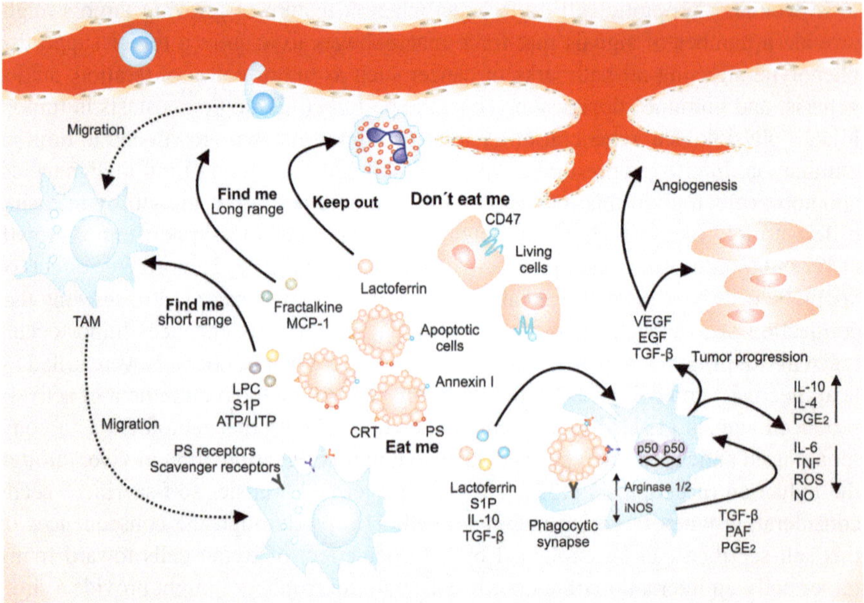

Fig. 9.5 Interaction of apoptotic cells with tumor-associated macrophages. TAMs are attracted to sites of apoptosis through find-me signals secreted from apoptotic cells. To ensure specific migration of mononuclear phagocytes, keep-out signals restrict granulocyte migration. Once at sites of apoptosis, specific receptors at the surface of differentiating monocytes or mature TAMs recognize eat-me signals expressed on apoptotic cells originating in the phagocytic synapse. This interaction together with soluble signals secreted from apoptotic cells programs macrophages toward a tumor-promoting phenotype, which is characterized by the secretion of growth factors and anti-inflammatory mediators. *CRT* calreticulin, *LPC* lysophosphatidylcholine, *S1P* sphingosine-1-phosphate, *PS* phosphatidylserine, *PGE*$_2$ prostaglandin E2, *TGF-β* transforming growth factor-β, *PAF* platelet activating factor, *iNOS* inducible nitric oxide synthase, *NO* nitric oxide, *ROS* radical oxygen species, *TNF* tumor necrosis factor, *VEGF* vascular endothelial growth factor, *EGF* epidermal growth factor

Apoptotic cells attract mononuclear phagocytes through the release of find-me signals such as lysophosphatidylcholine (LPC), the chemokine fractalkine (CX_3CL1), sphingosine-1-phosphate (S1P), the ribosomal protein S19, EMAPII, a fragment of human tyrosyl tRNA synthetase, and the nucleotides ATP and UTP [135, 136]. These factors exert short- and long-range actions, inducing the attraction of local resident and recruited macrophages, or the long-distance recruitment of blood monocytes, respectively. For instance, based on its mode of release, CX_3CL1 acts as a long-distance recruitment factor [137] with the potential to attract monocytes. Indeed, CX_3CL1 was already been implicated in TAM recruitment to promote tumor growth, survival, and metastasis [138]. Accordingly, expression of the fractalkine receptor CX_3CR1 was associated with poor prognosis, TAM infiltration, and metastasis [139]. During recruitment it is important to attract the right type of professional phagocyte. Mononuclear phagocytes but not granulocytes are able to remove apoptotic cells in the absence of a proinflammatory response.

Therefore, the attracting signals must be specific for mononuclear phagocytes. Chemotactic factors such as EMAPII and ATP are able to attract both macrophages and granulocytes, hence besides the above-mentioned find me signals, keep out signals are necessary for the recruitment of the adequate type of phagocytes. Keep out signals may present another macrophage-independent factor, relevant to how apoptotic tumor cells promote the survival of their viable kin, by limiting the influx of tumoricidal granulocytes. Lactoferrin, an iron transporting molecule, has been identified as such a keep out signal, which inhibits the migration of granulocytes to sites of apoptosis [140]. However, while the relevance of keep out signals in tumor immunity remains unclear, lactoferrin has certain immunoregulatory functions besides limiting granulocyte influx. These include limiting IL-6 and TNF-α production and inducing IL-10, IL-4, and TGF-β release by phagocytes [135]. Another putative find-me signal, the sphingolipid S1P, also showed immune-regulatory potential. S1P is produced by apoptotic tumor cells via sphingosine kinase 1 or 2 [141–143] to recruit and/or activate TAMs. S1P released from apoptotic tumor cells limited antitumor cytokine production, e.g., IL-12 [144], and downregulated major histocompatibility complex II (MHC-II) expression to suppress T cell activation by macrophages [145]. On the other hand, S1P promoted the expression of a multitude of protumor cytokines such as IL-10, IL-6, and VEGF, partly dependent on the action of heme oxygenase-1 or stabilization of the transcription factor HIF-1α [145–148]. Moreover, S1P induced the formation of the protumor prostaglandin PGE_2 by human antigen R (HuR)-mediated stabilization of cyclooxygenase-2 (COX-2) mRNA and microsomal prostaglandin E synthase-1 (mPGES-1) expression [6, 149]. In turn, formation of PGE_2 provoked a cAMP-increase in macrophages to limit CD80 expression (our unpublished observation), directly inhibited activated cytotoxic T cells [150], and promoted angiogenesis by supporting endothelial cell proliferation [6]. Consequently, depletion of sphingosine kinase 2 in tumor cells reduced tumor growth due to altered macrophage activation in a xenograft model [151]. Finally, S1P increased the resistance of macrophages against chemotherapeutic drugs [141], allowing the possibility that targeting S1P or its receptors on monocytes/macrophages might improve the efficacy of chemotherapy.

9.4.2 *Apoptotic Cell Surface Alterations: Recognition and TAM Generation*

Following recruitment and attraction of phagocytes to the site of apoptosis, intrinsic molecular signals enable specific apoptotic cell recognition. Especially plasma membrane alterations play a pivotal role in the distinction between viable and apoptotic cells, which leads to the clearance of the resulting apoptotic bodies. The plasma membrane of apoptotic cells is altered with respect to lipid, sugar, and protein composition [136]. Different proteins including calreticulin, annexin I, the large subunit of the eukaryotic translation initiation factor 3 (eIF3a), and the long pentraxin PTX3 are exposed on the surface of apoptotic cells and interact directly or indirectly with phagocytes [136]. Regarding the lipid architecture, most significant changes in the plasma membrane are the oxidation and redistribution of phosphatidylserine (PS) from the inner to the outer leaflet of the plasma membrane [152–154]. These molecules are considered as eat-me signals when their expression is increased at the surface of the cells. They are recognized, either directly by specific receptors that are highly expressed by phagocytes such as scavenger receptors (e.g., CD36), complement receptors, C-type lectin receptors, PS receptors (e.g., BAI1, TIM-4, and stabilin-2), and the prototypic pattern recognition receptor (PPR) CD14, or by phagocyte receptors such as Mer tyrosine kinase (MerTK) or the vitronectin receptor (VnR, $\alpha_v\beta_3$ integrin) via specific bridging molecules such as growth arrest-specific 6 (GAS6) and milk fat globule-EGF factor 8 (MFG-E8, lactadherin), respectively [135, 136, 155]. Numerous interactions of phagocyte receptors with eat-me signals provoke the formation of cell–cell junctions, known as the phagocytic synapse. Signals originating in the phagocytic synapse are involved in shaping a largely anti-inflammatory macrophage phenotype. Of note, not all receptor/ligand interactions in the phagocytic synapse suppress inflammation as shown with CD14, CD36, and $\alpha_v\beta_3$ integrin deficient mice [156, 157]. Rather PS recognition by its numerous receptors with or without bridging molecules is prominently associated with altered inflammatory macrophage function. In initial experiments, recognition of PS on apoptotic cells by macrophages induced autocrine signaling via TGF-β, platelet-activating factor (PAF) and PGE_2 to reduce the production of IL-1β, IL-8, and TNF-α. This was abolished by a neutralizing TGF-β-antibody, blocking COX enzymes with indomethacin, or by a PAF receptor antagonist [158]. Reduced cytokine expression was likely associated with inhibition of classical NF-kB activation (p65/p50 heterodimers), a major transcriptional regulator of proinflammatory signals [159]. Strikingly, signaling through nonclassical NF-kB pathways (e.g., p50/p50 homodimers) is also a common feature of certain TAM populations [160, 161]. In the case of apoptotic cells, PS recognition was coupled to defective classical NF-kB signaling. Cvetanovic et al. provided evidence that NF-kB binding to DNA was inactive, implying that coactivators/corepressors were involved [162]. Indeed, NF-kB inhibition resulted from the interplay with PPARγ and the nuclear receptor corepressor (NCoR). NCoR is bound to NF-kB sites under basal conditions, while coactivator/corepressor exchange is a common phenomenon switching between gene repression and activation. Apoptotic cells induced PPARγ sumoylation to attenuate the removal of NCoR, thereby blocking transactivation of NF-kB, which lowered proinflammatory cytokine production [163]. Activation of PPARγ by apoptotic cells was dependent on specific recognition of PS [164] and required sorting of 5-lipoxygenase (5-LO) into lipid rafts of apoptotic cells to synthesize ligands that in turn stimulated PPARγ in macrophages [165]. Since MerTK, recognizing PS via GAS6 was also associated with NF-kB inhibition in response to apoptotic cells as demonstrated by using MerTK knockdown cells and blocking antibodies [166], a signaling axis consisting of PS, MerTK, and PPARγ can be proposed to culminate in inhibition of NF-kB activity. Besides modulating cytokine production, PS recognition also impairs the ability of macrophages to produce reactive nitrogen and oxygen species. Reduced NO formation can result from increased arginase expression.

Arginase 1 and 2, metabolize L-arginine to urea and ornithine, thus competing with iNOS for the same substrate [44]. Apoptotic cells upregulated arginase 1 in macrophages, at the same time downregulating iNOS in a TGF-β- and PS-dependent manner [167], while soluble factors secreted by apoptotic cells upregulated arginase 2, which was TGF-β independent and left iNOS expression unaltered [168]. Likely, the rapid induction of arginase 2 by soluble apoptotic cell-derived molecules accounts for early responses, while the PS/TGF-β axis facilitates later phases of NO inhibition. The formation of ROS was blocked as well, presumably via oxidized PS inducing again PPARγ activation and inhibition of NADPH oxidase assembly [169, 170]. There is an apparent overlap between soluble mediators versus specific eat-me signals such as PS when looking at the consequences for macrophage function. This may be a way to ensure immunological silence even when apoptotic cell removal is not immediate. Alternatively, both pathways might be causatively linked. For instance, CX_3CL1 can induce expression of the bridging molecules such as MFG-E8, which is necessary for the interaction of VnR and PS [135], in turn inducing COX-2 expression in phagocytes [171].

There is emerging evidence that PS recognition indeed is involved in shaping TAM function. Therapeutic targeting PS induced inflammatory macrophage activation to suppress tumor growth and progression in prostate tumors [172]. Moreover, MerTK-deficiency protected mice with mammary carcinoma from metastasis, which occurred as a result of cell apoptosis as a means to promote mammary gland involution after pregnancy. In this model, MerTK deficiency prevented uptake of apoptotic tumor cells and production of TGF-β by tumor-associated macrophages [173].

Taken together, the evidence summarized earlier strongly suggests tumor cell apoptosis as a crucial driver of TAM polarization in the tumor microenvironment. However, more precise experimental evidence needs to be provided to identify specific molecules in the apoptotic cell/macrophage interaction that can be targeted for tumor therapy. Moreover, molecular signatures triggered by apoptotic cell phagocytosis may help to identify protumor TAM subpopulations that should be targeted preferentially. Resident (embryonic) macrophages may show a higher potential to ingest apoptotic cells in a noninflammatory manner, at least in the peritoneum [174]. Thus, it might be interesting to observe whether selectively depleting these cells will improve antitumor immunity in tumor entities. Interestingly, studying apoptotic cell/macrophage interaction may yield other mechanisms to interfere with tumor growth besides inhibiting the establishment of a protumor macrophage phenotype. In addition to the eat-me signals, the loss of don't eat-me signals such as CD31 and CD47 is necessary for the removal of apoptotic cells [136]. CD47 is ubiquitously expressed on normal viable cells, functioning as a don't eat-me signal that is downregulated to allow homeostatic phagocytosis of aged or damaged cells. Likewise, CD47 is overexpressed on viable tumor cells, allowing tumor cells to escape immune surveillance through inhibition of phagocytosis. Therapeutic targeting of CD47 induced phagocytosis of viable tumor cells apparently without triggering the negative consequences of apoptotic cell/macrophage interaction for antitumor therapy [175, 176].

As a summary, improving tumor therapy should not only kill malignant cells but also target the interaction of the host with apoptotic cancer cells, as this might be sufficient to limit the protumor action of apoptotic cells and boost the antitumor potential of macrophages. Leaving the apoptotic cell/macrophage interaction untouched might also limit the benefit of conventional tumor cell apoptosis-focused therapy since surviving tumor cells might receive overwhelming support by the wound healing response that apoptotic tumor cells will trigger in local macrophages, thereby enhancing tumor recurrence.

Acknowledgments We gratefully acknowledge the following funding organizations: Deutsche Forschungsgemeinschaft (SFB 815, SFB 1039, BR999, and Excellence Cluster Cardiopulmonary System), Deutsche Krebshilfe (110637, 111578), Sander Foundation (2013.036.1), Else Kröner-Fresenius-Graduiertenkolleg and Translational Research Innovation - Pharma (TRIP) (both funded by the Else Kröner-Fresenius-Stiftung).

We apologize to researchers whose primary observations that form the basis of current knowledge in the field have been acknowledged indirectly, by referring to current reviews.

References

1. Lavin Y, Merad M. Macrophages: gatekeepers of tissue integrity. Cancer Immunol Res. 2013;1(4):201–9. doi:10.1158/2326-6066.CIR-13-0117. pii: 1/4/201.
2. Wynn TA, Chawla A, Pollard JW. Macrophage biology in development, homeostasis and disease. Nature. 2013;496(7446):445–55. doi:10.1038/nature12034. pii: nature12034.
3. Epelman S, Lavine KJ, Randolph GJ. Origin and functions of tissue macrophages. Immunity. 2014;41(1):21–35. doi:10.1016/j.immuni.2014.06.013.
4. Murray PJ, Wynn TA. Protective and pathogenic functions of macrophage subsets. Nat Rev Immunol. 2011;11(11):723–37.
5. Sola A, Weigert A, Jung M, Vinuesa E, Brecht K, Weis N, Brune B, Borregaard N, Hotter G. Sphingosine-1-phosphate signalling induces the production of Lcn-2 by macrophages to promote kidney regeneration. J Pathol. 2011;225(4):597–608.
6. Brecht K, Weigert A, Hu J, Popp R, Fisslthaler B, Korff T, Fleming I, Geisslinger G, Brune B. Macrophages programmed by apoptotic cells promote angiogenesis via prostaglandin E2. FASEB J. 2011;25(7):2408–17.
7. Pollard JW. Trophic macrophages in development and disease. Nat Rev Immunol. 2009;9(4):259–70.
8. Kaufmann SH. Immunology's foundation: the 100-year anniversary of the Nobel Prize to Paul Ehrlich and Elie Metchnikoff. Nat Immunol. 2008;9(7):705–12.
9. van Furth R, Cohn ZA, Hirsch JG, Humphrey JH, Spector WG, Langevoort HL. The mononuclear phagocyte system: a new classification of macrophages, monocytes, and their precursor cells. Bull World Health Organ. 1972;46(6):845–52.
10. Volkman A. Disparity in origin of mononuclear phagocyte populations. J Reticuloendothel Soc. 1976;19(4):249–68.
11. van Furth R, Cohn ZA. The origin and kinetics of mononuclear phagocytes. J Exp Med. 1968;128(3):415–35.
12. Naito M, Umeda S, Yamamoto T, Moriyama H, Umezu H, Hasegawa G, Usuda H, Shultz LD, Takahashi K. Development, differentiation, and phenotypic heterogeneity of murine tissue macrophages. J Leukoc Biol. 1996;59(2):133–8.
13. Ginhoux F, Jung S. Monocytes and macrophages: developmental pathways and tissue homeostasis. Nat Rev Immunol. 2014;14(6):392–404. doi:10.1038/nri3671.

14. Schulz C, Gomez Perdiguero E, Chorro L, Szabo-Rogers H, Cagnard N, Kierdorf K, Prinz M, Wu B, Jacobsen SE, Pollard JW, Frampton J, Liu KJ, Geissmann F. A lineage of myeloid cells independent of Myb and hematopoietic stem cells. Science. 2012;336(6077):86–90.
15. Perdiguero EG, Klapproth K, Schulz C, Busch K, Azzoni E, Crozet L, Garner H, Trouillet C, de Bruijn MF, Geissmann F, Rodewald H. Tissue-resident macrophages originate from yolk-sac-derived erythro-myeloid progenitors. Nature. 2014;518:547–51. doi:10.1038/nature13989. pii: nature13989.
16. Ginhoux F, Greter M, Leboeuf M, Nandi S, See P, Gokhan S, Mehler MF, Conway SJ, Ng LG, Stanley ER, Samokhvalov IM, Merad M. Fate mapping analysis reveals that adult microglia derive from primitive macrophages. Science. 2010;330(6005):841–5. doi:10.1126/science.1194637.
17. Kierdorf K, Erny D, Goldmann T, Sander V, Schulz C, Perdiguero EG, Wieghofer P, Heinrich A, Riemke P, Holscher C, Muller DN, Luckow B, Brocker T, Debowski K, Fritz G, Opdenakker G, Diefenbach A, Biber K, Heikenwalder M, Geissmann F, Rosenbauer F, Prinz M. Microglia emerge from erythromyeloid precursors via Pu.1- and Irf8-dependent pathways. Nat Neurosci. 2013;16(3):273–80. doi:10.1038/nn.3318. pii: nn.3318.
18. McGovern N, Schlitzer A, Gunawan M, Jardine L, Shin A, Poyner E, Green K, Dickinson R, Wang XN, Low D, Best K, Covins S, Milne P, Pagan S, Aljefri K, Windebank M, Saavedra DM, Larbi A, Wasan PS, Duan K, Poidinger M, Bigley V, Ginhoux F, Collin M, Haniffa M. Human dermal CD14(+) cells are a transient population of monocyte-derived macrophages. Immunity. 2014;41(3):465–77. doi:10.1016/j.immuni.2014.08.006. pii: S1074-7613(14)00300-8.
19. Bain CC, Mowat AM. Macrophages in intestinal homeostasis and inflammation. Immunol Rev. 2014;260(1):102–17. doi:10.1111/imr.12192.
20. Heinz S, Benner C, Spann N, Bertolino E, Lin YC, Laslo P, Cheng JX, Murre C, Singh H, Glass CK. Simple combinations of lineage-determining transcription factors prime cis-regulatory elements required for macrophage and B cell identities. Mol Cell. 2010;38(4):576–89. doi:10.1016/j.molcel.2010.05.004. pii: S1097-2765(10)00366-7.
21. Jenkins SJ, Hume DA. Homeostasis in the mononuclear phagocyte system. Trends Immunol. 2014;35(8):358–67. doi:10.1016/j.it.2014.06.006. pii: S1471-4906(14)00111-2.
22. Lavin Y, Winter D, Blecher-Gonen R, David E, Keren-Shaul H, Merad M, Jung S, Amit I. Tissue-resident macrophage enhancer landscapes are shaped by the local microenvironment. Cell. 2014;159(6):1312–26. doi:10.1016/j.cell.2014.11.018. pii: S0092-8674(14)01449-4.
23. Haldar M, Kohyama M, So AY, Kc W, Wu X, Briseno CG, Satpathy AT, Kretzer NM, Arase H, Rajasekaran NS, Wang L, Egawa T, Igarashi K, Baltimore D, Murphy TL, Murphy KM. Heme-mediated SPI-C induction promotes monocyte differentiation into iron-recycling macrophages. Cell. 2014;156(6):1223–34. doi:10.1016/j.cell.2014.01.069. pii: S0092-8674(14)00276-1.
24. Okabe Y, Medzhitov R. Tissue-specific signals control reversible program of localization and functional polarization of macrophages. Cell. 2014;157(4):832–44. doi:10.1016/j.cell.2014.04.016. pii: S0092-8674(14)00494-2.
25. Rosas M, Davies LC, Giles PJ, Liao CT, Kharfan B, Stone TC, O'Donnell VB, Fraser DJ, Jones SA, Taylor PR. The transcription factor Gata6 links tissue macrophage phenotype and proliferative renewal. Science. 2014;344(6184):645–8. doi:10.1126/science.1251414. pii: science.1251414.
26. Schneider C, Nobs SP, Kurrer M, Rehrauer H, Thiele C, Kopf M. Induction of the nuclear receptor PPAR-gamma by the cytokine GM-CSF is critical for the differentiation of fetal monocytes into alveolar macrophages. Nat Immunol. 2014;15(11):1026–37. doi:10.1038/ni.3005. pii: ni.3005.
27. Butovsky O, Jedrychowski MP, Moore CS, Cialic R, Lanser AJ, Gabriely G, Koeglsperger T, Dake B, Wu PM, Doykan CE, Fanek Z, Liu L, Chen Z, Rothstein JD, Ransohoff RM, Gygi SP, Antel JP, Weiner HL. Identification of a unique TGF-beta-dependent molecular and functional signature in microglia. Nat Neurosci. 2014;17(1):131–43. doi:10.1038/nn.3599. pii: nn.3599.

28. Gosselin D, Link VM, Romanoski CE, Fonseca GJ, Eichenfield DZ, Spann NJ, Stender JD, Chun HB, Garner H, Geissmann F, Glass CK. Environment drives selection and function of enhancers controlling tissue-specific macrophage identities. Cell. 2014;159(6):1327–40. doi:10.1016/j.cell.2014.11.023. pii: S0092-8674(14)01500-1.
29. Brune B, Dehne N, Grossmann N, Jung M, Namgaladze D, Schmid T, von Knethen A, Weigert A. Redox control of inflammation in macrophages. Antioxid Redox Signal. 2013;19:595–637. doi:10.1089/ars.2012.4785.
30. Martinez FO, Helming L, Milde R, Varin A, Melgert BN, Draijer C, Thomas B, Fabbri M, Crawshaw A, Ho LP, Ten Hacken NH, Cobos Jimenez V, Kootstra NA, Hamann J, Greaves DR, Locati M, Mantovani A, Gordon S. Genetic programs expressed in resting and IL-4 alternatively activated mouse and human macrophages: similarities and differences. Blood. 2013;121(9):e57–69.
31. Xue J, Schmidt SV, Sander J, Draffehn A, Krebs W, Quester I, De Nardo D, Gohel TD, Emde M, Schmidleithner L, Ganesan H, Nino-Castro A, Mallmann MR, Labzin L, Theis H, Kraut M, Beyer M, Latz E, Freeman TC, Ulas T, Schultze JL. Transcriptome-based network analysis reveals a spectrum model of human macrophage activation. Immunity. 2014;40(2):274–88. doi:10.1016/j.immuni.2014.01.006.
32. Mantovani A, Sica A, Sozzani S, Allavena P, Vecchi A, Locati M. The chemokine system in diverse forms of macrophage activation and polarization. Trends Immunol. 2004;25(12):677–86.
33. Krausgruber T, Blazek K, Smallie T, Alzabin S, Lockstone H, Sahgal N, Hussell T, Feldmann M, Udalova IA. IRF5 promotes inflammatory macrophage polarization and TH1-TH17 responses. Nat Immunol. 2011;12(3):231–8. doi:10.1038/ni.1990. pii: ni.1990.
34. Noy R, Pollard JW. Tumor-associated macrophages: from mechanisms to therapy. Immunity. 2014;41(1):49–61. doi:10.1016/j.immuni.2014.06.010.
35. Pyonteck SM, Akkari L, Schuhmacher AJ, Bowman RL, Sevenich L, Quail DF, Olson OC, Quick ML, Huse JT, Teijeiro V, Setty M, Leslie CS, Oei Y, Pedraza A, Zhang J, Brennan CW, Sutton JC, Holland EC, Daniel D, Joyce JA. CSF-1R inhibition alters macrophage polarization and blocks glioma progression. Nat Med. 2013;19(10):1264–72. doi:10.1038/nm.3337. pii: nm.3337.
36. Franklin RA, Liao W, Sarkar A, Kim MV, Bivona MR, Liu K, Pamer EG, Li MO. The cellular and molecular origin of tumor-associated macrophages. Science. 2014;344:921–5.
37. Tymoszuk P, Evens H, Marzola V, Wachowicz K, Wasmer MH, Datta S, Muller-Holzner E, Fiegl H, Bock G, van Rooijen N, Theurl I, Doppler W. In situ proliferation contributes to accumulation of tumor-associated macrophages in spontaneous mammary tumors. Eur J Immunol. 2014;44(8):2247–62. doi:10.1002/eji.201344304.
38. Broz ML, Binnewies M, Boldajipour B, Nelson AE, Pollack JL, Erle DJ, Barczak A, Rosenblum MD, Daud A, Barber DL, Amigorena S, Van't Veer LJ, Sperling AI, Wolf DM, Krummel MF. Dissecting the tumor myeloid compartment reveals rare activating antigen-presenting cells critical for T cell immunity. Cancer Cell. 2014;26(5):638–52. doi:10.1016/j.ccell.2014.09.007. pii: S1535-6108(14)00370-5.
39. Laoui D, Movahedi K, Van Overmeire E, Van den Bossche J, Schouppe E, Mommer C, Nikolaou A, Morias Y, De Baetselier P, Van Ginderachter JA. Tumor-associated macrophages in breast cancer: distinct subsets, distinct functions. Int J Dev Biol. 2011;55(7-9):861–7.
40. Kuraishy A, Karin M, Grivennikov SI. Tumor promotion via injury- and death-induced inflammation. Immunity. 2011;35(4):467–77.
41. Weigert A, Jennewein C, Brune B. The liaison between apoptotic cells and macrophages—the end programs the beginning. Biol Chem. 2009;390(5-6):379–90.
42. Galdiero MR, Garlanda C, Jaillon S, Marone G, Mantovani A. Tumor associated macrophages and neutrophils in tumor progression. J Cell Physiol. 2013;228(7):1404–12. doi:10.1002/jcp.24260.
43. Sica A, Larghi P, Mancino A, Rubino L, Porta C, Totaro MG, Rimoldi M, Biswas SK, Allavena P, Mantovani A. Macrophage polarization in tumour progression. Semin Cancer Biol. 2008;18(5):349–55. doi:10.1016/j.semcancer.2008.03.004. pii: S1044-579X(08)00023-0.

44. Weigert A, Brune B. Nitric oxide, apoptosis and macrophage polarization during tumor progression. Nitric Oxide. 2008;19(2):95–102.
45. Dave SS, Wright G, Tan B, Rosenwald A, Gascoyne RD, Chan WC, Fisher RI, Braziel RM, Rimsza LM, Grogan TM, Miller TP, LeBlanc M, Greiner TC, Weisenburger DD, Lynch JC, Vose J, Armitage JO, Smeland EB, Kvaloy S, Holte H, Delabie J, Connors JM, Lansdorp PM, Ouyang Q, Lister TA, Davies AJ, Norton AJ, Muller-Hermelink HK, Ott G, Campo E, Montserrat E, Wilson WH, Jaffe ES, Simon R, Yang L, Powell J, Zhao H, Goldschmidt N, Chiorazzi M, Staudt LM. Prediction of survival in follicular lymphoma based on molecular features of tumor-infiltrating immune cells. N Engl J Med. 2004;351(21):2159–69.
46. Leek RD, Lewis CE, Whitehouse R, Greenall M, Clarke J, Harris AL. Association of macrophage infiltration with angiogenesis and prognosis in invasive breast carcinoma. Cancer Res. 1996;56(20):4625–9.
47. Lissbrant IF, Stattin P, Wikstrom P, Damber JE, Egevad L, Bergh A. Tumor associated macrophages in human prostate cancer: relation to clinicopathological variables and survival. Int J Oncol. 2000;17(3):445–51.
48. Nishie A, Ono M, Shono T, Fukushi J, Otsubo M, Onoue H, Ito Y, Inamura T, Ikezaki K, Fukui M, Iwaki T, Kuwano M. Macrophage infiltration and heme oxygenase-1 expression correlate with angiogenesis in human gliomas. Clin Cancer Res. 1999;5(5):1107–13.
49. Mantovani A, Allavena P, Sica A, Balkwill F. Cancer-related inflammation. Nature. 2008;454(7203):436–44.
50. Li J, Sejas DP, Zhang X, Qiu Y, Nattamai KJ, Rani R, Rathbun KR, Geiger H, Williams DA, Bagby GC, Pang Q. TNF-alpha induces leukemic clonal evolution ex vivo in Fanconi anemia group C murine stem cells. J Clin Invest. 2007;117(11):3283–95. doi:10.1172/JCI31772.
51. Balkwill FR, Mantovani A. Cancer-related inflammation: common themes and therapeutic opportunities. Semin Cancer Biol. 2012;22(1):33–40. doi:10.1016/j.semcancer.2011.12.005.
52. Deng J, Liu Y, Lee H, Herrmann A, Zhang W, Zhang C, Shen S, Priceman SJ, Kujawski M, Pal SK, Raubitschek A, Hoon DS, Forman S, Figlin RA, Liu J, Jove R, Yu H. S1PR1-STAT3 signaling is crucial for myeloid cell colonization at future metastatic sites. Cancer Cell. 2012;21(5):642–54. doi:10.1016/j.ccr.2012.03.039.
53. De Palma M, Lewis CE. Macrophage regulation of tumor responses to anticancer therapies. Cancer Cell. 2013;23(3):277–86.
54. Biswas SK, Allavena P, Mantovani A. Tumor-associated macrophages: functional diversity, clinical significance, and open questions. Semin Immunopathol. 2013;35(5):585–600. doi:10.1007/s00281-013-0367-7.
55. Biswas SK, Mantovani A. Orchestration of metabolism by macrophages. Cell Metab. 2012;15(4):432–7. doi:10.1016/j.cmet.2011.11.013.
56. Mantovani A, Sozzani S, Locati M, Allavena P, Sica A. Macrophage polarization: tumor-associated macrophages as a paradigm for polarized M2 mononuclear phagocytes. Trends Immunol. 2002;23(11):549–55.
57. Jung M, Mertens C, Brune B. Macrophage iron homeostasis and polarization in the context of cancer. Immunobiology. 2015;220(2):295–304. doi:10.1016/j.imbio.2014.09.011.
58. Starossom SC, Mascanfroni ID, Imitola J, Cao L, Raddassi K, Hernandez SF, Bassil R, Croci DO, Cerliani JP, Delacour D, Wang Y, Elyaman W, Khoury SJ, Rabinovich GA. Galectin-1 deactivates classically activated microglia and protects from inflammation-induced neurodegeneration. Immunity. 2012;37(2):249–63. doi:10.1016/j.immuni.2012.05.023.
59. Nishikawa K, Seo N, Torii M, Ma N, Muraoka D, Tawara I, Masuya M, Tanaka K, Takei Y, Shiku H, Katayama N, Kato T. Interleukin-17 induces an atypical M2-like macrophage subpopulation that regulates intestinal inflammation. PLoS One. 2014;9(9), e108494. doi:10.1371/journal.pone.0108494.
60. Jourdan T, Godlewski G, Cinar R, Bertola A, Szanda G, Liu J, Tam J, Han T, Mukhopadhyay B, Skarulis MC, Ju C, Aouadi M, Czech MP, Kunos G. Activation of the Nlrp3 inflammasome in infiltrating macrophages by endocannabinoids mediates beta cell loss in type 2 diabetes. Nat Med. 2013;19(9):1132–40. doi:10.1038/nm.3265.

61. Gazzaniga S, Bravo AI, Guglielmotti A, van Rooijen N, Maschi F, Vecchi A, Mantovani A, Mordoh J, Wainstok R. Targeting tumor-associated macrophages and inhibition of MCP-1 reduce angiogenesis and tumor growth in a human melanoma xenograft. J Invest Dermatol. 2007;127(8):2031–41.
62. Wu X, Schulte BC, Zhou Y, Haribhai D, Mackinnon AC, Plaza JA, Williams CB, Hwang ST. Depletion of M2-like tumor-associated macrophages delays cutaneous T-cell lymphoma development in vivo. J Invest Dermatol. 2014;134(11):2814–22. doi:10.1038/jid.2014.206.
63. Shen KY, Song YC, Chen IH, Chong P, Liu SJ. Depletion of tumor-associated macrophages enhances the anti-tumor immunity induced by a Toll-like receptor agonist-conjugated peptide. Hum Vaccin Immunother. 2014;10:3241–50. doi:10.4161/hv.29275.
64. Kiyota T, Takahashi Y, Watcharanurak K, Nishikawa M, Ohara S, Ando M, Watanabe Y, Takakura Y. Enhancement of anticancer effect of interferon-gamma gene transfer against interferon-gamma-resistant tumor by depletion of tumor-associated macrophages. Mol Pharm. 2014;11(5):1542–9. doi:10.1021/mp4007216.
65. Alzuguren P, Hervas-Stubbs S, Gonzalez-Aseguinolaza G, Poutou J, Fortes P, Mancheno U, Bunuales M, Olague C, Razquin N, Van Rooijen N, Enguita M, Hernandez-Alcoceba R. Transient depletion of specific immune cell populations to improve adenovirus-mediated transgene expression in the liver. Liver Int. 2015;35:1274–89. doi:10.1111/liv.12571.
66. Shmeeda H, Amitay Y, Tzemach D, Gorin J, Gabizon A. Liposome encapsulation of zoledronic acid results in major changes in tissue distribution and increase in toxicity. J Control Release. 2013;167(3):265–75. doi:10.1016/j.jconrel.2013.02.003.
67. Sabatino R, Antonelli A, Battistelli S, Schwendener R, Magnani M, Rossi L. Macrophage depletion by free bisphosphonates and zoledronate-loaded red blood cells. PLoS One. 2014;9(6), e101260. doi:10.1371/journal.pone.0101260.
68. Galmbacher K, Heisig M, Hotz C, Wischhusen J, Galmiche A, Bergmann B, Gentschev I, Goebel W, Rapp UR, Fensterle J. Shigella mediated depletion of macrophages in a murine breast cancer model is associated with tumor regression. PLoS One. 2010;5(3), e9572. doi:10.1371/journal.pone.0009572.
69. Melancon MP, Lu W, Huang Q, Thapa P, Zhou D, Ng C, Li C. Targeted imaging of tumor-associated M2 macrophages using a macromolecular contrast agent PG-Gd-NIR813. Biomaterials. 2010;31(25):6567–73. doi:10.1016/j.biomaterials.2010.05.001.
70. Hattori Y, Yamashita J, Sakaida C, Kawano K, Yonemochi E. Evaluation of antitumor effect of zoledronic acid entrapped in folate-linked liposome for targeting to tumor-associated macrophages. J Liposome Res. 2015;25:131–40. doi:10.3109/08982104.2014.954128.
71. Zhan X, Jia L, Niu Y, Qi H, Chen X, Zhang Q, Zhang J, Wang Y, Dong L, Wang C. Targeted depletion of tumour-associated macrophages by an alendronate-glucomannan conjugate for cancer immunotherapy. Biomaterials. 2014;35(38):10046–57. doi:10.1016/j.biomaterials.2014.09.007.
72. Huang Z, Zhang Z, Jiang Y, Zhang D, Chen J, Dong L, Zhang J. Targeted delivery of oligonucleotides into tumor-associated macrophages for cancer immunotherapy. J Control Release. 2012;158(2):286–92. doi:10.1016/j.jconrel.2011.11.013.
73. Luo Y, Zhou H, Krueger J, Kaplan C, Lee SH, Dolman C, Markowitz D, Wu W, Liu C, Reisfeld RA, Xiang R. Targeting tumor-associated macrophages as a novel strategy against breast cancer. J Clin Invest. 2006;116(8):2132–41.
74. Lewis JS, Landers RJ, Underwood JC, Harris AL, Lewis CE. Expression of vascular endothelial growth factor by macrophages is up-regulated in poorly vascularized areas of breast carcinomas. J Pathol. 2000;192(2):150–8. doi:10.1002/1096-9896(2000)9999:9999<::AID-PATH687>3.0.CO;2-G. pii: 10.1002/1096-9896(2000)9999:9999<::AID-PATH687>3.0.CO;2-G.
75. Mantovani A, Ming WJ, Balotta C, Abdeljalil B, Bottazzi B. Origin and regulation of tumor-associated macrophages: the role of tumor-derived chemotactic factor. Biochim Biophys Acta. 1986;865(1):59–67.
76. Murdoch C, Tazzyman S, Webster S, Lewis CE. Expression of Tie-2 by human monocytes and their responses to angiopoietin-2. J Immunol. 2007;178(11):7405–11. pii: 178/11/7405.

77. Lin EY, Nguyen AV, Russell RG, Pollard JW. Colony-stimulating factor 1 promotes progression of mammary tumors to malignancy. J Exp Med. 2001;193(6):727–40.
78. Ries CH, Cannarile MA, Hoves S, Benz J, Wartha K, Runza V, Rey-Giraud F, Pradel LP, Feuerhake F, Klaman I, Jones T, Jucknischke U, Scheiblich S, Kaluza K, Gorr IH, Walz A, Abiraj K, Cassier PA, Sica A, Gomez-Roca C, de Visser KE, Italiano A, Le Tourneau C, Delord JP, Levitsky H, Blay JY, Ruttinger D. Targeting tumor-associated macrophages with anti-CSF-1R antibody reveals a strategy for cancer therapy. Cancer Cell. 2014;25(6):846–59. doi:10.1016/j.ccr.2014.05.016.
79. Fend L, Accart N, Kintz J, Cochin S, Reymann C, Le Pogam F, Marchand JB, Menguy T, Slos P, Rooke R, Fournel S, Bonnefoy JY, Preville X, Haegel H. Therapeutic effects of anti-CD115 monoclonal antibody in mouse cancer models through dual inhibition of tumor-associated macrophages and osteoclasts. PLoS One. 2013;8(9), e73310. doi:10.1371/journal.pone.0073310.
80. Milke L, Schulz K, Weigert A, Sha W, Schmid T, Brune B. Depletion of tristetraprolin in breast cancer cells increases interleukin-16 expression and promotes tumor infiltration with monocytes/macrophages. Carcinogenesis. 2013;34(4):850–7. doi:10.1093/carcin/bgs387.
81. Kee JY, Ito A, Hojo S, Hashimoto I, Igarashi Y, Tsuneyama K, Tsukada K, Irimura T, Shibahara N, Takasaki I, Inujima A, Nakayama T, Yoshie O, Sakurai H, Saiki I, Koizumi K. CXCL16 suppresses liver metastasis of colorectal cancer by promoting TNF-alpha-induced apoptosis by tumor-associated macrophages. BMC Cancer. 2014;14(1):949. doi:10.1186/1471-2407-14-949.
82. Germano G, Frapolli R, Belgiovine C, Anselmo A, Pesce S, Liguori M, Erba E, Uboldi S, Zucchetti M, Pasqualini F, Nebuloni M, van Rooijen N, Mortarini R, Beltrame L, Marchini S, Fuso Nerini I, Sanfilippo R, Casali PG, Pilotti S, Galmarini CM, Anichini A, Mantovani A, D'Incalci M, Allavena P. Role of macrophage targeting in the antitumor activity of trabectedin. Cancer Cell. 2013;23(2):249–62.
83. Sonda N, Simonato F, Peranzoni E, Cali B, Bortoluzzi S, Bisognin A, Wang E, Marincola FM, Naldini L, Gentner B, Trautwein C, Sackett SD, Zanovello P, Molon B, Bronte V. miR-142-3p prevents macrophage differentiation during cancer-induced myelopoiesis. Immunity. 2013;38(6):1236–49. doi:10.1016/j.immuni.2013.06.004.
84. Campbell MJ, Tonlaar NY, Garwood ER, Huo D, Moore DH, Khramtsov AI, Au A, Baehner F, Chen Y, Malaka DO, Lin A, Adeyanju OO, Li S, Gong C, McGrath M, Olopade OI, Esserman LJ. Proliferating macrophages associated with high grade, hormone receptor negative breast cancer and poor clinical outcome. Breast Cancer Res Treat. 2011;128(3):703–11.
85. Mitchem JB, Brennan DJ, Knolhoff BL, Belt BA, Zhu Y, Sanford DE, Belaygorod L, Carpenter D, Collins L, Piwnica-Worms D, Hewitt S, Udupi GM, Gallagher WM, Wegner C, West BL, Wang-Gillam A, Goedegebuure P, Linehan DC, DeNardo DG. Targeting tumor-infiltrating macrophages decreases tumor-initiating cells, relieves immunosuppression, and improves chemotherapeutic responses. Cancer Res. 2013;73(3):1128–41. doi:10.1158/0008-5472.CAN-12-2731.
86. Linde N, Lederle W, Depner S, van Rooijen N, Gutschalk CM, Mueller MM. Vascular endothelial growth factor-induced skin carcinogenesis depends on recruitment and alternative activation of macrophages. J Pathol. 2012;227(1):17–28. doi:10.1002/path.3989.
87. Zhu Y, Knolhoff BL, Meyer MA, Nywening TM, West BL, Luo J, Wang-Gillam A, Goedegebuure SP, Linehan DC, DeNardo DG. CSF1/CSF1R blockade reprograms tumor-infiltrating macrophages and improves response to T-cell checkpoint immunotherapy in pancreatic cancer models. Cancer Res. 2014;74(18):5057–69. doi:10.1158/0008-5472.CAN-13-3723.
88. Chittezhath M, Dhillon MK, Lim JY, Laoui D, Shalova IN, Teo YL, Chen J, Kamaraj R, Raman L, Lum J, Thamboo TP, Chiong E, Zolezzi F, Yang H, Van Ginderachter JA, Poidinger M, Wong AS, Biswas SK. Molecular profiling reveals a tumor-promoting phenotype of monocytes and macrophages in human cancer progression. Immunity. 2014;41(5):815–29. doi:10.1016/j.immuni.2014.09.014.
89. Murray PJ, Allen JE, Biswas SK, Fisher EA, Gilroy DW, Goerdt S, Gordon S, Hamilton JA, Ivashkiv LB, Lawrence T, Locati M, Mantovani A, Martinez FO, Mege JL, Mosser DM,

Natoli G, Saeij JP, Schultze JL, Shirey KA, Sica A, Suttles J, Udalova I, van Ginderachter JA, Vogel SN, Wynn TA. Macrophage activation and polarization: nomenclature and experimental guidelines. Immunity. 2014;41(1):14–20. doi:10.1016/j.immuni.2014.06.008.

90. Guiducci C, Vicari AP, Sangaletti S, Trinchieri G, Colombo MP. Redirecting in vivo elicited tumor infiltrating macrophages and dendritic cells towards tumor rejection. Cancer Res. 2005;65(8):3437–46. doi:10.1158/0008-5472.CAN-04-4262.
91. Chatterjee S, Mookerjee A, Basu JM, Chakraborty P, Ganguly A, Adhikary A, Mukhopadhyay D, Ganguli S, Banerjee R, Ashraf M, Biswas J, Das PK, Sa G, Chatterjee M, Das T, Choudhuri SK. A novel copper chelate modulates tumor associated macrophages to promote anti-tumor response of T cells. PLoS One. 2009;4(9), e7048. doi:10.1371/journal.pone.0007048.
92. Chakraborty P, Chatterjee S, Ganguly A, Saha P, Adhikary A, Das T, Chatterjee M, Choudhuri SK. Reprogramming of TAM toward proimmunogenic type through regulation of MAP kinases using a redox-active copper chelate. J Leukoc Biol. 2012;91(4):609–19. doi:10.1189/jlb.0611287.
93. Hagemann T, Lawrence T, McNeish I, Charles KA, Kulbe H, Thompson RG, Robinson SC, Balkwill FR. "Re-educating" tumor-associated macrophages by targeting NF-kappaB. J Exp Med. 2008;205(6):1261–8.
94. Cieslewicz M, Tang J, Yu JL, Cao H, Zavaljevski M, Motoyama K, Lieber A, Raines EW, Pun SH. Targeted delivery of proapoptotic peptides to tumor-associated macrophages improves survival. Proc Natl Acad Sci U S A. 2013;110(40):15919–24. doi:10.1073/pnas.1312197110.
95. Kroemer G, Galluzzi L, Kepp O, Zitvogel L. Immunogenic cell death in cancer therapy. Annu Rev Immunol. 2013;31:51–72. doi:10.1146/annurev-immunol-032712-100008.
96. Apetoh L, Ghiringhelli F, Tesniere A, Obeid M, Ortiz C, Criollo A, Mignot G, Maiuri MC, Ullrich E, Saulnier P, Yang H, Amigorena S, Ryffel B, Barrat FJ, Saftig P, Levi F, Lidereau R, Nogues C, Mira JP, Chompret A, Joulin V, Clavel-Chapelon F, Bourhis J, Andre F, Delaloge S, Tursz T, Kroemer G, Zitvogel L. Toll-like receptor 4-dependent contribution of the immune system to anticancer chemotherapy and radiotherapy. Nat Med. 2007;13(9):1050–9.
97. Krysko DV, Garg AD, Kaczmarek A, Krysko O, Agostinis P, Vandenabeele P. Immunogenic cell death and DAMPs in cancer therapy. Nat Rev Cancer. 2012;12(12):860–75. doi:10.1038/nrc3380.
98. Kodumudi KN, Woan K, Gilvary DL, Sahakian E, Wei S, Djeu JY. A novel chemoimmunomodulating property of docetaxel: suppression of myeloid-derived suppressor cells in tumor bearers. Clin Cancer Res. 2010;16(18):4583–94. doi:10.1158/1078-0432.CCR-10-0733.
99. Rolny C, Mazzone M, Tugues S, Laoui D, Johansson I, Coulon C, Squadrito ML, Segura I, Li X, Knevels E, Costa S, Vinckier S, Dresselaer T, Akerud P, De Mol M, Salomaki H, Phillipson M, Wyns S, Larsson E, Buysschaert I, Botling J, Himmelreich U, Van Ginderachter JA, De Palma M, Dewerchin M, Claesson-Welsh L, Carmeliet P. HRG inhibits tumor growth and metastasis by inducing macrophage polarization and vessel normalization through downregulation of PlGF. Cancer Cell. 2011;19(1):31–44. doi:10.1016/j.ccr.2010.11.009.
100. Klug F, Prakash H, Huber PE, Seibel T, Bender N, Halama N, Pfirschke C, Voss RH, Timke C, Umansky L, Klapproth K, Schakel K, Garbi N, Jager D, Weitz J, Schmitz-Winnenthal H, Hammerling GJ, Beckhove P. Low-dose irradiation programs macrophage differentiation to an iNOS(+)/M1 phenotype that orchestrates effective T cell immunotherapy. Cancer Cell. 2013;24(5):589–602. doi:10.1016/j.ccr.2013.09.014.
101. Luo Y, Knudson MJ. Mycobacterium bovis bacillus Calmette-Guerin-induced macrophage cytotoxicity against bladder cancer cells. Clin Dev Immunol. 2010;2010:357591. doi:10.1155/2010/357591.
102. Mathsyaraja H, Thies K, Taffany DA, Deighan C, Liu T, Yu L, Fernandez SA, Shapiro C, Otero J, Timmers C, Lustberg MB, Chalmers J, Leone G, Ostrowski MC. CSF1-ETS2-induced microRNA in myeloid cells promote metastatic tumor growth. Oncogene. 2015;34:3651–61. doi:10.1038/onc.2014.294.
103. Esquela-Kerscher A, Slack FJ. Oncomirs—microRNAs with a role in cancer. Nat Rev Cancer. 2006;6(4):259–69. doi:10.1038/nrc1840.

104. Zonari E, Pucci F, Saini M, Mazzieri R, Politi LS, Gentner B, Naldini L. A role for miR-155 in enabling tumor-infiltrating innate immune cells to mount effective antitumor responses in mice. Blood. 2013;122(2):243–52. doi:10.1182/blood-2012-08-449306.
105. Wang P, Hou J, Lin L, Wang C, Liu X, Li D, Ma F, Wang Z, Cao X. Inducible microRNA-155 feedback promotes type I IFN signaling in antiviral innate immunity by targeting suppressor of cytokine signaling 1. J Immunol. 2010;185(10):6226–33. doi:10.4049/jimmunol.1000491.
106. Nazari-Jahantigh M, Wei Y, Noels H, Akhtar S, Zhou Z, Koenen RR, Heyll K, Gremse F, Kiessling F, Grommes J, Weber C, Schober A. MicroRNA-155 promotes atherosclerosis by repressing Bcl6 in macrophages. J Clin Invest. 2012;122(11):4190–202. doi:10.1172/JCI61716.
107. Martinez-Nunez RT, Louafi F, Sanchez-Elsner T. The interleukin 13 (IL-13) pathway in human macrophages is modulated by microRNA-155 via direct targeting of interleukin 13 receptor alpha1 (IL13Ralpha1). J Biol Chem. 2011;286(3):1786–94. doi:10.1074/jbc.M110.169367.
108. Cai X, Yin Y, Li N, Zhu D, Zhang J, Zhang CY, Zen K. Re-polarization of tumor-associated macrophages to pro-inflammatory M1 macrophages by microRNA-155. J Mol Cell Biol. 2012;4(5):341–3. doi:10.1093/jmcb/mjs044.
109. He M, Xu Z, Ding T, Kuang DM, Zheng L. MicroRNA-155 regulates inflammatory cytokine production in tumor-associated macrophages via targeting C/EBPbeta. Cell Mol Immunol. 2009;6(5):343–52. doi:10.1038/cmi.2009.45.
110. Hanahan D, Weinberg RA. Hallmarks of cancer: the next generation. Cell. 2011;144(5):646–74.
111. Yang J, Zhang Z, Chen C, Liu Y, Si Q, Chuang TH, Li N, Gomez-Cabrero A, Reisfeld RA, Xiang R, Luo Y. MicroRNA-19a-3p inhibits breast cancer progression and metastasis by inducing macrophage polarization through downregulated expression of Fra-1 proto-oncogene. Oncogene. 2014;33(23):3014–23. doi:10.1038/onc.2013.258.
112. Madsen DH, Leonard D, Masedunskas A, Moyer A, Jurgensen HJ, Peters DE, Amornphimoltham P, Selvaraj A, Yamada SS, Brenner DA, Burgdorf S, Engelholm LH, Behrendt N, Holmbeck K, Weigert R, Bugge TH. M2-like macrophages are responsible for collagen degradation through a mannose receptor-mediated pathway. J Cell Biol. 2013;202(6):951–66. doi:10.1083/jcb.201301081.
113. Squadrito ML, Pucci F, Magri L, Moi D, Gilfillan GD, Ranghetti A, Casazza A, Mazzone M, Lyle R, Naldini L, De Palma M. miR-511-3p modulates genetic programs of tumor-associated macrophages. Cell Rep. 2012;1(2):141–54. doi:10.1016/j.celrep.2011.12.005.
114. Cubillos-Ruiz JR, Baird JR, Tesone AJ, Rutkowski MR, Scarlett UK, Camposeco-Jacobs AL, Anadon-Arnillas J, Harwood NM, Korc M, Fiering SN, Sempere LF, Conejo-Garcia JR. Reprogramming tumor-associated dendritic cells in vivo using miRNA mimetics triggers protective immunity against ovarian cancer. Cancer Res. 2012;72(7):1683–93. doi:10.1158/0008-5472.CAN-11-3160.
115. Ortega RA, Barham WJ, Kumar B, Tikhomirov O, McFadden ID, Yull FE, Giorgio TD. Biocompatible mannosylated endosomal-escape nanoparticles enhance selective delivery of short nucleotide sequences to tumor associated macrophages. Nanoscale. 2014;7(2):500–10. doi:10.1039/c4nr03962a.
116. Yang M, Chen J, Su F, Yu B, Lin L, Liu Y, Huang JD, Song E. Microvesicles secreted by macrophages shuttle invasion-potentiating microRNAs into breast cancer cells. Mol Cancer. 2011;10:117. doi:10.1186/1476-4598-10-117.
117. Elmen J, Lindow M, Schutz S, Lawrence M, Petri A, Obad S, Lindholm M, Hedtjarn M, Hansen HF, Berger U, Gullans S, Kearney P, Sarnow P, Straarup EM, Kauppinen S. LNA-mediated microRNA silencing in non-human primates. Nature. 2008;452(7189):896–9. doi:10.1038/nature06783.
118. Lanford RE, Hildebrandt-Eriksen ES, Petri A, Persson R, Lindow M, Munk ME, Kauppinen S, Orum H. Therapeutic silencing of microRNA-122 in primates with chronic hepatitis C virus infection. Science. 2010;327(5962):198–201. doi:10.1126/science.1178178.
119. Sica A, Mantovani A. Macrophage plasticity and polarization: in vivo veritas. J Clin Invest. 2012;122(3):787–95. doi:10.1172/JCI59643.

120. Ivashkiv LB. Inflammatory signaling in macrophages: transitions from acute to tolerant and alternative activation states. Eur J Immunol. 2011;41(9):2477–81. doi:10.1002/eji.201141783.
121. Satoh T, Takeuchi O, Vandenbon A, Yasuda K, Tanaka Y, Kumagai Y, Miyake T, Matsushita K, Okazaki T, Saitoh T, Honma K, Matsuyama T, Yui K, Tsujimura T, Standley DM, Nakanishi K, Nakai K, Akira S. The Jmjd3-Irf4 axis regulates M2 macrophage polarization and host responses against helminth infection. Nat Immunol. 2010;11(10):936–44. doi:10.1038/ni.1920.
122. Chen X, Barozzi I, Termanini A, Prosperini E, Recchiuti A, Dalli J, Mietton F, Matteoli G, Hiebert S, Natoli G. Requirement for the histone deacetylase Hdac3 for the inflammatory gene expression program in macrophages. Proc Natl Acad Sci U S A. 2012;109(42):E2865–74. doi:10.1073/pnas.1121131109.
123. Shakespear MR, Halili MA, Irvine KM, Fairlie DP, Sweet MJ. Histone deacetylases as regulators of inflammation and immunity. Trends Immunol. 2011;32(7):335–43. doi:10.1016/j.it.2011.04.001.
124. Kerr JF, Wyllie AH, Currie AR. Apoptosis: a basic biological phenomenon with wide-ranging implications in tissue kinetics. Br J Cancer. 1972;26(4):239–57.
125. Li F, Huang Q, Chen J, Peng Y, Roop DR, Bedford JS, Li CY. Apoptotic cells activate the "phoenix rising" pathway to promote wound healing and tissue regeneration. Sci Signal. 2010;3(110):ra13.
126. Weidenbusch M, Anders HJ. Tissue microenvironments define and get reinforced by macrophage phenotypes in homeostasis or during inflammation, repair and fibrosis. J Innate Immun. 2012;4(5-6):463–77. doi:10.1159/000336717.
127. Buckley CD, Gilroy DW, Serhan CN, Stockinger B, Tak PP. The resolution of inflammation. Nat Rev Immunol. 2013;13(1):59–66.
128. Dvorak HF. Tumors: wounds that do not heal. Similarities between tumor stroma generation and wound healing. N Engl J Med. 1986;315(26):1650–9.
129. Shalini S, Dorstyn L, Dawar S, Kumar S. Old, new and emerging functions of caspases. Cell Death Differ. 2015;22:526–39. doi:10.1038/cdd.2014.216.
130. Revesz L. Effect of tumour cells killed by x-rays upon the growth of admixed viable cells. Nature. 1956;178(4547):1391–2.
131. Revesz L. Effect of lethally damaged tumor cells upon the development of admixed viable cells. J Natl Cancer Inst. 1958;20(6):1157–86.
132. Huang Q, Li F, Liu X, Li W, Shi W, Liu FF, O'Sullivan B, He Z, Peng Y, Tan AC, Zhou L, Shen J, Han G, Wang XJ, Thorburn J, Thorburn A, Jimeno A, Raben D, Bedford JS, Li CY. Caspase 3-mediated stimulation of tumor cell repopulation during cancer radiotherapy. Nat Med. 2011;17(7):860–6.
133. Willems JJ, Arnold BP, Gregory CD. Sinister self-sacrifice: the contribution of apoptosis to malignancy. Front Immunol. 2014;5:299. doi:10.3389/fimmu.2014.00299.
134. Gough MJ, Melcher AA, Ahmed A, Crittenden MR, Riddle DS, Linardakis E, Ruchatz AN, Emiliusen LM, Vile RG. Macrophages orchestrate the immune response to tumor cell death. Cancer Res. 2001;61(19):7240–7.
135. Gregory CD, Pound JD. Cell death in the neighbourhood: direct microenvironmental effects of apoptosis in normal and neoplastic tissues. J Pathol. 2011;223(2):177–94.
136. Gregory CD, Pound JD. Microenvironmental influences of apoptosis in vivo and in vitro. Apoptosis. 2010;15(9):1029–49. doi:10.1007/s10495-010-0485-9.
137. Truman LA, Ford CA, Pasikowska M, Pound JD, Wilkinson SJ, Dumitriu IE, Melville L, Melrose LA, Ogden CA, Nibbs R, Graham G, Combadiere C, Gregory CD. CX3CL1/fractalkine is released from apoptotic lymphocytes to stimulate macrophage chemotaxis. Blood. 2008;112(13):5026–36. doi:10.1182/blood-2008-06-162404. pii: blood-2008-06-162404.
138. Lee HW, Choi HJ, Ha SJ, Lee KT, Kwon YG. Recruitment of monocytes/macrophages in different tumor microenvironments. Biochim Biophys Acta. 2013;1835(2):170–9. doi:10.1016/j.bbcan.2012.12.007.
139. Zheng J, Yang M, Shao J, Miao Y, Han J, Du J. Chemokine receptor CX3CR1 contributes to macrophage survival in tumor metastasis. Mol Cancer. 2013;12(1):141. doi:10.1186/1476-4598-12-141. pii: 1476-4598-12-141.

140. Bournazou I, Pound JD, Duffin R, Bournazos S, Melville LA, Brown SB, Rossi AG, Gregory CD. Apoptotic human cells inhibit migration of granulocytes via release of lactoferrin. J Clin Invest. 2009;119(1):20–32. doi:10.1172/JCI36226. pii: 36226.
141. Weigert A, Johann AM, von Knethen A, Schmidt H, Geisslinger G, Brune B. Apoptotic cells promote macrophage survival by releasing the antiapoptotic mediator sphingosine-1-phosphate. Blood. 2006;108(5):1635–42.
142. Gude DR, Alvarez SE, Paugh SW, Mitra P, Yu J, Griffiths R, Barbour SE, Milstien S, Spiegel S. Apoptosis induces expression of sphingosine kinase 1 to release sphingosine-1-phosphate as a "come-and-get-me" signal. FASEB J. 2008;22(8):2629–38.
143. Weigert A, Cremer S, Schmidt MV, von Knethen A, Angioni C, Geisslinger G, Brune B. Cleavage of sphingosine kinase 2 by caspase-1 provokes its release from apoptotic cells. Blood. 2010;115(17):3531–40.
144. Weigert A, Tzieply N, von Knethen A, Johann AM, Schmidt H, Geisslinger G, Brune B. Tumor cell apoptosis polarizes macrophages role of sphingosine-1-phosphate. Mol Biol Cell. 2007;18(10):3810–9.
145. Weis N, Weigert A, von Knethen A, Brune B. Heme oxygenase-1 contributes to an alternative macrophage activation profile induced by apoptotic cell supernatants. Mol Biol Cell. 2009;20(5):1280–8.
146. Herr B, Zhou J, Werno C, Menrad H, Namgaladze D, Weigert A, Dehne N, Brune B. The supernatant of apoptotic cells causes transcriptional activation of hypoxia-inducible factor-1alpha in macrophages via sphingosine-1-phosphate and transforming growth factor-beta. Blood. 2009;114(10):2140–8.
147. Ley S, Weigert A, Heriche JK, Mille-Baker B, Janssen RA, Brune B. RNAi screen in apoptotic cancer cell-stimulated human macrophages reveals co-regulation of IL-6/IL-10 expression. Immunobiology. 2013;218(1):40–51.
148. Ley S, Weigert A, Weichand B, Henke N, Mille-Baker B, Janssen RA, Brune B. The role of TRKA signaling in IL-10 production by apoptotic tumor cell-activated macrophages. Oncogene. 2013;32(5):631–40.
149. Johann AM, Weigert A, Eberhardt W, Kuhn AM, Barra V, von Knethen A, Pfeilschifter JM, Brune B. Apoptotic cell-derived sphingosine-1-phosphate promotes HuR-dependent cyclooxygenase-2 mRNA stabilization and protein expression. J Immunol. 2008;180(2):1239–48.
150. Sha W, Olesch C, Hanaka H, Radmark O, Weigert A, Brune B. Necrosis in DU145 prostate cancer spheroids induces COX-2/mPGES-1-derived PGE2 to promote tumor growth and to inhibit T cell activation. Int J Cancer. 2013;133(7):1578–88.
151. Weigert A, Schiffmann S, Sekar D, Ley S, Menrad H, Werno C, Grosch S, Geisslinger G, Brune B. Sphingosine kinase 2 deficient tumor xenografts show impaired growth and fail to polarize macrophages towards an anti-inflammatory phenotype. Int J Cancer. 2009;125(9):2114–21.
152. Fadok VA, Voelker DR, Campbell PA, Cohen JJ, Bratton DL, Henson PM. Exposure of phosphatidylserine on the surface of apoptotic lymphocytes triggers specific recognition and removal by macrophages. J Immunol. 1992;148(7):2207–16.
153. Matsura T, Serinkan BF, Jiang J, Kagan VE. Phosphatidylserine peroxidation/externalization during staurosporine-induced apoptosis in HL-60 cells. FEBS Lett. 2002;524(1-3):25–30. pii: S0014579302029903.
154. Gardai SJ, Bratton DL, Ogden CA, Henson PM. Recognition ligands on apoptotic cells: a perspective. J Leukoc Biol. 2006;79(5):896–903. doi:10.1189/jlb.1005550. pii: 79/5/896.
155. Sancho D, Reis e Sousa C. Sensing of cell death by myeloid C-type lectin receptors. Curr Opin Immunol. 2013;25(1):46–52. doi:10.1016/j.coi.2012.12.007. pii: S0952-7915(12)00197-5.
156. Devitt A, Parker KG, Ogden CA, Oldreive C, Clay MF, Melville LA, Bellamy CO, Lacy-Hulbert A, Gangloff SC, Goyert SM, Gregory CD. Persistence of apoptotic cells without autoimmune disease or inflammation in CD14-/- mice. J Cell Biol. 2004;167(6):1161–70.
157. Lucas M, Stuart LM, Zhang A, Hodivala-Dilke K, Febbraio M, Silverstein R, Savill J, Lacy-Hulbert A. Requirements for apoptotic cell contact in regulation of macrophage responses. J Immunol. 2006;177(6):4047–54.

158. Fadok VA, Bratton DL, Konowal A, Freed PW, Westcott JY, Henson PM. Macrophages that have ingested apoptotic cells in vitro inhibit proinflammatory cytokine production through autocrine/paracrine mechanisms involving TGF-beta, PGE2, and PAF. J Clin Invest. 1998;101(4):890–8.
159. Bonizzi G, Karin M. The two NF-kappaB activation pathways and their role in innate and adaptive immunity. Trends Immunol. 2004;25(6):280–8.
160. Saccani A, Schioppa T, Porta C, Biswas SK, Nebuloni M, Vago L, Bottazzi B, Colombo MP, Mantovani A, Sica A. p50 nuclear factor-kappaB overexpression in tumor-associated macrophages inhibits M1 inflammatory responses and antitumor resistance. Cancer Res. 2006;66(23):11432–40.
161. Biswas SK, Gangi L, Paul S, Schioppa T, Saccani A, Sironi M, Bottazzi B, Doni A, Vincenzo B, Pasqualini F, Vago L, Nebuloni M, Mantovani A, Sica A. A distinct and unique transcriptional program expressed by tumor-associated macrophages (defective NF-kappaB and enhanced IRF-3/STAT1 activation). Blood. 2006;107(5):2112–22. doi:10.1182/blood-2005-01-0428.
162. Cvetanovic M, Ucker DS. Innate immune discrimination of apoptotic cells: repression of proinflammatory macrophage transcription is coupled directly to specific recognition. J Immunol. 2004;172(2):880–9.
163. Jennewein C, Kuhn AM, Schmidt MV, Meilladec-Jullig V, von Knethen A, Gonzalez FJ, Brune B. Sumoylation of peroxisome proliferator-activated receptor gamma by apoptotic cells prevents lipopolysaccharide-induced NCoR removal from kappaB binding sites mediating transrepression of proinflammatory cytokines. J Immunol. 2008;181(8):5646–52.
164. Ramos GC, Fernandes D, Charao CT, Souza DG, Teixeira MM, Assreuy J. Apoptotic mimicry: phosphatidylserine liposomes reduce inflammation through activation of peroxisome proliferator-activated receptors (PPARs) in vivo. Br J Pharmacol. 2007;151(6):844–50. doi:10.1038/sj.bjp.0707302.
165. von Knethen A, Sha LK, Kuchler L, Heeg AK, Fuhrmann D, Heide H, Wittig I, Maier TJ, Steinhilber D, Brune B. 5-Lipoxygenase contributes to PPARgamma activation in macrophages in response to apoptotic cells. Cell Signal. 2013;25(12):2762–8. doi:10.1016/j.cellsig.2013.08.045.
166. Sen P, Wallet MA, Yi Z, Huang Y, Henderson M, Mathews CE, Earp HS, Matsushima G, Baldwin Jr AS, Tisch RM. Apoptotic cells induce Mer tyrosine kinase-dependent blockade of NF-{kappa}B activation in dendritic cells. Blood J. 2007;109:653–60.
167. Freire-de-Lima CG, Xiao YQ, Gardai SJ, Bratton DL, Schiemann WP, Henson PM. Apoptotic cells, through transforming growth factor-beta, coordinately induce anti-inflammatory and suppress pro-inflammatory eicosanoid and NO synthesis in murine macrophages. J Biol Chem. 2006;281(50):38376–84.
168. Johann AM, Barra V, Kuhn AM, Weigert A, von Knethen A, Brune B. Apoptotic cells induce arginase II in macrophages, thereby attenuating NO production. FASEB J. 2007;21(11):2704–12.
169. Serinkan BF, Gambelli F, Potapovich AI, Babu H, Di Giuseppe M, Ortiz LA, Fabisiak JP, Kagan VE. Apoptotic cells quench reactive oxygen and nitrogen species and modulate TNF-alpha/TGF-beta1 balance in activated macrophages: involvement of phosphatidylserine-dependent and -independent pathways. Cell Death Differ. 2005;12(8):1141–4.
170. Johann AM, von Knethen A, Lindemann D, Brune B. Recognition of apoptotic cells by macrophages activates the peroxisome proliferator-activated receptor-gamma and attenuates the oxidative burst. Cell Death Differ. 2006;13(9):1533–40.
171. Freire-de-Lima CG, Nascimento DO, Soares MB, Bozza PT, Castro-Faria-Neto HC, de Mello FG, DosReis GA, Lopes MF. Uptake of apoptotic cells drives the growth of a pathogenic trypanosome in macrophages. Nature. 2000;403(6766):199–203.
172. Yin Y, Huang X, Lynn KD, Thorpe PE. Phosphatidylserine-targeting antibody induces M1 macrophage polarization and promotes myeloid-derived suppressor cell differentiation. Cancer Immunol Res. 2013;1(4):256–68. doi:10.1158/2326-6066.CIR-13-0073. pii: 2326-6066.CIR-13-0073.

173. Stanford JC, Young C, Hicks D, Owens P, Williams A, Vaught DB, Morrison MM, Lim J, Williams M, Brantley-Sieders DM, Balko JM, Tonetti D, Earp III HS, Cook RS. Efferocytosis produces a prometastatic landscape during postpartum mammary gland involution. J Clin Invest. 2014;124(11):4737–52. doi:10.1172/JCI76375.
174. Uderhardt S, Herrmann M, Oskolkova OV, Aschermann S, Bicker W, Ipseiz N, Sarter K, Frey B, Rothe T, Voll R, Nimmerjahn F, Bochkov VN, Schett G, Kronke G. 12/15-Lipoxygenase orchestrates the clearance of apoptotic cells and maintains immunologic tolerance. Immunity. 2012;36(5):834–46. doi:10.1016/j.immuni.2012.03.010.
175. Chao MP, Weissman IL, Majeti R. The CD47-SIRPalpha pathway in cancer immune evasion and potential therapeutic implications. Curr Opin Immunol. 2012;24(2):225–32. doi:10.1016/j.coi.2012.01.010. pii: S0952-7915(12)00012-X.
176. Willingham SB, Volkmer JP, Gentles AJ, Sahoo D, Dalerba P, Mitra SS, Wang J, Contreras-Trujillo H, Martin R, Cohen JD, Lovelace P, Scheeren FA, Chao MP, Weiskopf K, Tang C, Volkmer AK, Naik TJ, Storm TA, Mosley AR, Edris B, Schmid SM, Sun CK, Chua MS, Murillo O, Rajendran P, Cha AC, Chin RK, Kim D, Adorno M, Raveh T, Tseng D, Jaiswal S, Enger PO, Steinberg GK, Li G, So SK, Majeti R, Harsh GR, van de Rijn M, Teng NN, Sunwoo JB, Alizadeh AA, Clarke MF, Weissman IL. The CD47-signal regulatory protein alpha (SIRPa) interaction is a therapeutic target for human solid tumors. Proc Natl Acad Sci U S A. 2012;109(17):6662–7. doi:10.1073/pnas.1121623109. pii: 1121623109.

About the Author

Christopher D. Gregory is Professor of Inflammatory Cell Biology and the Deputy Director of the MRC Centre for Inflammation Research at the University of Edinburgh. Professor Gregory is an active research group leader in the fields of immunology, inflammation, and cancer with a particular interest in apoptosis. He has published over 100 scientific papers and has received over 6950 citations.

C.D. Gregory (ed.), *Apoptosis in Cancer Pathogenesis and Anti-cancer Therapy*, Advances in Experimental Medicine and Biology 930,
DOI 10.1007/978-3-319-39406-0

Index

C.D. Gregory (ed.), *Apoptosis in Cancer Pathogenesis and Anti-cancer Therapy*, Advances in Experimental Medicine and Biology 930,
DOI 10.1007/978-3-319-39406-0

MIX
Papier aus verantwortungsvollen Quellen
Paper from responsible sources
FSC® C105338
FSC
www.fsc.org

If you have any concerns about our products,
you can contact us on
ProductSafety@springernature.com

In case Publisher is established outside the EU,
the EU authorized representative is:
Springer Nature Customer Service Center GmbH
Europaplatz 3, 69115 Heidelberg, Germany

Printed by Libri Plureos GmbH
in Hamburg, Germany